T0324471

An Introduction to Lagrangian Mechanics

2nd Edition

An Introduction to Lagrangian Mechanics

2nd Edition

Alain J. Brizard

Saint Michael's College, USA

World Scientific

NEW JERSEY · LONDON · SINGAPORE · BEIJING · SHANGHAI · HONG KONG · TAIPEI · CHENNAI

Published by

World Scientific Publishing Co. Pte. Ltd.

5 Toh Tuck Link, Singapore 596224

USA office: 27 Warren Street, Suite 401-402, Hackensack, NJ 07601

UK office: 57 Shelton Street, Covent Garden, London WC2H 9HE

Library of Congress Cataloging-in-Publication Data
Brizard, Alain Jean.
  An introduction to Lagrangian mechanics / Alain J Brizard, Saint Michael's College, USA. --
2nd edition.
    pages cm
  Includes bibliographical references and index.
  ISBN 978-9814623612 (hardcover : alk. paper) -- ISBN 981462361X (hardcover : alk. paper) --
ISBN 978-9814623629 (pbk. : alk. paper) -- ISBN 9814623628 (pbk. : alk. paper)
  1. Lagrangian functions--Textbooks. 2. Hamiltonian systems--Textbooks. I. Title.
QA315.B84 2015
515'.39--dc23

                                                                    2014042298

British Library Cataloguing-in-Publication Data
A catalogue record for this book is available from the British Library.

Printed in Singapore

To the memory of my parents
Yvon Brizard (1929-2007)
Berthe (St-Laurent) Brizard (1927-2010)

Preface to the Second Edition

The Second Edition of *An Introduction to Lagrangian Mechanics* includes a larger selection of examples and problems (with hints) in each chapter and the reduction of the unavoidable typos that crept into the First Edition. The Second Edition also continues the strong emphasis of the First Edition on the development and application of mathematical methods (mostly calculus) to the solution of problems in Classical Mechanics.

New material has been added to most chapters. In Chapter 1, the Frenet-Serret formulas are given for a light ray propagating in an arbitrary medium and a new derivation of the Noether theorem for discrete Lagrangian systems is given in Chapter 2. In Chapter 3, the motion of a particle in the Morse potential is given as another example of motion in a potential that exhibits a separatrix solution separating bounded motion from unbounded motion. Sections have been rearranged in Chapter 5 and a modified Rutherford scattering problem is solved exactly to show that the total scattering cross section σ_T associated with scattering by a confined potential (i.e., which vanishes beyond a certain radius R) yields the hard-sphere result $\sigma_T = \pi R^2$. In Chapter 6, the Frenet-Serret formulas for the Coriolis-corrected projectile motion are presented and the Frenet-Serret torsion is shown to be directly related to the Coriolis deflection. In Chapter 7, complete solutions of the body-frame and space-frame precession motions are given and a new solution of the sleeping-top problem is given, while in Chapter 8, the normal-mode stability analysis of the sleeping top is presented in a two-dimensional configuration space. Lastly, Appendix A presents a simple solution for the roots of a general cubic polynomial and a compendium of integral formulas evaluated by the trigonometric and hyperbolic-trigonometric substitution methods.

Alain Jean Brizard (2014)

Preface to the First Edition

The structure of the present lecture notes on the Lagrangian mechanics of particles and fields is based on achieving several goals. As a first goal, I wanted to model these notes after the wonderful monograph of Landau and Lifschitz on *Mechanics* [13], which is often thought to be too concise for most undergraduate students. One of the many positive characteristics of Landau and Lifschitz's *Mechanics*, however, is that Lagrangian mechanics is introduced in its first chapter and not in later chapters as is usually done in more standard textbooks used at the sophomore/junior undergraduate level.[1] Consequently, the Lagrangian method becomes the centerpiece of the present course and provides a continuous thread throughout the text. This course has been taught at Dartmouth College and Saint Michael's College in approximately the same format proposed in these lecture notes.

As a second goal, the lecture notes introduce several numerical investigations of dynamical equations appearing throughout the text. These numerical investigations present an interactive pedagogical approach, which should enable students to begin their own numerical investigations. As a third goal, an attempt was made to introduce historical facts (whenever appropriate) about the pioneers of Classical Mechanics. Much of the historical information included in the Notes is taken from excellent books by René Dugas [4], Wolfgang Yourgrau and Stanley Mandelstam [21], and Cornelius Lanczos [12]. In fact, from a pedagogical point of view, this historical perspective helps educating undergraduate students in establishing the deep connections between Classical and Quantum Mechanics, which are often ignored or even inverted (as can be observed when students are surprised

---

[1] The reader is invited to read *A call to action* by E. F. Taylor [Am. J. Phys. **71**, 423–425 (2003)], which promotes a reorganization of undergraduate physics education that includes an early introduction of Lagrangian Mechanics (the Principle of Least Action) into the physics curriculum.

to learn that Hamiltonians have an independent classical existence). As a fourth and final goal, I wanted to keep the scope of these notes limited to a one-semester course in contrast to standard textbooks, which often include an extensive review of Newtonian Mechanics as well as additional material such as Hamiltonian chaos.

It is expected that students taking this course will have had a one-year calculus-based introductory physics course followed by a one-semester course in Modern Physics. Ideally, students should have completed their full calculus sequence and, perhaps, have taken a course on ordinary differential equations. On the other hand, this course should be taken before a rigorous course in Quantum Mechanics in order to provide students with a sound historical perspective involving the connection between Classical Physics and Quantum Physics. Hence, the fall semester of the junior year provides a perfect niche for this course. Topics identified with an asterisk can also be included in a more advanced course.

The standard topics covered in these notes are: The Calculus of Variations (Chapter 1), Lagrangian Mechanics (Chapter 2), Hamiltonian Mechanics (Chapter 3), Motion in a Central Field (Chapter 4), Collisions and Scattering Theory (Chapter 5), Motion in a Non-Inertial Frame (Chapter 6), Rigid Body Motion (Chapter 7), Normal-Mode Analysis (Chapter 8), and Continuous Lagrangian Systems (Chapter 9). Each chapter contains a set of problems with variable level of difficulty. Lastly, in order to ensure a self-contained presentation, a summary of mathematical methods associated with linear algebra and numerical analysis is presented in Appendix A. Appendix B presents a brief introduction to the applications of the Jacobi and Weierstrass elliptic functions in Classical Mechanics; see Whittaker's textbook [20] for many more applications. Lastly, Appendix C presents a brief summary of differential geometric methods in the modern formulation of Hamiltonian mechanics and perturbation theory.

Several innovative topics not normally discussed in standard undergraduate textbooks are included throughout the notes. In Chapter 1, a complete discussion of Fermat's Principle of Least Time is presented, from which a generalization of Snell's Law for light refraction through a nonuniform medium is derived and the equations of geometric optics are obtained [3]. We note that Fermat's Principle proves to be an ideal introduction to variational methods in the undergraduate physics curriculum since students are already familiar with Snell's Law of light refraction.

In Chapter 2, we establish the connection between Fermat's Principle of Least Time and Maupertuis-Jacobi's Principle of Least Action.

In particular, Jacobi's Principle introduces a geometric representation of single-particle dynamics that establishes a clear pre-relativistic connection between Geometry and Physics. Next, the nature of mechanical forces (e.g., active versus passive forces) is discussed within the context of d'Alembert's Principle, which is based on a dynamical generalization of the Principle of Virtual Work. Lastly, the fundamental link between the energy-momentum conservation laws and the symmetries of the Lagrangian function is first discussed through Noether's Theorem and then Routh's procedure to eliminate ignorable coordinates is applied to a Lagrangian with symmetries.

In Chapter 3, we present a brief discussion of Hamiltonian optics and the wave-particle duality that established the connection between Classical Physics and Quantum Physics. The problem of charged-particle motion in an electromagnetic field is also investigated by the Lagrangian method in the three-dimensional configuration space and the Hamiltonian method in six-dimensional phase space. This important physical example presents a clear link between the Lagrangian and Hamiltonian methods. In Chapter 4, we discuss the role of the Laplace-Runge-Lenz vector invariant in determining the shape of the Kepler bounded orbit. We also use the Laplace-Runge-Lenz vector to study the precession of a perturbed Keplerian orbit. In Chapter 5, we present a complete solution of the soft-sphere scattering problem as well as the problem of elastic scattering by a hard surface. In Chapter 9, we present the variational derivations of the Schroedinger equation and the Euler equations for a perfect fluid. Using the Noether method, we also derive their respective conservation laws.

In Appendix B, we present an introduction to the applications of the Jacobi and Weierstrass elliptic functions in Classical Mechanics. These interesting functions used to be part of the standard curriculum in Classical Mechanics [13, 20] and have now all but disappeared from modern textbooks [7, 15]. For the Jacobi elliptic function, we consider the problems of motion in a quartic potential, while for the Weierstrass elliptic function, we consider the problem of motion in a cubic potential. The problem of the planar pendulum is used to establish the connection between the Jacobi and Weierstrass elliptic functions. Lastly, in Appendix C, we present a brief introduction to noncanonical Hamiltonian mechanics and canonical Hamiltonian perturbation theory.

My interest in Lagrangian Mechanics was awakened more than 30 years ago when I was an undergraduate student at the Collège Militaire Royal de Saint Jean (Canada). One of my professors (Fernand Ledoyen) bravely taught me Lagrangian Mechanics with Landau and Lifschitz [13] and Arnold

[1] as our constant companions. I remember being immediately struck by the beauty of Lagrangian Mechanics and the power of its methods. I have used Lagrangian methods in my own research in plasma physics for the past 20 years. I would like to thank my *Lagrangian* collaborators Allan N. Kaufman (University of California at Berkeley) and Eugene (Gene) R. Tracy (College of William and Mary) for their friendship and support during this time.

Lastly, I owe a great debt of love and gratitude to my wife (Dinah Larsen) and son (Peter Brizard Larsen) and I thank them for their patience and understanding during the arduous process of writing this book.

Alain Jean Brizard (2007)

Contents

Preface to the Second Edition   vii

Preface to the First Edition   ix

1. The Calculus of Variations   1

    1.1   Foundations of the Calculus of Variations   1
            1.1.1   A Simple Minimization Problem   1
            1.1.2   Methods of the Calculus of Variations   3
            1.1.3   Path of Shortest Distance and Geodesic Equation   8
    1.2   Classical Variational Problems   10
            1.2.1   Isoperimetric Problem   11
            1.2.2   Brachistochrone Problem   13
    1.3   Fermat's Principle of Least Time   15
            1.3.1   Light Propagation in a Nonuniform Medium . . .   17
            1.3.2   Snell's Law   19
            1.3.3   Application of Fermat's Principle   20
    1.4   Geometric Formulation of Ray Optics*   21
            1.4.1   General Geometric Optics   21
            1.4.2   Light-ray Frenet-Serret Equations   23
            1.4.3   Light Propagation in Spherical Geometry   24
            1.4.4   Geodesic Representation of Light Propagation . .   26
            1.4.5   Wavefront Representation   29
    1.5   Summary   31
    1.6   Problems   31

2. Lagrangian Mechanics   37

    2.1   Maupertuis-Jacobi Principle of Least Action   38

		2.1.1	Maupertuis' Principle	38
		2.1.2	Jacobi's Principle	39
	2.2	d'Alembert's Principle		41
		2.2.1	Principle of Virtual Work	41
		2.2.2	Lagrange's Equations from d'Alembert's Principle	42
	2.3	Hamilton's Principle		45
		2.3.1	Constraint Forces	45
		2.3.2	Generalized Coordinates in Configuration Space	47
		2.3.3	Constrained Motion on a Surface	48
		2.3.4	Euler-Lagrange Equations	50
		2.3.5	Four-step Lagrangian Method	51
		2.3.6	Lagrangian Mechanics in Curvilinear Coordinates*	52
	2.4	Lagrangian Mechanics in Configuration Space		53
		2.4.1	Example I: Pendulum	53
		2.4.2	Example II: Bead on a Rotating Hoop	54
		2.4.3	Example III: Rotating Pendulum	56
		2.4.4	Example IV: Compound Atwood Machine	57
		2.4.5	Example V: Pendulum with Oscillating Fulcrum	59
	2.5	Symmetries and Conservation Laws		61
		2.5.1	Energy Conservation Law	62
		2.5.2	Momentum Conservation Laws	62
		2.5.3	Invariance Properties of a Lagrangian	63
		2.5.4	Lagrangian Mechanics with Symmetries	64
		2.5.5	Routh's Procedure	65
	2.6	Lagrangian Mechanics in the CM Frame		67
	2.7	Summary		69
	2.8	Problems		70
3.	Hamiltonian Mechanics			77
	3.1	Hamilton's Canonical Equations		77
	3.2	Legendre Transformation*		79
	3.3	Hamiltonian Optics and Wave-Particle Duality*		80
	3.4	Motion in an Electromagnetic Field		82
		3.4.1	Euler-Lagrange Equations	82
		3.4.2	Gauge Invariance	84
		3.4.3	Canonical Hamilton's Equations	84
		3.4.4	Maupertuis' Principle for Particle-Beam Optics	85
	3.5	One-degree-of-freedom Hamiltonian Dynamics		86
		3.5.1	Energy Method	86

3.5.2   Simple Harmonic Oscillator   88

3.5.3   Morse Potential   89

3.5.4   Pendulum   93

3.5.5   Constrained Motion on the Surface of a Cone . .   98

3.6   Summary   100

3.7   Problems   100

4.   **Motion in a Central-Force Field**   **105**

4.1   Motion in a Central-Force Field   105

4.1.1   Lagrangian Formalism   106

4.1.2   Hamiltonian Formalism   109

4.2   Homogeneous Central Potentials*   110

4.2.1   The Virial Theorem   110

4.2.2   General Properties of Homogeneous Potentials . .   112

4.3   Kepler Problem   113

4.3.1   Bounded Keplerian Orbits   114

4.3.2   Unbounded Keplerian Orbits   117

4.3.3   Laplace-Runge-Lenz Vector*   118

4.4   Isotropic Simple Harmonic Oscillator   121

4.5   Internal Reflection inside a Well   122

4.6   Summary   124

4.7   Problems   125

5.   **Collisions and Scattering Theory**   **133**

5.1   Two-Particle Collisions in the LAB Frame   134

5.2   Two-Particle Collisions in the CM Frame   137

5.3   Connection between the CM and LAB Frames   138

5.4   Scattering Cross Sections   139

5.4.1   Definitions   140

5.4.2   Cross Sections in CM and LAB Frames   141

5.5   Hard-Sphere Scattering   143

5.6   Rutherford Scattering   145

5.6.1   Classical Rutherford Scattering   145

5.6.2   Modified Rutherford Scattering   147

5.7   Soft-Sphere Scattering   150

5.8   Elastic Scattering by a Hard Surface   152

5.9   Summary   154

5.10   Problems   155

6. Motion in a Non-Inertial Frame   161

    6.1   Time Derivatives in Rotating Frames 161
    6.2   Accelerations in Rotating Frames 163
    6.3   Lagrangian Formulation of Non-Inertial Motion 164
    6.4   Motion Relative to Earth 166
          6.4.1   Coriolis-corrected Projectile Motion 168
          6.4.2   Frenet-Serret-Coriolis Formulas 170
          6.4.3   Free-Fall Problem Revisited 172
          6.4.4   Foucault Pendulum 172
    6.5   Summary 175
    6.6   Problems 176

7. Rigid Body Motion   181

    7.1   Inertia Tensor of a Rigid Body 181
          7.1.1   Discrete Particle Distribution 181
          7.1.2   Parallel-Axes Theorem 183
          7.1.3   Continuous Particle Distribution 184
          7.1.4   Principal Axes of Inertia 185
    7.2   Eulerian Rigid-Body Dynamics 189
          7.2.1   Euler Equations 189
          7.2.2   Euler Equations for a Torque-free Symmetric Top   190
          7.2.3   Euler Equations for a Free Asymmetric Top . . .   194
    7.3   Lagrangian Rigid-Body Dynamics   197
          7.3.1   Eulerian Angles as Generalized Coordinates . . . 197
          7.3.2   Angular Velocity in Terms of Eulerian Angles . . 198
          7.3.3   Rotational Kinetic Energy of a Symmetric Top . . 199
          7.3.4   Space-frame Precession and Space-cone Solutions   202
    7.4   Symmetric Top with One Fixed Point   204
          7.4.1   Nutation   206
          7.4.2   Slow and Fast Precession   209
          7.4.3   The Sleeping Top   209
    7.5   Summary 212
    7.6   Problems 213

8. Normal-Mode Analysis   219

    8.1   Stability of Equilibrium Points 219
          8.1.1   Bead on a Rotating Hoop 219
          8.1.2   Circular Orbits in Central-Force Fields 220

8.2   Small Oscillations about Stable Equilibria 221
8.3   Normal-Mode Analysis of Coupled Oscillations 223
      8.3.1   Normal-Mode Analysis 223
      8.3.2   Coupled Simple Harmonic Oscillators 226
      8.3.3   Coupled Nonlinear Oscillators 228
      8.3.4   Stability of the Sleeping Top 231
8.4   Summary 235
8.5   Problems 235

9. Continuous Lagrangian Systems   241

9.1   Waves on a Stretched String 241
      9.1.1   Wave Equation 241
      9.1.2   Lagrangian Formulation 242
9.2   Variational Principle for Field Theory* 243
      9.2.1   Lagrangian Formulation 244
      9.2.2   Noether Method and Conservation Laws 245
9.3   Schroedinger's Equation 248
9.4   Euler Equations for a Perfect Fluid 250
      9.4.1   Lagrangian Formulation 251
      9.4.2   Energy-Momentum Conservation Laws 252
9.5   Summary 253
9.6   Problems 254

Appendix A   Basic Mathematical Methods   255

A.1   Roots of a General Cubic Polynomial 255
A.2   Integration by Trigonometric Substitution 257
      A.2.1   Trigonometric Functions 257
      A.2.2   Hyperbolic-Trigonometric Functions 258
A.3   Frenet-Serret Formulas 259
      A.3.1   Frenet Frame 259
      A.3.2   Darboux Frame 261
      A.3.3   Example: Seiffert Spiral on the Unit Sphere . . . 262
      A.3.4   Frenet-Serret Formulas for Helical Path 264
A.4   Linear Algebra 265
      A.4.1   Matrix Algebra 266
      A.4.2   Eigenvalue Analysis of a 2×2 Matrix 268
A.5   Numerical Analysis 272

Appendix B   Elliptic Functions and Integrals*   275

B.1   Jacobi Elliptic Functions 275
      B.1.1   Definitions and Notation 275
      B.1.2   Motion in a Quartic Potential 279
B.2   Weierstrass Elliptic Functions 280
      B.2.1   Definitions and Notation 280
      B.2.2   Motion in a Cubic Potential 285
B.3   Connection between Elliptic Functions 287

Appendix C   Noncanonical Hamiltonian Mechanics*   291

C.1   Differential Geometry 291
C.2   Lagrange and Poisson Tensors 293
C.3   Hamiltonian Perturbation Theory 295

Bibliography   301

Index   303

Chapter 1

The Calculus of Variations

A wide range of equations in physics, from quantum field and superstring theories to general relativity, from fluid dynamics to plasma physics and condensed-matter theory, are derived from action (variational) principles [2, 17]. The purpose of this Chapter is to introduce the methods of the Calculus of Variations that figure prominently in the formulation of action principles in physics.

1.1   Foundations of the Calculus of Variations

1.1.1   *A Simple Minimization Problem*

It is a well-known fact that the shortest distance between two points in (Euclidean) space is calculated along a straight line joining the two points. Although this fact is intuitively obvious, we begin our discussion of the problem of minimizing certain integrals in mathematics and physics with a search for an explicit proof. In particular, we prove that the straight line $y_0(x) = mx$ yields a path of shortest distance between the two points $(0,0)$ and $(1, m)$ on the (x, y)-plane as follows.

First, we consider the length integral

$$\mathcal{L}[y] = \int_0^1 \sqrt{1 + (y')^2}\, dx, \tag{1.1}$$

where $y' = y'(x)$ is the slope of the function y at point x. The notation $\mathcal{L}[y]$ is used to denote the fact that the value of the integral (1.1) depends on the choice we make for the function $y(x)$; thus, $\mathcal{L}[y]$ is called a *functional* of y. We insist, however, that the function $y(x)$ satisfy the boundary conditions $y(0) = 0$ and $y(1) = m$. For example, for $y_0(x) = mx$, we find $\mathcal{L}[y_0] =$

1

$\sqrt{1 + m^2}$, while for $y_1(x) = m\,x^2$, we find

$$\mathcal{L}[y_1] = \int_0^1 \sqrt{1 + 4\,m^2\,x^2}\,dx \;=\; \frac{1}{2m} \int_0^{\text{arcsinh}\,2m} \cosh^2 z\,dz$$
$$= \frac{1}{4\,m} \left[2m\,\sqrt{1 + 4\,m^2} \;+\; \text{arcsinh}(2\,m) \right],$$

where we used the hyperbolic-trigonometric substitution[1] $2m\,x = \sinh z$. We can readily show that $\mathcal{L}[y_0] < \mathcal{L}[y_1]$ for all non-vanishing m-values.

Next, we introduce the modified function

$$y(x; \epsilon) \;=\; y_0(x) \;+\; \epsilon\,\delta y(x),$$

where $y_0(x) = mx$ (i.e., the solution to our problem) and the variation function $\delta y(x)$ is required to satisfy the prescribed boundary conditions $\delta y(0) = 0 = \delta y(1)$. We thus define the modified length integral

$$\mathcal{L}[y_0 + \epsilon\,\delta y] \;=\; \int_0^1 \sqrt{1 + (m + \epsilon\,\delta y')^2}\,dx$$

as a function of ϵ and a functional of δy. We now show that the function $y_0(x) = mx$ minimizes the integral (1.1) by evaluating the following derivatives

$$\left(\frac{d}{d\epsilon} \mathcal{L}[y_0 + \epsilon\,\delta y] \right)_{\epsilon=0} = \frac{m}{\sqrt{1 + m^2}} \int_0^1 \delta y'\,dx$$
$$= \frac{m}{\sqrt{1 + m^2}} \left[\delta y(1) - \delta y(0) \right] \;=\; 0,$$

and

$$\left(\frac{d^2}{d\epsilon^2} \mathcal{L}[y_0 + \epsilon\,\delta y] \right)_{\epsilon=0} = \int_0^1 \frac{(\delta y')^2}{(1 + m^2)^{3/2}}\,dx \;>\; 0,$$

which holds for a fixed value of m and all variations $\delta y(x)$ that satisfy the conditions $\delta y(0) = 0 = \delta y(1)$. Hence, we have shown that $y(x) = mx$ minimizes the length integral (1.1) since the first derivative (with respect to ϵ) vanishes at $\epsilon = 0$, while its second derivative is positive at $\epsilon = 0$. We note, however, that our task was made easier by our knowledge of the actual minimizing function $y_0(x) = mx$; without this knowledge, we would be required to choose a trial function $y_0(x)$ and test for all variations $\delta y(x)$ that vanish at the integration boundaries.

Another way to tackle this minimization problem is to find a way to characterize the function $y_0(x)$ that minimizes the length integral (1.1), for

---

[1]The trigonometric and hyperbolic-trigonometric substitutions are used extensively in this textbook and reviewed in App. A.

all variations $\delta y(x)$, without actually solving for $y(x)$. For example, the characteristic property of a straight line $y(x)$ is that its second derivative vanishes for all values of x. The methods of the Calculus of Variations introduced in this Chapter present a mathematical procedure for transforming the problem of minimizing an integral to the problem of finding the solution to an ordinary differential equation for $y(x)$. The mathematical foundations of the Calculus of Variations were developed by Leonhard Euler (1707-1783) and Joseph-Louis Lagrange (1736-1813), who developed the mathematical method for finding curves that minimize (or maximize) certain integrals.

1.1.2 *Methods of the Calculus of Variations*

1.1.2.1 *Euler's First Equation*

The methods of the Calculus of Variations transform the problem of minimizing (or maximizing) an integral of the form

$$\mathcal{F}[y] = \int_a^b F(y, y'; x)\, dx \qquad (1.2)$$

(with fixed boundary points a and b) into the solution of a differential equation for $y(x)$ expressed in terms of derivatives of the integrand $F(y, y'; x)$, which is assumed to be a smooth function of $y(x)$ and its first derivative $y'(x)$, with a possible explicit dependence on x. See problem 1 at the end of this Chapter for a generalization of Eq. (1.2).

The problem of *extremizing* the integral (1.2) will be treated in analogy with the problem of finding the extremal value of any (smooth) function $f(x)$, i.e., finding the value x_0 such that

$$f'(x_0) = \lim_{\epsilon \to 0} \frac{1}{\epsilon}\left(f(x_0 + \epsilon) - f(x_0) \right) \equiv \frac{1}{h}\left(\frac{d}{d\epsilon} f(x_0 + \epsilon\, h) \right)_{\epsilon=0} = 0,$$

where $h \neq 0$ is an arbitrary constant factor.[2] First, we introduce the first-

---

[2] An *extremum* point refers to either the minimum or maximum of a one-variable function. A *critical* point, on the other hand, refers to a point where the gradient of a multi-variable function vanishes. Critical points include minima and maxima as well as saddle points (where the function exhibits maxima in some directions and minima in other directions). A function $y(x)$ is said to be a *stationary* solution of the functional (1.2) if the first variation (1.3) vanishes for all variations δy that satisfy the boundary conditions.

order *functional variation* $\delta\mathcal{F}[y; \delta y]$ defined as

$$\delta\mathcal{F}[y; \delta y] \equiv \left(\frac{d}{d\epsilon}\mathcal{F}[y + \epsilon\,\delta y]\right)_{\epsilon=0}$$

$$= \left[\frac{d}{d\epsilon}\left(\int_a^b F\left(y + \epsilon\,\delta y, y' + \epsilon\,\delta y', x\right)\,dx\right)\right]_{\epsilon=0}, \quad (1.3)$$

where $\delta y(x)$ is an arbitrary smooth variation of the path $y(x)$ subject to the boundary conditions $\delta y(a) = 0 = \delta y(b)$, i.e., the end points of the path are not affected by the variation (see Fig. 1.1). By performing the

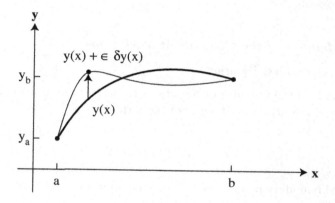

Fig. 1.1   Virtual displacement $\delta y(x)$ for the functional variation (1.3).

ϵ-derivatives in the functional variation (1.3), which involves partial derivatives of $F(y, y', x)$ with respect to y and y', we find

$$\delta\mathcal{F}[y; \delta y] = \int_a^b \left[\delta y(x)\,\frac{\partial F}{\partial y(x)} + \delta y'(x)\,\frac{\partial F}{\partial y'(x)}\right]dx.$$

When the second term is integrated by parts, we obtain

$$\delta\mathcal{F}[y; \delta y] = \int_a^b \delta y \left[\frac{\partial F}{\partial y} - \frac{d}{dx}\left(\frac{\partial F}{\partial y'}\right)\right]dx$$

$$+ \left[\delta y_b \left(\frac{\partial F}{\partial y'}\right)_b - \delta y_a \left(\frac{\partial F}{\partial y'}\right)_a\right]. \quad (1.4)$$

Here, since the variation $\delta y(x)$ vanishes at the integration boundaries ($\delta y_b = 0 = \delta y_a$), the last terms involving δy_b and δy_a vanish explicitly and Eq. (1.4) becomes

$$\delta\mathcal{F}[y; \delta y] = \int_a^b \delta y \left[\frac{\partial F}{\partial y} - \frac{d}{dx}\left(\frac{\partial F}{\partial y'}\right)\right]dx \equiv \int_a^b \delta y\,\frac{\delta\mathcal{F}}{\delta y}\,dx, \quad (1.5)$$

where $\delta\mathcal{F}/\delta y$ is called the *functional derivative* of $\mathcal{F}[y]$ with respect to the function y. The stationarity condition

$$\delta\mathcal{F}[y; \delta y] = 0 \tag{1.6}$$

for all variations δy yields Euler's First equation

$$\frac{d}{dx}\left(\frac{\partial F}{\partial y'}\right) \equiv y''\,\frac{\partial^2 F}{\partial y'\,\partial y'} + y'\,\frac{\partial^2 F}{\partial y\,\partial y'} + \frac{\partial^2 F}{\partial x\,\partial y'} = \frac{\partial F}{\partial y}, \tag{1.7}$$

which represents a second-order ordinary differential equation for $y(x)$. According to the Calculus of Variations, the solution $y(x)$ to this ordinary differential equation, subject to the boundary conditions $y(a) = y_a$ and $y(b) = y_b$, yields a solution to the problem of minimizing (or maximizing) the integral (1.2). Lastly, we note that Lagrange's variation operator δ, while analogous to the derivative operator d, commutes with the integral operator, i.e.,

$$\delta\int_a^b P(y(x))\,dx = \int_a^b \delta y(x)\,P'(y(x))\,dx,$$

for any smooth function P.

1.1.2.2 *Extremal Values of an Integral*

Euler's First Equation (1.7), which results from the stationarity condition (1.6), does not necessarily imply that the Euler path $y(x)$, in fact, minimizes the integral (1.2). To investigate whether the path $y(x)$ actually minimizes Eq. (1.2), we must evaluate the second-order functional variation

$$\delta^2\mathcal{F}[y; \delta y] \equiv \left(\frac{d^2}{d\epsilon^2}\mathcal{F}[y + \epsilon\,\delta y]\right)_{\epsilon=0},$$

and investigate its sign. By following steps similar to the derivation of Eq. (1.5), the second-order variation is expressed as

$$\delta^2\mathcal{F}[y; \delta y] = \int_a^b \left\{ \delta y^2 \left[\frac{\partial^2 F}{\partial y^2} - \frac{d}{dx}\left(\frac{\partial^2 F}{\partial y \partial y'}\right)\right] + (\delta y')^2\,\frac{\partial^2 F}{\partial(y')^2} \right\} dx, \tag{1.8}$$

after integration by parts was performed. The necessary and sufficient condition for a minimum is $\delta^2\mathcal{F} > 0$ and, thus, the sufficient conditions for a minimal integral are

$$\frac{\partial^2 F}{\partial y^2} - \frac{d}{dx}\left(\frac{\partial^2 F}{\partial y \partial y'}\right) > 0 \quad \text{and} \quad \frac{\partial^2 F}{(\partial y')^2} > 0, \tag{1.9}$$

for all smooth variations $\delta y(x)$. For a small enough interval (a, b), however, the $(\delta y')^2$-term will normally dominate over the $(\delta y)^2$-term and the sufficient condition becomes $\partial^2 F/(\partial y')^2 > 0$ (Legendre's Condition [6]).

Because variational problems often involve finding the minima or maxima of certain integrals, the methods of the Calculus of Variations enable us to find extremal solutions $y_0(x)$ for which the integral $\mathcal{F}[y]$ is stationary (i.e., $\delta \mathcal{F}[y_0] = 0$), without specifying whether the second-order variation is positive-definite (corresponding to a minimum), negative-definite (corresponding to a maximum), or with indefinite sign (i.e., when the coefficients of $(\delta y)^2$ and $(\delta y')^2$ have opposite signs).

1.1.2.3 *Jacobi Equation**

> Warning: Material identified by an asterisk is not meant to be covered in an undergraduate-level course and can be skipped without loss of continuity.

Carl Gustav Jacobi (1804-1851) derived a useful differential equation describing the deviation $u(x) = \overline{y}(x) - y(x)$ between two extremal curves that solve Euler's First Equation (1.7) for a given function $F(x, y, y')$. Upon Taylor expanding Euler's First Equation (1.7) for $\overline{y} = y + u$ and keeping only linear terms in u (which is assumed to be small in the interval (a, b)), we easily obtain the linear ordinary differential equation

$$\frac{d}{dx}\left(u' \frac{\partial^2 F}{(\partial y')^2} + u \frac{\partial^2 F}{\partial y \partial y'} \right) = u \frac{\partial^2 F}{\partial y^2} + u' \frac{\partial^2 F}{\partial y' \partial y}. \tag{1.10}$$

By performing the x-derivative on the second term on the left side, we obtain a partial cancellation with the second term on the right side and obtain the Jacobi equation [6]

$$\frac{d}{dx}\left(\frac{\partial^2 F}{(\partial y')^2} \frac{du}{dx} \right) = u \left[\frac{\partial^2 F}{\partial y^2} - \frac{d}{dx}\left(\frac{\partial^2 F}{\partial y \partial y'} \right) \right]. \tag{1.11}$$

We immediately see that the extremal properties (1.9) of the solutions of Euler's First Equation (1.7) are intimately connected to the behavior of the deviation $u(x)$ between two nearby extremal curves.

We note that the differential equation (1.10) may be derived from the variational principle $\delta \int J(u, u')\, dx = 0$ as the Jacobi-Euler equation

$$\frac{d}{dx}\left(\frac{\partial J}{\partial u'} \right) = \frac{\partial J}{\partial u}, \tag{1.12}$$

where the Jacobi function $J(u, u'; x)$ is defined as

$$J(u, u') \equiv \frac{1}{2} \left(\frac{d^2}{d\epsilon^2} F(y + \epsilon u, y' + \epsilon u') \right)_{\epsilon=0}$$

$$\equiv \frac{u^2}{2} \frac{\partial^2 F}{\partial y^2} + u u' \frac{\partial^2 F}{\partial y \partial y'} + \frac{u'^2}{2} \frac{\partial^2 F}{(\partial y')^2}. \tag{1.13}$$

For example, for $F(y, y') = \sqrt{1 + (y')^2}$, we find $\partial^2 F/\partial y^2 = 0 = \partial^2 F/\partial y \partial y'$ and $\partial^2 F/\partial (y')^2 = \Lambda^{-3}$, where $\Lambda \equiv \sqrt{1 + m^2}$ for the extremal solution $y(x) = m x$. The Jacobi function (1.13) for this case is $J(u, u') = \frac{1}{2} (u')^2/\Lambda^3$ and the Jacobi equation (1.11) becomes $(\Lambda^{-3} u')' = 0$, or $u'' = 0$ (i.e., deviations between straight lines diverge linearly).

Lastly, the second functional variation (1.8) can be combined with the Jacobi equation (1.11) to yield the expression [6]

$$\delta^2 \mathcal{F}[y; \delta y] = \int_a^b \frac{\partial^2 F}{(\partial y')^2} \left(\delta y' - \delta y \frac{u'}{u} \right)^2 dx, \tag{1.14}$$

where $u(x)$ is a solution of the Jacobi equation (1.11). We note that the minimum condition $\delta^2 \mathcal{F} > 0$ is now clearly associated with the Legendre condition $\partial^2 F/\partial (y')^2 > 0$. Furthermore, we note that the Jacobi equation describing space-time geodesic deviations plays a fundamental role in Einstein's Theory of General Relativity.[3] We shall return to the Jacobi equation (1.11) in Sec. 1.4, where we briefly discuss Fermat's Principle of Least Time and its applications to the general theory of geometric optics.

1.1.2.4 *Euler's Second Equation*

Whenever the function F appearing in Eq. (1.2) satisfies the condition $\partial F/\partial x \equiv 0$, we may obtain a partial solution to Euler's First Equation (1.7) as follows. First, we write the exact x-derivative of $F(y, y'; x)$ as

$$\frac{dF}{dx} = \frac{\partial F}{\partial x} + y' \frac{\partial F}{\partial y} + y'' \frac{\partial F}{\partial y'},$$

and substitute Euler's First Equation (1.7) for $\partial F/\partial y$ in order to combine the last two terms, so that we obtain Euler's Second equation

$$\frac{d}{dx} \left(F - y' \frac{\partial F}{\partial y'} \right) = \frac{\partial F}{\partial x}. \tag{1.15}$$

This equation is especially useful when the integrand $F(y, y')$ in Eq. (1.2) is independent of x (i.e., $\partial F/\partial x \equiv 0$), for which Eq. (1.15) yields the solution

$$F(y, y') - y' \frac{\partial F}{\partial y'}(y, y') = \alpha, \tag{1.16}$$

---

[3]See, for example, S. Weinberg, *Gravitation and Cosmology: Principles and Applications of the General Theory of Relativity* (Wiley 1972).

where the constant α is determined from the conditions $y(x_0) = y_0$ and $y'(x_0) = y'_0$. Here, Eq. (1.16) is a *partial* solution (in some sense) of Eq. (1.7), since we have reduced the derivative order from a second-order derivative $y''(x)$ in Eq. (1.7) to a first-order derivative $y'(x)$ in Eq. (1.16) on the solution $y(x)$. Hence, Euler's Second Equation has produced an equation of the form

$$G(y, y'; \alpha) \equiv F(y, y') - y' \frac{\partial F}{\partial y'}(y, y') - \alpha = 0,$$

which can often be integrated by *quadrature* (as we shall see later) by solving for y' as a function of y and x. For example, for $F(y, y') = \sqrt{1 + (y')^2}$, we find $F(y, y') - y' \, \partial F(y, y')/\partial y' = 1/\sqrt{1 + (y')^2} = \alpha$, which yields $y' = \pm \alpha^{-1} \sqrt{1 - \alpha^2}$.

1.1.3   *Path of Shortest Distance and Geodesic Equation*

We now return to the problem of minimizing the length integral (1.1), with the integrand written as $F(y, y') = \sqrt{1 + (y')^2}$. Here, Euler's First Equation (1.7) yields

$$\frac{d}{dx}\left(\frac{\partial F}{\partial y'}\right) = \frac{y''}{[1 + (y')^2]^{3/2}} = \frac{\partial F}{\partial y} = 0,$$

so that the function $y(x)$ that minimizes the length integral (1.1) is the solution of the differential equation $y''(x) = 0$ subject to the boundary conditions $y(0) = 0$ and $y(1) = m$, i.e., the extremal solution is $y(x) = m\,x$. Note that the integrand $F(y, y')$ also satisfies the sufficient minimum conditions (1.9) so that the path $y(x) = m\,x$ is indeed the path of shortest distance between two points on the plane.

1.1.3.1   *Geodesic Equation**

We generalize the problem of finding the path of shortest distance on the Euclidean plane (x, y) to the problem of finding *geodesic* paths in arbitrary geometry because it introduces important geometric concepts in Classical Mechanics needed in later chapters. For this purpose, let us consider a path in n-dimensional space from point $\mathbf{x}_A$ to point $\mathbf{x}_B$ parameterized by the continuous parameter σ: $\mathbf{x}(\sigma)$ such that $\mathbf{x}(A) = \mathbf{x}_A$ and $\mathbf{x}(B) = \mathbf{x}_B$. The length integral from point A to B is

$$\mathcal{L}[\mathbf{x}] = \int_A^B \left(g_{ij}\frac{dx^i}{d\sigma}\frac{dx^j}{d\sigma}\right)^{1/2} d\sigma, \tag{1.17}$$

where the space metric g_{ij} is defined so that the squared infinitesimal length element is $ds^2 \equiv g_{ij}(\mathbf{x})\,dx^i\,dx^j$ (summation over repeated indices is implied throughout the text).

Next, using the definition (1.3), the first-order variation $\delta\mathcal{L}[\mathbf{x}]$ is given as

$$\delta\mathcal{L}[\mathbf{x}] = \frac{1}{2}\int_A^B \left[\frac{\partial g_{ij}}{\partial x^k}\,\delta x^k\,\frac{dx^i}{d\sigma}\frac{dx^j}{d\sigma} + 2\,g_{ij}\,\frac{d\delta x^i}{d\sigma}\frac{dx^j}{d\sigma}\right]\frac{d\sigma}{ds/d\sigma}$$

$$= \frac{1}{2}\int_a^b \left[\frac{\partial g_{ij}}{\partial x^k}\,\delta x^k\,\frac{dx^i}{ds}\frac{dx^j}{ds} + 2\,g_{ij}\,\frac{d\delta x^i}{ds}\frac{dx^j}{ds}\right]ds,$$

where $a = s(A)$ and $b = s(B)$ and we have performed a parameterization change: $\mathbf{x}(\sigma) \rightarrow \mathbf{x}(s)$. By integrating the second term by parts (with $\delta\mathbf{x}$ vanishing at the end points), we obtain

$$\delta\mathcal{L}[\mathbf{x}] = -\int_a^b \left[\frac{d}{ds}\left(g_{ij}\frac{dx^j}{ds}\right) - \frac{1}{2}\frac{\partial g_{jk}}{\partial x^i}\frac{dx^j}{ds}\frac{dx^k}{ds}\right]\delta x^i\,ds \qquad (1.18)$$

$$= -\int_a^b \left[g_{ij}\frac{d^2 x^j}{ds^2} + \left(\frac{\partial g_{ij}}{\partial x^k} - \frac{1}{2}\frac{\partial g_{jk}}{\partial x^i}\right)\frac{dx^j}{ds}\frac{dx^k}{ds}\right]\delta x^i\,ds.$$

We now note that, using symmetry properties under interchange of the j-k indices, the second term in Eq. (1.18) can also be written as

$$\left(\frac{\partial g_{ij}}{\partial x^k} - \frac{1}{2}\frac{\partial g_{jk}}{\partial x^i}\right)\frac{dx^j}{ds}\frac{dx^k}{ds} = \frac{1}{2}\left(\frac{\partial g_{ij}}{\partial x^k} + \frac{\partial g_{ik}}{\partial x^j} - \frac{\partial g_{jk}}{\partial x^i}\right)\frac{dx^j}{ds}\frac{dx^k}{ds}$$

$$= \Gamma_{i|jk}\,\frac{dx^j}{ds}\frac{dx^k}{ds},$$

using the definition of the Christoffel symbol

$$\Gamma^\ell_{jk} = g^{\ell i}\,\Gamma_{i|jk} = \frac{1}{2}g^{\ell i}\left(\frac{\partial g_{ij}}{\partial x^k} + \frac{\partial g_{ik}}{\partial x^j} - \frac{\partial g_{jk}}{\partial x^i}\right) \equiv \Gamma^i_{kj}, \qquad (1.19)$$

where g^{ij} denotes a component of the inverse metric (i.e., $g^{ij}\,g_{jk} = \delta^i{}_k$). Hence, the first-order variation (1.18) can be expressed as

$$\delta\mathcal{L}[\mathbf{x}] = \int_a^b \left[\frac{d^2 x^i}{ds^2} + \Gamma^i_{jk}\frac{dx^j}{ds}\frac{dx^k}{ds}\right]g_{i\ell}\,\delta x^\ell\,ds. \qquad (1.20)$$

The stationarity condition $\delta\mathcal{L} = 0$ for arbitrary variations δx^ℓ yields an equation for the path $\mathbf{x}(s)$ of shortest distance known as the *geodesic* equation

$$\frac{d^2 x^i}{ds^2} + \Gamma^i_{jk}\frac{dx^j}{ds}\frac{dx^k}{ds} = 0. \qquad (1.21)$$

Returning to two-dimensional Euclidean geometry, where the components of the metric tensor are constants (i.e., $ds^2 = dx^2 + dy^2$), the geodesic equations are $x''(s) = 0 = y''(s)$, which once again leads to a straight line.

1.1.3.2   *Geodesic Equation on a Sphere*

We now consider the example of geodesic curves on the surface of a sphere of radius R, which are known to be great circles as we will now show. According to Eq. (1.17), they are extremal curves of the length functional

$$\mathcal{L}[\varphi] = \int R\sqrt{1 + \sin^2\theta \left(\frac{d\varphi}{d\theta}\right)^2} \, d\theta \equiv R \int L(\varphi', \theta) \, d\theta, \qquad (1.22)$$

where the azimuthal angle $\varphi(\theta)$ is an arbitrary function of the polar angle θ. Since the function $L(\varphi', \theta)$ in Eq. (1.22) is independent of the azimuthal angle φ, its corresponding Euler's First Equation (1.7) yields the partial solution

$$\frac{\partial L}{\partial \varphi'} = \frac{\sin^2\theta\, \varphi'}{\sqrt{1 + \sin^2\theta\, (\varphi')^2}} = \sin\alpha,$$

where α is an arbitrary constant angle. Solving for φ' we find

$$\varphi'(\theta) = \frac{\sin\alpha}{\sin\theta \sqrt{\sin^2\theta - \sin^2\alpha}},$$

which can, thus, be integrated to give

$$\varphi - \beta = \int \frac{\sin\alpha \, d\theta}{\sin\theta \sqrt{\sin^2\theta - \sin^2\alpha}} = -\int \frac{\tan\alpha \, du}{\sqrt{1 - u^2 \tan^2\alpha}},$$

where β is another constant angle and we used the change of variable $u = \cot\theta$. A simple trigonometric substitution finally yields

$$\cos(\varphi - \beta) = \tan\alpha \cot\theta, \qquad (1.23)$$

which describes a great circle on the surface of the sphere. We easily verify this statement by converting Eq. (1.23) into the equation for a plane that passes through the origin:

$$z \sin\alpha = x \cos\alpha \cos\beta + y \cos\alpha \sin\beta.$$

The intersection of this plane with the unit sphere is expressed in terms of the coordinate functions $x = \sin\alpha \cos\beta$, $y = \sin\alpha \sin\beta$, and $z = \cos\alpha$.

1.2   Classical Variational Problems

The development of the Calculus of Variations led to the resolution of several classical optimization problems in mathematics and physics [2, 17]. In this Section, we present two classical variational problems that are connected to its original development. First, in the isoperimetric problem, we show how Lagrange modified Euler's formulation of the Calculus of Variations by allowing constraints to be imposed on the search for finding extremal values of certain integrals. Next, in the brachistochrone problem, we show how the Calculus of Variations is used to find the path of *quickest* descent for a bead sliding along a frictionless wire under the action of gravity.

1.2.1 *Isoperimetric Problem*

Isoperimetric problems represent some of the earliest applications of the variational approach to solving mathematical optimization problems. Pappus (ca. 290-350) was among the first to recognize that among all the isoperimetric closed planar curves (i.e., closed curves that have the same perimeter length), the circle encloses the greatest area.[4] The variational formulation of the (planar) isoperimetric problem requires that we maximize the area integral $A = \int y(x)\,dx$ while keeping the perimeter length integral $L = \int \sqrt{1 + (y')^2}\,dx$ constant.

The isoperimetric problem falls in a class of variational problems called *constrained* variational principles, where a certain functional $\int f(y, y', x)\,dx$ is to be optimized under the constraint that another functional $\int g(y, y', x)\,dx$ be held constant (say at value G). The constrained variational principle is then expressed in terms of the functional

$$\mathcal{F}_\lambda[y] = \int f(y, y', x)\,dx \ + \ \lambda \left(G \ - \ \int g(y, y', x)\,dx \right)$$
$$= \int \left[f(y, y', x) \ - \ \lambda\,g(y, y', x) \right]\,dx \ + \ \lambda\,G, \qquad (1.24)$$

where the parameter λ is called a Lagrange *multiplier*. Note that the functional $\mathcal{F}_\lambda[y]$ is chosen, on the one hand, so that the derivative

$$\frac{d\mathcal{F}_\lambda[y]}{d\lambda} \ = \ G \ - \ \int g(y, y', x)\,dx \ = \ 0$$

enforces the constraint for all curves $y(x)$. On the other hand, the stationarity condition $\delta \mathcal{F}_\lambda = 0$ for the functional (1.24) with respect to arbitrary

---

[4]Such results are normally described in terms of the so-called isoperimetric inequalities $4\pi A \le L^2$, where A denotes the area enclosed by a closed curve of perimeter length L; here, equality is satisfied by the circle, where $L = 2\pi a$ and $A = \pi a^2$.

variations $\delta y(x)$ (which vanish at the integration boundaries) yields Euler's First Equation:

$$\frac{d}{dx}\left(\frac{\partial f}{\partial y'} - \lambda\,\frac{\partial g}{\partial y'}\right) = \frac{\partial f}{\partial y} - \lambda\,\frac{\partial g}{\partial y}. \qquad (1.25)$$

Here, we assume that this second-order differential equation is to be solved (at fixed λ) subject to the conditions $y(x_0) = y_0$ and $y'(x_0) = 0$; the solution $y(x; \lambda)$ of Eq. (1.25) is, however, parameterized by the Lagrange multiplier λ.

If the integrands $f(y, y')$ and $g(y, y')$ in Eq. (1.24) are both explicitly independent of x, then Euler's Second Equation (1.16) for the functional (1.24) becomes

$$\frac{d}{dx}\left[\left(f - y'\,\frac{\partial f}{\partial y'}\right) - \lambda\left(g - y'\,\frac{\partial g}{\partial y'}\right)\right] = 0. \qquad (1.26)$$

By integrating this equation we obtain

$$\left(f - y'\,\frac{\partial f}{\partial y'}\right) - \lambda\left(g - y'\,\frac{\partial g}{\partial y'}\right) = f_0 - \lambda\,g_0 = 0,$$

where the constant of integration on the right is chosen from the conditions $y(x_0) = y_0$ and $y'(x_0) \equiv 0$ (i.e., x_0 is an extremum point of $y(x)$), so that the value of the constant Lagrange multiplier is now defined as $\lambda = f(y_0, 0)/g(y_0, 0) \equiv f_0/g_0$. Hence, the solution $y(x; \lambda)$ of the constrained variational problem (1.24) is now uniquely determined.

We return to the isoperimetric problem now represented in terms of the constrained functional

$$\mathcal{A}_\lambda[y] = \int y\,dx + \lambda\left(L - \int \sqrt{1 + (y')^2}\,dx\right)$$

$$= \int \left[y - \lambda\,\sqrt{1 + (y')^2}\right]\,dx + \lambda\,L, \qquad (1.27)$$

where L denotes the value of the constant-length constraint. From Eq. (1.25), the stationarity of the functional (1.27) with respect to arbitrary variations $\delta y(x)$ yields

$$\frac{d}{dx}\left(-\frac{\lambda\,y'}{\sqrt{1 + (y')^2}}\right) = 1,$$

which can be integrated to give

$$-\frac{\lambda\,y'}{\sqrt{1 + (y')^2}} = x - x_0, \qquad (1.28)$$

where x_0 denotes a constant of integration associated with $y'(x_0) = 0$. Since the integrands $f(y, y') = y$ and $g(y, y') = \sqrt{1 + (y')^2}$ are both explicitly independent of x, then Euler's Second Equation (1.26) applies, and we obtain

$$\frac{d}{dx}\left(y - \frac{\lambda}{\sqrt{1 + (y')^2}}\right) = 0,$$

which can be integrated to give

$$\frac{\lambda}{\sqrt{1 + (y')^2}} = y, \qquad (1.29)$$

where we chose $y(x_0) = \lambda$ with $y'(x_0) = 0$.

Lastly, by combining Eqs. (1.28) and (1.29), we obtain $y\,y' + (x - x_0) = 0$, which can be integrated to give $y^2(x) = \lambda^2 - (x - x_0)^2$. We immediately recognize that the maximal isoperimetric curve $y(x)$ is a circle of radius $r = \lambda$, centered at $(x, y) = (x_0, 0)$, with perimeter length $L = 2\pi\lambda$ and maximal enclosed area $A = \pi\lambda^2 = L^2/4\pi$.

1.2.2 Brachistochrone Problem

The brachistochrone problem is a *least-time* variational problem, which was first solved in 1696 by Jean (Johann) Bernoulli (1667-1748). The problem can be stated as follows. A bead is released from rest (at the origin in Fig. 1.2) and slides down a frictionless wire that connects the origin to a given point (x_f, y_f). The question posed by the brachistochrone problem is to determine the shape $y(x)$ of the wire for which the frictionless descent of the bead under gravity takes the shortest amount of time.

Using the (x, y)-coordinates set up in Fig. 1.2, the speed of the bead after it has fallen a vertical distance x along the wire is $v = \sqrt{2gx}$ (where g denotes the gravitational acceleration) and, thus, the time integral

$$\mathcal{T}[y] = \int \frac{ds}{v} = \int_0^{x_f} \sqrt{\frac{1 + (y')^2}{2\,gx}}\,dx = \int_0^{x_f} F(y, y', x)\,dx, \qquad (1.30)$$

is a functional of the path $y(x)$. Note that, in the absence of friction, the bead's mass does not enter into the problem. Since the integrand of Eq. (1.30) is independent of the y-coordinate ($\partial F/\partial y = 0$), Euler's First Equation (1.7) simply yields

$$\frac{d}{dx}\left(\frac{\partial F}{\partial y'}\right) = 0 \quad \rightarrow \quad \frac{\partial F}{\partial y'} = \frac{y'}{\sqrt{2\,gx}\,[1 + (y')^2]} = \alpha,$$

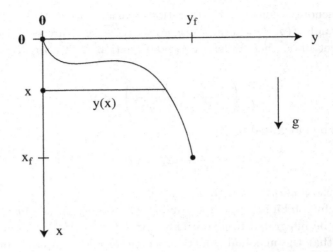

Fig. 1.2 Brachistochrone problem.

where α is a constant, which can be rewritten in terms of the scale length $\ell = (2\alpha^2 g)^{-1}$ as

$$\frac{(y')^2}{1 + (y')^2} = \frac{x}{\ell} \;\rightarrow\; (y')^2 = \frac{x}{(\ell - x)}.$$

Integration by quadrature yields the solution

$$y(x) = \int_0^x \sqrt{\frac{s}{\ell - s}}\, ds,$$

subject to the initial condition $y(x = 0) = 0$. Using the trigonometric substitution (with $\ell = 2a$)

$$s = 2a \sin^2(\theta/2) = a\,(1 - \cos\theta),$$

we obtain the parametric solution

$$x_{\mathrm{B}}(\theta) = a\,(1 - \cos\theta) \tag{1.31}$$

and

$$y_{\mathrm{B}}(\theta) = \int_0^\theta \sqrt{\frac{1 - \cos\theta}{1 + \cos\theta}}\, a\,\sin\theta\, d\theta$$

$$= a \int_0^\theta (1 - \cos\theta)\, d\theta = a\,(\theta - \sin\theta). \tag{1.32}$$

This solution yields a parametric representation of the *cycloid* (Fig. 1.3) where the bead is placed on a rolling hoop of radius a.

Fig. 1.3 Brachistochrone solution.

Using the parametric solution (1.31)-(1.32) for the brachistochrone problem, we calculate the time taken to go from the origin $(0,0)$ to the end point $(2a, \pi a)$ to be

$$\mathcal{T}[x_B, y_B] = \int_0^\pi \sqrt{\frac{(x_B')^2 + (y_B')^2}{2g\, x_B}}\, d\theta = \int_0^\pi \sqrt{\frac{a^2\,[\sin^2\theta + (1 - \cos\theta)^2]}{2\,ag\,(1 - \cos\theta)}}\, d\theta$$

$$= \sqrt{\frac{a}{g}} \int_0^\pi d\theta = \pi\,\sqrt{\frac{a}{g}} \equiv T_B. \tag{1.33}$$

By comparison, we note that the time taken by following a straight-line path $(y_L = \pi\, x/2)$ between these two points would be

$$\mathcal{T}[y_L] = \int_0^{2a} \sqrt{\frac{1 + (\pi/2)^2}{2g\, x}}\, dx = \sqrt{\pi^2 + 4}\,\sqrt{\frac{a}{g}} > T_B,$$

which is longer than the minimal time (1.33).

1.3 Fermat's Principle of Least Time

Several *minimum* principles have been invoked throughout the history of Physics to explain the behavior of light and particles. In one of its earliest form, Hero of Alexandria (ca. 75 AD) stated that *light travels in a straight line* and that light follows a path of shortest distance when it is reflected. In 1657, Pierre de Fermat (1601-1665) stated the Principle of *Least Time*, whereby light travels between two points along a path that minimizes the travel time, to explain Snell's Law (Willebrord Snell, 1591-1626) associated with light refraction in a stratified medium. Using the index of refraction $n_0 \geq 1$ of the uniform medium, the speed of light in the medium is expressed as $v_0 = c/n_0 \leq c$, where c is the speed of light in vacuum. This straight-line

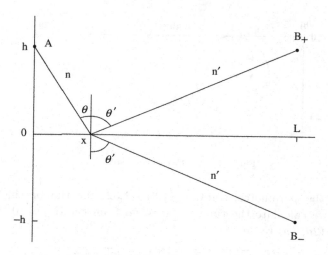

Fig. 1.4   Reflection path (AxB_+) and refraction path (AxB_-) for light propagating in a stratified medium $(n' < n)$.

path in a uniform medium is not only a path of shortest distance but also a path of least time.

The laws of reflection and refraction as light propagates in uniform media separated by sharp boundaries (see segments AxB_+ and AxB_- in Fig. 1.4) are easily formulated as minimization problems as follows. The time taken by light to go from point $A = (0, h)$ to point $B_\pm = (L, \pm h)$ after being reflected or refracted at point $(x, 0)$ is given by

$$T_{AB}(x) = c^{-1} \left[n \sqrt{x^2 + h^2} + n' \sqrt{(L - x)^2 + h^2} \right],$$

where n and n' denote the indices of refraction of the medium along path Ax and $xB_\pm$, respectively. We easily evaluate the derivative of $T_{AB}(x)$ to find

$$\frac{dT_{AB}(x)}{dx} = c^{-1} \left[n \frac{x}{\sqrt{x^2 + h^2}} - n' \frac{(L - x)}{\sqrt{(L - x)^2 + h^2}} \right]$$

$$\equiv c^{-1} \left(n \sin \theta - n' \sin \theta' \right), \tag{1.34}$$

where the angles θ and θ' are defined in Fig. 1.4. Here, the law of reflection ($n' = n$ and $B = B_+$ in Fig. 1.4) is expressed in terms of the extremum condition $T'_{AB}(x) = 0$, which implies that the path of least time is obtained when the reflected angle θ' is equal to the incidence angle θ (or $x = L/2$ in Fig. 1.4).

Next, the extremum condition $T'_{AB}(x) = 0$ for refraction ($n' \neq n$ and $B = B_-$ in Fig. 1.4) yields Snell's law of refraction

$$n \sin\theta = n' \sin\theta'. \tag{1.35}$$

Note that Snell's law implies that the refracted light ray bends toward the medium with the largest index of refraction ($n > n'$ in Fig. 1.4). In what follows, we generalize Snell's law to describe the case of light refraction in a continuous nonuniform medium.

Before proceeding with this general case, we note that the second derivative is strictly positive:

$$\frac{d^2 T_{AB}(x)}{dx^2} = \frac{1}{hc} \left(n \cos^3\theta + n' \cos^3\theta' \right) > 0,$$

which proves that the paths of light reflection and refraction in a flat stratified medium are indeed paths of minimal optical lengths. For some curved reflecting surfaces, however, the reflected path corresponds to a path of maximum optical length (see problem 16). This example emphasizes the fact that Fermat's Principle is in fact a principle of *stationary* time.

1.3.1 *Light Propagation in a Nonuniform Medium*

According to Fermat's Principle, light propagates in a nonuniform medium by traveling along a path that *minimizes* the travel time between an initial point A (where a light ray is launched) and a final point B (where the light ray is received). The time taken by a light ray following a path γ from point A to point B (parameterized by σ) is [3]

$$\mathcal{T}[\mathbf{x}] = \int c^{-1} n(\mathbf{x}) \left| \frac{d\mathbf{x}}{d\sigma} \right| d\sigma = c^{-1} \mathcal{L}_n[\mathbf{x}], \tag{1.36}$$

where $\mathcal{L}_n[\mathbf{x}]$ represents the length of the *optical* path taken by light as it travels in a nonuniform medium with refractive index $n(\mathbf{x})$, and

$$\left| \frac{d\mathbf{x}}{d\sigma} \right| = \sqrt{\left(\frac{dx}{d\sigma} \right)^2 + \left(\frac{dy}{d\sigma} \right)^2 + \left(\frac{dz}{d\sigma} \right)^2}. \tag{1.37}$$

Fermat's Principle of Least Time states that light traveling in a nonuniform medium follows an optical path $\mathbf{x}(\sigma)$ that is a stationary solution of the variational principle

$$\delta \mathcal{T}[\mathbf{x}] \equiv 0. \tag{1.38}$$

We now consider ray propagation in two dimensions (x, y), with the index of refraction $n(y)$, so that the optical length

$$\mathcal{L}_n[y] = \int_a^b n(y) \sqrt{1 + (y')^2} \, dx \qquad (1.39)$$

is a *functional* of $y(x)$. We shall return to the general properties of ray propagation in Sec. 1.4.

By applying the variational principle (1.38) for the case where $F(y, y') = n(y) \sqrt{1 + (y')^2}$, we find

$$\frac{\partial F}{\partial y'} = \frac{n(y) \, y'}{\sqrt{1 + (y')^2}} \quad \text{and} \quad \frac{\partial F}{\partial y} = n'(y) \sqrt{1 + (y')^2},$$

so that Euler's First Equation (1.7) becomes

$$n(y) \, y'' = n'(y) \left[1 + (y')^2 \right]. \qquad (1.40)$$

Although the solution of this (nonlinear) second-order ordinary differential equation is difficult to obtain for general functions $n(y)$, we can nonetheless obtain a qualitative picture of its solution by noting that y'' has the same sign as $n'(y)$. Hence, when $n'(y) = 0$ for all y (i.e., the medium is spatially uniform), the solution $y'' = 0$ yields the straight line $y(x; \varphi_0) = \tan \varphi_0 \, x$, where φ_0 denotes the initial launch angle (as measured from the horizontal axis). The case where $n'(y) > 0$ (or < 0), on the other hand, yields a light path which is concave upward, i.e., $y'' > 0$ (or downward, i.e., $y'' < 0$), as will be shown below.

Note that the sufficient conditions (1.9) for a minimal optical path are expressed as

$$\frac{\partial^2 F}{(\partial y')^2} = \frac{n}{[1 + (y')^2]^{3/2}} > 0,$$

which is satisfied for all refractive media, and

$$\frac{\partial^2 F}{\partial y^2} - \frac{d}{dx} \left(\frac{\partial^2 F}{\partial y \partial y'} \right) = n'' \sqrt{1 + (y')^2} - \frac{d}{dx} \left(\frac{n' \, y'}{\sqrt{1 + (y')^2}} \right)$$

$$= \frac{1}{F} \left(n \, n'' - (n')^2 \right) = \frac{n^2}{F} \frac{d^2 \ln n}{dy^2},$$

whose sign is indefinite. Hence, the sufficient condition for a minimal optical length for light traveling in a nonuniform refractive medium is $d^2 \ln n / dy^2 > 0$; note, however, that only the stationarity of the optical path is physically meaningful and, thus, we shall not discuss the minimal properties of light paths in what follows.

Since the function $F(y, y') = n(y) \sqrt{1 + (y')^2}$ is explicitly independent of x, Euler's Second Equation yields

$$F - y' \frac{\partial F}{\partial y'} = \frac{n(y)}{\sqrt{1 + (y')^2}} = \text{constant},$$

and, thus, the partial solution of Eq. (1.40) is

$$n(y) = N \sqrt{1 + (y')^2}, \tag{1.41}$$

where N is a constant determined from the initial conditions of the light ray. We note that Eq. (1.41) states that as a light ray enters a region of increased (decreased) refractive index, the slope of its path also increases (decreases). In particular, by substituting Eq. (1.40) into Eq. (1.41), we find $N^2 y'' = n(y) n'(y)$, and, hence, the path of a light ray is concave upward (downward) where $n'(y)$ is positive (negative), as previously discussed. Eq. (1.41) can be integrated by *quadrature* to give the integral solution

$$x(y) = \int_0^y \frac{N \, ds}{\sqrt{[n(s)]^2 - N^2}}, \tag{1.42}$$

subject to the condition $x(y = 0) = 0$. From the explicit dependence of the index of refraction $n(y)$, we may be able to perform the integration in Eq. (1.42) to obtain $x(y)$ and, thus, obtain an explicit solution $y(x)$ by inverting $x(y)$.

1.3.2 Snell's Law

We now show that the partial solution (1.41) corresponds to Snell's Law for light refraction in a nonuniform medium. Consider a light ray traveling in the (x, y)-plane launched from the initial position $(0, 0)$ at an initial tangent angle $\varphi_0 \leq \pi/2$ (measured from the x-axis) so that $y'(0) = \tan \varphi_0$ is the slope at $x = 0$. The constant N is then simply determined from Eq. (1.41) as $N = n_0 \cos \varphi_0$, where $n_0 = n(0)$ is the refractive index at $y(0) = 0$. Next, let $y'(x) = \tan \varphi(x)$ be the slope of the light ray at $(x, y(x))$, so that $\sqrt{1 + (y')^2} = \sec \varphi$ and Eq. (1.41) becomes $n(y) \cos \varphi = n_0 \cos \varphi_0$. Lastly, when we substitute the complementary angle $\theta = \pi/2 - \varphi$ (measured from the vertical y-axis), we obtain the *local* form of Snell's Law of refraction

$$n[y(x)] \sin \theta(x) = n_0 \sin \theta_0, \tag{1.43}$$

properly generalized to include a light path in a nonuniform refractive medium. Note that Snell's Law (1.43) does not tell us anything about the actual light path $y(x)$; this solution must come from solving Eq. (1.42).

1.3.3   *Application of Fermat's Principle*

In order to obtain an explicit solution (1.42) of Fermat's Principle in two
dimensions, we consider the propagation of a light ray in a medium with
linear refractive index

$$n(y) = n_0 (1 - \beta y) \tag{1.44}$$

exhibiting a constant gradient $n'(y) = - n_0 \beta$. Substituting this profile into
the optical-path solution (1.42), we find

$$x(y) = \int_0^y \frac{\cos \varphi_0 \, ds}{\sqrt{(1 - \beta s)^2 - \cos^2 \varphi_0}}. \tag{1.45}$$

Next, we use the trigonometric substitution

$$y(\varphi) = \frac{1}{\beta} \left(1 - \frac{\cos \varphi_0}{\cos \varphi} \right), \tag{1.46}$$

with $\varphi = \varphi_0$ at $(x, y) = (0, 0)$, so that Eq. (1.45) becomes

$$x(\varphi) = - \frac{\cos \varphi_0}{\beta} \ln \left(\frac{\sec \varphi + \tan \varphi}{\sec \varphi_0 + \tan \varphi_0} \right). \tag{1.47}$$

Fig. 1.5   Parametric solution (1.46)-(1.47) for $\varphi_0 = \pi/4$ and $\beta = 1$ (solid), $\beta = 1/2$
(dashed), and $\beta = 0$ (dotted). The plots are shown for (x, y) from $(0, 0)$ to $(\bar{x}, \bar{y})$, except
for $\beta = 0$, for which $y = x \tan \varphi_0$.

The *parametric* solution (1.46)-(1.47) for the optical path in a linear
medium (see Fig. 1.5) shows that the path reaches a maximum height $\bar{y} =
y(0)$ at a distance $\bar{x} = x(0)$ when the tangent angle φ is zero:

$$\bar{x} = \frac{\cos \varphi_0}{\beta} \ln(\sec \varphi_0 + \tan \varphi_0) \quad \text{and} \quad \bar{y} = \frac{1 - \cos \varphi_0}{\beta}.$$

The time taken for the light path parameterized by Eqs. (1.46)-(1.47) in going from the origin $(0, 0)$ to the end point $(\overline{x}, \overline{y})$ is a function of the initial angle φ_0:

$$
\begin{aligned}
\mathcal{T}(\varphi_0) &= \frac{n_0}{c} \int_0^{\varphi_0} \left(1 - \beta\, y(\varphi)\right) \sqrt{(x')^2 + (y')^2}\, d\varphi \\
&= \frac{n_0}{\beta c} \cos^2 \varphi_0 \int_0^{\varphi_0} \sec^3 \varphi\, d\varphi \\
&= \frac{n_0}{2\,\beta c} \left[\sin \varphi_0 + \cos^2 \varphi_0 \ln \left(\sec \varphi_0 + \tan \varphi_0 \right) \right].
\end{aligned}
\tag{1.48}
$$

Lastly, we note that by expressing $y(x; \beta)$ as a function of x, Eq. (1.46) becomes

$$
y(x; \beta) = \frac{1}{\beta} \left[1 - \cos \varphi_0 \cosh \left(\beta \sec \varphi_0 \left(x - \overline{x}\right) \right) \right].
\tag{1.49}
$$

In the uniform limit ($\beta = 0$), we use L'Hospital's rule on Eq. (1.49) to find the straight-line equation $y(x; 0) \equiv x \tan \varphi_0$.

1.4 Geometric Formulation of Ray Optics*

1.4.1 General Geometric Optics

We now return to the general formulation for light-ray propagation based on the time integral (1.36), where the integrand is

$$
F\left(\mathbf{x}, \frac{d\mathbf{x}}{d\sigma}\right) = n(\mathbf{x}) \left| \frac{d\mathbf{x}}{d\sigma} \right|,
$$

and light rays are allowed to travel in a three-dimensional refractive medium with a general index of refraction $n(\mathbf{x})$. Euler's First equation in this case is

$$
\frac{d}{d\sigma} \left(\frac{\partial F}{\partial (d\mathbf{x}/d\sigma)} \right) = \frac{\partial F}{\partial \mathbf{x}},
\tag{1.50}
$$

where

$$
\frac{\partial F}{\partial (d\mathbf{x}/d\sigma)} = \frac{n}{\Lambda} \frac{d\mathbf{x}}{d\sigma} \quad \text{and} \quad \frac{\partial F}{\partial \mathbf{x}} = \Lambda \, \nabla n,
$$

with $\Lambda = |d\mathbf{x}/d\sigma|$ given by Eq. (1.37). Euler's First Equation (1.50), therefore, yields the Euler-Fermat equation

$$
\frac{d}{d\sigma} \left(\frac{n}{\Lambda} \frac{d\mathbf{x}}{d\sigma} \right) = \Lambda \, \nabla n.
\tag{1.51}
$$

By choosing the ray parameterization $\Lambda = ds/d\sigma \equiv n$, the Euler-Fermat equation (1.51) becomes $d^2\mathbf{x}/d\sigma^2 = n\nabla n = \frac{1}{2}\nabla n^2$, which implies that the light ray is *accelerated* toward regions of higher index of refraction (see Fig. 1.6).

Euler's Second Equation, on the other hand, states that

$$H(\sigma) \equiv F\left(\mathbf{x}, \frac{d\mathbf{x}}{d\sigma}\right) - \frac{d\mathbf{x}}{d\sigma} \cdot \frac{\partial F}{\partial(d\mathbf{x}/d\sigma)} = 0$$

is a constant of motion. Note that, while Euler's Second Equation (1.41) proved very useful in providing an explicit solution (Snell's Law) to finding the optical path in a nonuniform medium with index of refraction $n(y)$, it appears that Euler's Second Equation $H(\sigma) \equiv 0$ now reveals no information about the optical path. Where did the information go? To answer this question, we apply the Euler-Fermat equation (1.51) to the two-dimensional case where $\sigma = x$ and $\Lambda = \sqrt{1 + (y')^2}$ with $\nabla n = n'(y)\,\hat{\mathbf{y}}$. Hence, the Euler-Fermat equation (1.51) becomes

$$\frac{d}{dx}\left[\frac{n}{\Lambda}\left(\hat{\mathbf{x}} + y'\,\hat{\mathbf{y}}\right)\right] = \Lambda n'\,\hat{\mathbf{y}},$$

from which we immediately conclude that Euler's Second Equation (1.41), $n = N\Lambda$, now appears as a constant of the motion $d(n/\Lambda)/dx = 0$ associated with a symmetry of the optical medium (i.e., the optical properties of the medium are invariant under translation along the x-axis). The association of symmetries with constants of the motion will later be discussed in terms of Noether's Theorem (see problem 17 and Sec. 2.5).

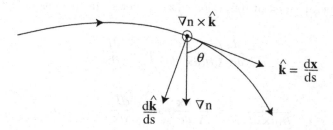

Fig. 1.6   Light-ray curvature $\kappa \equiv \hat{\mathbf{n}} \cdot d\hat{k}/ds$ and the Frenet-Serret frame $(\hat{k}, \hat{\mathbf{n}}, \hat{\mathbf{b}})$ following a light ray. The vectors $(\nabla n, \hat{k}, \hat{\mathbf{n}})$ lie on the same plane (i.e., the surface of the page), while the vector $\hat{\mathbf{b}} \equiv -\kappa^{-1}\,\nabla\ln n \times \hat{k}$ is perpendicular to that plane (i.e., into the page).

1.4.2 *Light-ray Frenet-Serret Equations*

Next, by choosing the ray parametrization $d\sigma = ds$ (so that $\Lambda = 1$), the Euler-Fermat equation (1.51) becomes

$$\frac{d}{ds}\left(n\,\frac{d\mathbf{x}}{ds}\right) = \nabla n. \tag{1.52}$$

Since the ray *velocity* $d\mathbf{x}/ds = \widehat{\mathbf{k}}$ is a unit vector, which defines the direction of the wave vector $\mathbf{k}$, Eq. (1.52) yields the *light-curvature* equation [see Eq. (A.24)]

$$\frac{d\widehat{\mathbf{k}}}{ds} = \widehat{\mathbf{k}} \times \left(\nabla\ln n \times \widehat{\mathbf{k}}\right) \equiv \kappa\,\widehat{\mathbf{n}}, \tag{1.53}$$

where $\widehat{\mathbf{n}}$ defines the principal normal unit vector and the light-ray *Frenet-Serret* curvature is

$$\kappa = |\nabla\ln n \times \widehat{\mathbf{k}}| = \widehat{\mathbf{n}}\cdot\nabla\ln n; \tag{1.54}$$

see App. A for a review of the Frenet-Serret formulas for a general spatial curve. Note that for the one-dimensional problem discussed in Sec. 1.3.1, the curvature $\kappa = |n'|/(n\,\Lambda) = |y''|/\Lambda^3$ is in agreement with the standard Frenet-Serret curvature.

The light-ray Frenet-Serret torsion is calculated as follows. First, using the definition of the binormal unit vector $\widehat{\mathbf{b}} \equiv \widehat{\mathbf{k}} \times \widehat{\mathbf{n}}$, we use Eq. (1.53) to find

$$\kappa\widehat{\mathbf{b}} = \widehat{\mathbf{k}} \times \frac{d\widehat{\mathbf{k}}}{ds} = -\nabla\ln n \times \widehat{\mathbf{k}}, \tag{1.55}$$

which shows that $\widehat{\mathbf{b}}\cdot\nabla n \equiv 0$ (see Fig. 1.6). Next, we introduce the remaining Frenet-Serret equations [see Eqs. (A.28)-(A.30)]

$$\frac{d\widehat{\mathbf{n}}}{ds} = \tau\,\widehat{\mathbf{b}} - \kappa\,\widehat{\mathbf{k}} \quad\text{and}\quad \frac{d\widehat{\mathbf{b}}}{ds} = -\tau\,\widehat{\mathbf{n}}, \tag{1.56}$$

where κ and τ denote the curvature and torsion of the light ray. By taking the s-derivative of Eq. (1.55) and then taking the dot product with the normal unit vector $\widehat{\mathbf{n}}$, we obtain the light-ray Frenet-Serret torsion

$$\tau = \frac{\widehat{\mathbf{n}}}{\kappa}\cdot\left[\left(\frac{d}{ds}\nabla\ln n\right)\times\widehat{\mathbf{k}}\right] = \frac{\widehat{\mathbf{b}}}{\kappa}\cdot\left(\frac{d}{ds}\nabla\ln n\right). \tag{1.57}$$

Lastly, a light wave is characterized by a polarization unit vector $\widehat{\mathbf{e}}$ (defined in terms of the wave's electric field $\mathbf{E} \equiv |\mathbf{E}|\,\widehat{\mathbf{e}}$) that is perpendicular to the wavevector direction $\widehat{k}$. We may, thus, write the polarization vector as

$$\widehat{e} \equiv \cos\varphi\,\widehat{n} + \sin\varphi\,\widehat{b}, \tag{1.58}$$

where φ denotes the polarization angle measured from the local normal $\widehat{n}$-axis. Using the Frenet-Serret equations (1.56), we obtain the evolution equation for the polarization unit vector along a light ray:

$$\frac{d\widehat{e}}{ds} = -\kappa\cos\varphi\,\widehat{k} + \left(\frac{d\varphi}{ds} + \tau\right)\frac{\partial\widehat{e}}{\partial\varphi}, \tag{1.59}$$

which is expressed in terms of the Frenet-Serret curvature and torsion (κ, τ): $\kappa_g \equiv \kappa\cos\varphi$ denotes the geodesic curvature and $\tau_r \equiv \tau + d\varphi/ds$ denotes the relative torsion, which are derived in Eq. (A.34).

1.4.3 *Light Propagation in Spherical Geometry*

We now explore the case where the medium is spherically symmetric. By using the general ray-orbit equation (1.53), we can also show that, for a spherically-symmetric nonuniform medium with index of refraction $n(r)$, the light-ray orbit $\mathbf{r}(s)$ satisfies the conservation law

$$\frac{d}{ds}\left(\mathbf{r} \times n(r)\,\frac{d\mathbf{r}}{ds}\right) = \mathbf{r} \times \frac{d}{ds}\left(n(r)\,\frac{d\mathbf{r}}{ds}\right) = \mathbf{r} \times \nabla n(r) = 0. \tag{1.60}$$

Next, we use the fact that the light-ray path is planar (i.e., the torsion is zero, $\tau = 0$) and, thus, we write

$$\mathbf{r} \times \frac{d\mathbf{r}}{ds} = r\left(\sin\phi\cos\theta - \cos\phi\sin\theta\right)\widehat{z} = r\sin\varphi\,\widehat{z}, \tag{1.61}$$

where $\varphi \equiv \phi - \theta$ denotes the angle between the position vector $\mathbf{r} = r\left(\cos\theta\,\widehat{x} + \sin\theta\,\widehat{y}\right)$ and the tangent vector $d\mathbf{r}/ds = \cos\phi\,\widehat{x} + \sin\phi\,\widehat{y} \equiv \widehat{k}$ (see Fig. 1.7). The Frenet-Serret equation $d\widehat{k}/ds = \kappa\left(-\sin\phi\,\widehat{x} + \cos\phi\,\widehat{y}\right)$ yields the curvature $\kappa \equiv d\phi/ds$.

Using Eq. (1.61), the conservation law (1.60) for ray orbits in a spherically-symmetric medium can, therefore, be expressed as

$$n(r)\,r\,\sin\varphi(r) = N\,a, \tag{1.62}$$

which is known as Bouguer's formula (Pierre Bouguer, 1698-1758), where N and a are constants (see Fig. 1.7); note that the condition $n(r)\,r \geq N\,a$ must also be satisfied since $\sin\varphi(r) \leq 1$. This conservation law is analogous to the conservation law of angular momentum for particles moving in a central-force potential (see Chap. 4).

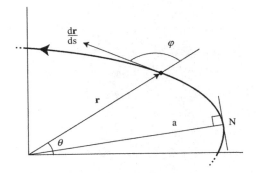

Fig. 1.7   Light path in a nonuniform medium with spherical symmetry.

An explicit expression for the ray orbit $r(\theta)$ is then obtained as follows. First, since $d\mathbf{r}/ds$ is a unit vector, we find

$$\frac{d\mathbf{r}}{ds} = \frac{d\theta}{ds}\left(r\,\widehat{\theta} + \frac{dr}{d\theta}\,\widehat{r}\right) = \frac{r\,\widehat{\theta} + (dr/d\theta)\,\widehat{r}}{\sqrt{r^2 + (dr/d\theta)^2}},$$

so that

$$\frac{d\theta}{ds} = \frac{1}{\sqrt{r^2 + (dr/d\theta)^2}}$$

and Eq. (1.61) yields

$$\mathbf{r} \times \frac{d\mathbf{r}}{ds} = r\sin\varphi\,\widehat{z} = r^2\frac{d\theta}{ds}\,\widehat{z} \quad \rightarrow \quad \sin\varphi = \frac{r}{\sqrt{r^2 + (dr/d\theta)^2}} = \frac{Na}{nr},$$

where we made use of Bouguer's formula (1.62). Next, integration by quadrature yields

$$\theta(r) = Na \int_{r_0}^{r} \frac{d\rho}{\rho\sqrt{n^2(\rho)\rho^2 - N^2 a^2}},$$

where we choose r_0 so that $\theta(r_0) = 0$. Lastly, a change of integration variable $\eta = Na/\rho$ yields

$$\theta(r) = \int_{Na/r}^{Na/r_0} \frac{d\eta}{\sqrt{\overline{n}^2(\eta) - \eta^2}}, \tag{1.63}$$

where $\overline{n}(\eta) \equiv n(Na/\eta)$. Hence, for a spherically-symmetric medium with index of refraction $n(r)$, we can compute the light-ray orbit $r(\theta)$ by inverting the integral (1.63) for $\theta(r)$.

Consider, for example, the spherically-symmetric refractive index $n(r) = n_0\sqrt{2 - (r/R)^2}$, where $n_0 = n(R)$ denotes the refractive index

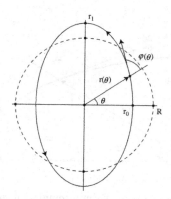

Fig. 1.8   Elliptical light path in a spherically-symmetric refractive medium.

at $r = R$. Introducing the dimensional parameter $\epsilon = a/R$ and the transformation $\sigma = \eta^2$, Eq. (1.63) becomes

$$\theta(r) = \int_{Na/r}^{Na/r_0} \frac{\eta \, d\eta}{\sqrt{n_0^2 \left(2 \eta^2 - N^2 \mathsf{e}^2\right) - \eta^4}}$$

$$= \frac{1}{2} \int_{(Na/r)^2}^{(Na/r_0)^2} \frac{d\sigma}{\sqrt{n_0^4 \, \mathsf{e}^2 - (\sigma - n_0^2)^2}},$$

where $\mathsf{e} = \sqrt{1 - N^2\epsilon^2/n_0^2}$ (assuming that $n_0 > N\epsilon$). Next, using the trigonometric substitution $\sigma = n_0^2 \left(1 + \mathsf{e} \cos \chi \right)$, we find $\theta(r) = \frac{1}{2} \chi(r)$ or

$$r(\theta) \;=\; \frac{r_0 \sqrt{1 + \mathsf{e}}}{\sqrt{1 + \mathsf{e} \cos 2\theta}}, \tag{1.64}$$

which represents an ellipse (see Fig. 1.8)

$$\left(\frac{x}{R \sqrt{1 - \mathsf{e}}}\right)^2 + \left(\frac{y}{R \sqrt{1 + \mathsf{e}}}\right)^2 = 1$$

with semi-major and semi-minor axes $r_1 = R \left(1 + \mathsf{e}\right)^{1/2}$ and $r_0 = R \left(1 - \mathsf{e}\right)^{1/2}$, respectively. This example shows that, surprisingly, it is possible to trap light!

1.4.4   *Geodesic Representation of Light Propagation*

We now investigate the geodesic properties of light propagation in a nonuniform refractive medium. For this purpose, let us consider a path AB in space from point A to point B parameterized by the continuous parameter

σ, i.e., $\mathbf{x}(\sigma)$ such that $\mathbf{x}(A) = \mathbf{x}_A$ and $\mathbf{x}(B) = \mathbf{x}_B$. The time taken by light in propagating from A to B is

$$\mathcal{T}[\mathbf{x}] = \int_A^B \frac{dt}{d\sigma} \, d\sigma = \int_A^B \frac{n}{c} \left(g_{ij} \frac{dx^i}{d\sigma} \frac{dx^j}{d\sigma} \right)^{1/2} d\sigma, \qquad (1.65)$$

where $dt = n \, ds/c$ denotes the infinitesimal time interval taken by light in moving an infinitesimal distance ds in a medium with refractive index n and the space metric is denoted by g_{ij}. The geodesic properties of light propagation are investigated with the *vacuum* metric g_{ij} or the *medium-modified* metric $\overline{g}_{ij} = n^2 \, g_{ij}$.

1.4.4.1 Vacuum-metric Case

We begin with the vacuum-metric case and consider the light-curvature equation (1.53). First, we define the vacuum-metric tensor $g_{ij} = \mathbf{e}_i \cdot \mathbf{e}_j$ in terms of the basis vectors $(\mathbf{e}_1, \mathbf{e}_2, \mathbf{e}_3)$, so that the ray velocity is

$$\widehat{\mathbf{k}} = \frac{d\mathbf{x}}{ds} = \frac{dx^i}{ds} \mathbf{e}_i.$$

Second, using the definition for the Christoffel symbol (1.19) and the relations

$$\frac{d\mathbf{e}_j}{ds} \equiv \Gamma^i_{jk} \frac{dx^k}{ds} \mathbf{e}_i,$$

we find

$$\frac{d\widehat{\mathbf{k}}}{ds} \equiv \frac{d^2\mathbf{x}}{ds^2} = \frac{d^2x^i}{ds^2} \mathbf{e}_i + \frac{dx^j}{ds} \frac{d\mathbf{e}_j}{ds} = \left(\frac{d^2x^i}{ds^2} + \Gamma^i_{jk} \frac{dx^j}{ds} \frac{dx^k}{ds} \right) \mathbf{e}_i.$$

By combining these relations, the light-ray curvature equation (1.53) becomes

$$\frac{d^2x^i}{ds^2} + \Gamma^i_{jk} \frac{dx^j}{ds} \frac{dx^k}{ds} = \left(g^{ij} - \frac{dx^i}{ds} \frac{dx^j}{ds} \right) \frac{\partial \ln n}{\partial x^j}. \qquad (1.66)$$

This equation shows that the path of a light ray departs from a vacuum geodesic line as a result of a refractive-index gradient projected along the tensor

$$h^{ij} \equiv g^{ij} - \frac{dx^i}{ds} \frac{dx^j}{ds}$$

which, by construction, is perpendicular to the ray velocity $d\mathbf{x}/ds$ (i.e., $h^{ij} \, dx_j/ds = 0$).

1.4.4.2   *Medium-metric Case*

Next, we investigate the geodesic propagation of a light ray associated with the medium-modified (conformal) metric $\overline{g}_{ij} = n^2\, g_{ij}$, where $c^2 dt^2 = n^2 ds^2 = \overline{g}_{ij}\, dx^i dx^j$. The derivation follows a variational formulation similar to that found in Sec. 1.1.3. Hence, the first-order variation $\delta\mathcal{T}[\mathbf{x}]$ is expressed as

$$\delta\mathcal{T}[\mathbf{x}] \;=\; \int_{t_A}^{t_B}\left[\frac{d^2 x^i}{dt^2} + \overline{\Gamma}^i_{jk}\,\frac{dx^j}{dt}\,\frac{dx^k}{dt}\right]\overline{g}_{i\ell}\,\delta x^\ell\,\frac{dt}{c^2},\tag{1.67}$$

where the medium-modified Christoffel symbol $\overline{\Gamma}^i_{jk}$ includes the effects of the gradient in the refractive index $n(\mathbf{x})$. We, therefore, find that the light path $\mathbf{x}(t)$ is a solution of the geodesic equation

$$\frac{d^2 x^i}{dt^2} + \overline{\Gamma}^i_{jk}\,\frac{dx^j}{dt}\,\frac{dx^k}{dt} \;=\; 0,\tag{1.68}$$

which is also the path of least time for which $\delta\mathcal{T}[\mathbf{x}] = 0$. By using the substitution $d/dt = (c/n)\, d/ds$ in Eq. (1.68), we find $d\mathbf{x}/dt = c\widehat{\mathbf{k}}/n$ and $d^2\mathbf{x}/dt^2 = (c/n)^2\,(d\widehat{\mathbf{k}}/ds - \widehat{\mathbf{k}}\, d\ln n/ds)$.

When using Cartesian coordinates (where $\overline{g}_{ij} = n^2\,\delta_{ij}$ and $\overline{g}^{ij} = n^{-2}\delta^{ij}$), for example, the medium-modified Christoffel symbol

$$\overline{\Gamma}^i_{jk} \;=\; \delta^i_j\,\partial_k \ln n \,+\, \delta^i_k\,\partial_j \ln n \,-\, \delta_{jk}\,\delta^{i\ell}\,\partial_\ell \ln n\tag{1.69}$$

is expressed in terms of gradient-components of the logarithm of the refraction index n. By inserting Eq. (1.69) into Eq. (1.68), we readily recover the light-ray curvature equation (1.53).

1.4.4.3   *Jacobi Equation for Light Propagation*

Lastly, we point out that the Jacobi equation for the deviation $\boldsymbol{\xi}(\sigma) = \overline{\mathbf{x}}(\sigma) - \mathbf{x}(\sigma)$ between two rays that satisfy the Euler-Fermat ray equation (1.51) can be obtained from the Jacobi function

$$J(\boldsymbol{\xi}, d\boldsymbol{\xi}/d\sigma) \equiv \frac{1}{2}\left[\frac{d^2}{d\epsilon^2}\left(n(\mathbf{x} + \epsilon\,\boldsymbol{\xi})\left|\frac{d\mathbf{x}}{d\sigma} + \epsilon\,\frac{d\boldsymbol{\xi}}{d\sigma}\right|\right)\right]_{\epsilon=0}\tag{1.70}$$

$$\equiv \frac{n}{2\Lambda^3}\left|\frac{d\boldsymbol{\xi}}{d\sigma}\times\frac{d\mathbf{x}}{d\sigma}\right|^2 + \frac{\boldsymbol{\xi}\cdot\nabla n}{\Lambda}\,\frac{d\boldsymbol{\xi}}{d\sigma}\cdot\frac{d\mathbf{x}}{d\sigma} + \frac{\Lambda}{2}\,\boldsymbol{\xi}\boldsymbol{\xi} : \nabla\nabla n,$$

where the Euler-Fermat ray equation (1.51) was taken into account and the exact σ-derivative, which cancels out upon integration, is omitted. Hence, the Jacobi equation describing light-ray deviation is expressed as the Jacobi-Euler-Fermat equation

$$\frac{d}{d\sigma}\left(\frac{\partial J}{\partial(d\boldsymbol{\xi}/d\sigma)}\right) = \frac{\partial J}{\partial\boldsymbol{\xi}},$$

which yields

$$\frac{d}{d\sigma}\left[\frac{n}{\Lambda^3}\frac{d\mathbf{x}}{d\sigma}\times\left(\frac{d\boldsymbol{\xi}}{d\sigma}\times\frac{d\mathbf{x}}{d\sigma}\right)\right] = \Lambda\,\boldsymbol{\xi}\cdot\nabla\nabla n\cdot\left(\mathbf{I} - \frac{1}{\Lambda^2}\frac{d\mathbf{x}}{d\sigma}\frac{d\mathbf{x}}{d\sigma}\right) \qquad (1.71)$$
$$+ \left[\frac{d\boldsymbol{\xi}}{d\sigma} - (\boldsymbol{\xi}\cdot\nabla\ln n)\frac{d\mathbf{x}}{d\sigma}\right]\times\left(\frac{\nabla n}{\Lambda}\times\frac{d\mathbf{x}}{d\sigma}\right).$$

The Jacobi equation (1.71) describes the property of nearby rays to converge or diverge in a nonuniform refractive medium. Note, here, that the terms involving $\Lambda^{-1}\nabla n\times d\mathbf{x}/d\sigma$ in Eq. (1.71) can be written in terms of the Euler-Fermat ray equation (1.51) as

$$\frac{\nabla n}{\Lambda}\times\frac{d\mathbf{x}}{d\sigma} = \frac{1}{\Lambda^2}\frac{d}{d\sigma}\left(\frac{n}{\Lambda}\frac{d\mathbf{x}}{d\sigma}\right)\times\frac{d\mathbf{x}}{d\sigma} = \frac{n}{\Lambda^3}\left(\frac{d^2\mathbf{x}}{d\sigma^2}\times\frac{d\mathbf{x}}{d\sigma}\right),$$

which, thus, involve the Frenet-Serret ray curvature [see Eq. (1.55)].

1.4.5 *Wavefront Representation*

The complementary picture of rays propagating in a nonuniform medium was proposed by Christiaan Huygens (1629-1695) in terms of the picture of propagating wavefronts. Here, a wavefront is defined as the surface that is locally perpendicular to a ray. Hence, the index of refraction itself (for an isotropic medium) can be written as

$$n = |\nabla S| = \frac{ck}{\omega} \quad \text{or} \quad \nabla S = n\frac{d\mathbf{x}}{ds} = \frac{c\mathbf{k}}{\omega}, \qquad (1.72)$$

where S is called the *eikonal* function (i.e., a wavefront is defined by the surface $S = $ constant; see Fig. 1.9). To show that this definition is consistent with Eq. (1.53), we easily check that

$$\frac{d}{ds}\left(n\frac{d\mathbf{x}}{ds}\right) = \frac{d\nabla S}{ds} = \frac{d\mathbf{x}}{ds}\cdot\nabla\nabla S = \frac{1}{n}\,\nabla S\cdot\nabla\nabla S$$
$$= \frac{1}{2n}\,\nabla|\nabla S|^2 = \frac{1}{2n}\,\nabla n^2 = \nabla n.$$

This definition, therefore, implies that the wavevector $\mathbf{k}$ is curl-free:

$$\nabla\times\mathbf{k} = \nabla\times\nabla\left(\frac{\omega}{c}S\right) \equiv 0, \qquad (1.73)$$

where we used the fact that the wave frequency ω is unchanged by refraction. Hence, we find that $\nabla \times \widehat{\mathbf{k}} = \widehat{\mathbf{k}} \times \nabla \ln n \equiv \kappa \, \widehat{\mathbf{b}}$, from which we obtain the light-curvature equation (1.53). Note also that because $\mathbf{k}$ is curl-free, we easily apply Stokes' Theorem to find that the closed contour integral $\oint_{\partial A} \mathbf{k} \cdot d\mathbf{x} = 0$ along the boundary ∂A of an open surface A vanishes, i.e., the path integral $\int \mathbf{k} \cdot d\mathbf{x}$ is path-independent.

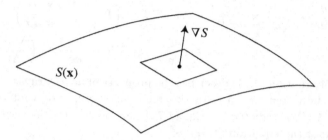

Fig. 1.9   Wavefront surface.

Lastly, in the absence of sources and sinks, the light energy flux entering a finite volume bounded by a closed surface is equal to the light energy flux leaving the volume and, thus, the intensity of light I satisfies the conservation law

$$0 = \nabla \cdot (I \, \nabla S) = I \, \nabla^2 S + \nabla S \cdot \nabla I. \tag{1.74}$$

Using the definition $\nabla S \cdot \nabla \equiv n \, \partial/\partial s$, we find the intensity *evolution* equation

$$\frac{\partial \ln I}{\partial s} = -n^{-1} \, \nabla^2 S,$$

whose solution is expressed as

$$I = I_0 \, \exp\left(-\int_0^s \nabla^2 S \, \frac{d\sigma}{n}\right), \tag{1.75}$$

where I_0 is the light intensity at position $s = 0$ along a ray. This equation, therefore, determines whether light intensity increases ($\nabla^2 S < 0$) or decreases ($\nabla^2 S > 0$) along a ray depending on the sign of $\nabla^2 S$. In a refractive medium with spherical symmetry, with $S'(r) = n(r)$ and $\widehat{\mathbf{k}} = \widehat{r}$, the conservation law (1.74) becomes

$$0 = \frac{1}{r^2} \frac{d}{dr} \left(r^2 \, I \, n\right),$$

which implies that the light intensity satisfies the generalized inverse-square law:

$$I(r) n(r) \, r^2 = I_0 n_0 \, r_0^2. \tag{1.76}$$

Table 1.1 Summary of Chapter 1: The Calculus of Variations.

Topic	Equation
Euler's First Equation	(1.7)
Euler's Second Equation	(1.16)
Constrained Variational Principle	(1.24)-(1.25)
Brachistochrone Problem	(1.30)-(1.32)
Fermat's Principle of Least Time	(1.36)
Euler-Fermat Equation	(1.51)
Frenet-Serret Formulas for Light Rays	(1.53)-(1.57)

1.5 Summary

Chapter 1 presented the mathematical foundations of the Calculus of Variations, which will form the basis upon which the Lagrangian method (introduced in Chapter 2) will be built. The brachistochrone problem and Fermat's Principle of Least Time are two examples that were discussed extensively. Table 1.1 presents a summary of the important topics of Chapter 1.

1.6 Problems

1. Consider the problem of finding the extremal solution $y(x)$ of the integral

$$\mathcal{F}[y] = \int_a^b F(y, y', y'') \, dx,$$

where $F(y, y', y'')$ is a smooth function of its arguments.

(a) Show that Euler's First Equation for this problem is

$$\frac{\partial F}{\partial y} = \frac{d}{dx}\left(\frac{\partial F}{\partial y'}\right) - \frac{d^2}{dx^2}\left(\frac{\partial F}{\partial y''}\right).$$

(b) Find Euler's Second Equation and state whether an additional set of boundary conditions for $\delta y'(a)$ and $\delta y'(b)$ are necessary.

2. Find the curve joining two points (x_1, y_1) and (x_2, y_2) that yields a surface of revolution (about the x-axis) of minimum area by minimizing the integral

$$\mathcal{A}[y] = \int_{x_1}^{x_2} y \sqrt{1 + (y')^2} \, dx.$$

3. Use the Jacobi equation (1.11) to obtain Eq. (1.14) for $\delta^2 \mathcal{F}$.

4. This problem deals with finding the equation for geodesics on a cone represented by $z(\phi) = \rho(\phi) \cot \alpha$, for which the infinitesimal length element ds is defined as

$$ds^2 = d\rho^2(\phi) + \rho^2(\phi)\, d\phi^2 + dz^2(\phi) = \left[\rho^2 + \csc^2\alpha\, (\rho')^2\right] d\phi^2.$$

The length integral is

$$\mathcal{L}[\rho] = \int \sqrt{\rho^2 + \csc^2\alpha\, (\rho')^2}\; d\phi \equiv \int F(\rho, \rho')\, d\phi,$$

and Euler's First and Second equations are

$$\frac{d}{d\phi}\left(\frac{\partial F}{\partial \rho'}\right) = \frac{\partial F}{\partial \rho} \quad \text{and} \quad \frac{d}{d\phi}\left(F - \rho'\,\frac{\partial F}{\partial \rho'}\right) = \frac{\partial F}{\partial \phi} \equiv 0.$$

(a) Show that Euler's Second equation for $\rho(\phi)$ can be written as

$$\frac{\rho^2 \sin\alpha}{\sqrt{\rho^2 \sin^2\alpha + (\rho')^2}} = \rho_0,$$

where $\rho_0 \equiv \rho(\phi_0)$ and $\rho'(\phi_0) \equiv 0$.

(b) Solve Euler's Second equation for $\rho(\phi)$ and show that the equation for geodesics on a cone is

$$\rho(\phi) = \rho_0 \sec\left[\sin\alpha\, (\phi - \phi_0)\right] \equiv \rho_0 \sec(\chi).$$

(c) With the solution given in part (b), show that

$$\frac{\partial F}{\partial \rho} = \cos(\chi) \quad \text{and} \quad \frac{\partial F}{\partial \rho'} = \frac{\sin(\chi)}{\sin\alpha} \;\to\; \frac{d}{d\phi}\left(\frac{\partial F}{\partial \rho'}\right) = \frac{\partial F}{\partial \rho}$$

5. Show that the time required for a particle to move without friction from the point (x_0, y_0) parametrized by the angle θ_0 to the minimum point $(\pi a, 2a)$ of the cycloid solution of the brachistochrone problem is

$$\int_{\theta_0}^{\pi} \sqrt{\frac{1 - \cos\theta}{\cos\theta_0 - \cos\theta}}\; d\theta = \pi \sqrt{\frac{a}{g}},$$

where we used $x(\theta) = \sqrt{2g\left[x(\theta) - x_0\right]}$.

6. A thin rope of mass m (and uniform density) is attached to two vertical poles of height H separated by a horizontal distance $2D$; the coordinates of the pole tops are set at $(\pm D, H)$. If the length L of the rope is greater than $2D$, it will sag under the action of gravity and its lowest point (at its midpoint) will be at a height $y(x = 0) = y_0$. The shape of the rope, subject to the boundary conditions $y(\pm D) = H$, is obtained by minimizing the gravitational potential energy of the rope expressed in terms of the functional

$$\mathcal{U}[y] = \int_D^{-D} mg\, y\, \sqrt{1 + (y')^2}\, dx.$$

Show that the extremal curve $y(x)$ (known as the *catenary* curve) for this problem is

$$y(x) = c \cosh\left(\frac{x - b}{c}\right),$$

where $b = 0$ and $c = y_0$.

7. Show that the parametric solution given by Eqs. (1.46)-(1.47) for the linear refractive medium can be expressed as Eq. (1.49).

8. A light ray travels in a medium with refractive index

$$n(y) = n_0 \exp(-\beta y),$$

where n_0 is the refractive index at $y = 0$ and β is a positive constant.

(a) Using Eq. (1.42), with the transformation $\exp(-\beta s) = \cos\varphi_0 \sec\theta$, show that the path of the light ray is expressed as

$$y(x; \beta) = \frac{1}{\beta} \ln\left[\frac{\cos(\beta x - \varphi_0)}{\cos\varphi_0}\right], \tag{1.77}$$

where the light ray is initially traveling upwards from $(x, y) = (0, 0)$ at an angle φ_0.

(b) Using the appropriate mathematical techniques, show that we recover the expected result $\lim_{\beta \to 0} y(x; \beta) = (\tan\varphi_0)\, x$ from Eq. (1.77).

(c) Show that the light ray reaches a maximum height $\bar{y} = \beta^{-1} \ln(\sec\varphi_0)$ at $x = \varphi_0/\beta$.

9. Consider the path associated with the index of refraction $n(y) = H/y$, where the height H is a constant and $0 < y < H\alpha^{-1} \equiv R$ to ensure that, according to Eq. (1.41), $n(y) > \alpha$. Show that the light path has the simple semi-circular form:

$$(R - x)^2 + y^2 = R^2 \quad \to \quad y(x) = \sqrt{x(2R - x)}.$$

10.[*] Using the parametric solutions (1.46)-(1.47) of the optical path in a linear refractive medium, calculate the Frenet-Serret curvature coefficient

$$\kappa(\varphi) = \frac{|\mathbf{r}''(\varphi) \times \mathbf{r}'(\varphi)|}{|\mathbf{r}'(\varphi)|^3},$$

and show that it is equal to $|\widehat{k} \times \nabla \ln n|$, where

$$\widehat{k} = \frac{d\mathbf{r}}{ds} = \mathbf{r}'(\varphi) \left(\frac{ds}{d\varphi}\right)^{-1} = \frac{x'\widehat{x} + y'\widehat{y}}{\sqrt{(x')^2 + (y')^2}},$$

and $\nabla \ln n(y) = \widehat{y}\, n'(y)/n(y)$.

11. Assuming that the refractive index $n(z)$ in a nonuniform medium is a function of z only, show that the Euler-Fermat equations (1.53) for the components (α, β, γ) of the unit vector $\widehat{k} = \alpha\widehat{x} + \beta\widehat{y} + \gamma\widehat{z}$ are

$$\alpha' = -\alpha\gamma\, n'/n,$$
$$\beta' = -\beta\gamma\, n'/n,$$
$$\gamma' = (1 - \gamma^2)\, n'/n = (\alpha^2 + \beta^2)\, n'/n.$$

12. In Fig. 1.8, show that the angle $\varphi(\theta)$ defined from Eq. (1.64) is expressed as

$$\varphi(\theta) = \arcsin\left[\frac{r}{\sqrt{r^2 + (dr/d\theta)^2}}\right] = \arcsin\left[\frac{1 + e\,\cos 2\theta}{\sqrt{1 + e^2 + 2e\,\cos 2\theta}}\right],$$

so that $\varphi = \pi/2$ at $\theta = 0$ and $\pi/2$, as expected for an ellipse.

13. Consider the light-path trajectory $r(\theta)$ for a spherically-symmetric medium, with index of refraction $n(r) = n_0\,(b/r)^2$, where b is an arbitrary constant and $n_0 = n(b)$.

(a) Using Eq. (1.63), show that the light-path trajectory is

$$r(\theta) = r_0\,\cos\theta + \sqrt{R^2 - r_0^2}\,\sin\theta,$$

where $r_0 = r(\theta = 0)$ and $R = (n_0/N)\,a^2/r_0$.

(b) Using the vector $\mathbf{r} \equiv r(\theta)\,(\sin\theta\,\widehat{x} + \cos\theta\,\widehat{z}) \equiv r(\theta)\,\widehat{r}(\theta)$, show that this solution satisfies the Euler-Fermat equation

$$\frac{1}{R^2}\frac{d}{d\theta}\left(n(r)\,\frac{d\mathbf{r}}{d\theta}\right) = \nabla n = -2\,n_0\,b^2\,\frac{\widehat{r}}{r^3},$$

where $ds = d\theta \sqrt{r^2 + (r')^2} \equiv R \, d\theta$.

14.* Derive the Jacobi equation (1.71) for two-dimensional light propagation in a nonuniform medium with index of refraction $n(y)$; *Hint*: choose $\sigma = x$. Compare your Jacobi equation with that obtained from Eq. (1.11).

15.* Lagrange showed in 1760 that a surface $z(x, y)$ has minimal area if it satisfies the partial differential equation

$$\left(1 + q^2\right) \frac{\partial^2 z}{\partial x^2} + \left(1 + p^2\right) \frac{\partial^2 z}{\partial y^2} - 2pq \frac{\partial^2 z}{\partial x \, \partial y} = 0, \qquad (1.78)$$

where $(p, q) \equiv (\partial z / \partial x, \ \partial z / \partial y)$.

(a) Derive Eq. (1.78) by minimizing the surface integral

$$I[z] = \int \int \sqrt{1 + p^2 + q^2} \, dx \, dy.$$

(b) Show that the surface $z(x, y) = \cosh^{-1}(\sqrt{x^2 + y^2})$ has minimal area.

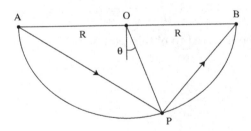

Fig. 1.10   Problem 16.

16. (a) Show that the optical length followed by a light ray along the path APB in Fig. 1.10 is $L(\theta) = 2 \sqrt{2} \, R \, \cos(\theta/2)$, where R is the radius of the circle.

(b) Show that the optical length $L(\theta)$ has a *maximum* for $\theta = 0$.

17. We now consider light propagation in axially-symmetric cylindrical geometry, where the index of refraction $n(\rho)$ is a function of the cylindrical radius ρ (measured from the z-axis). If we use the z-coordinate as the ray parameter, Fermat's Principle of Least Time (1.38) becomes

$$\delta \int_a^b n(\rho) \, \Lambda(\rho, \rho', \theta') \, dz \equiv \delta \int_a^b n(\rho) \sqrt{1 + (\rho')^2 + \rho^2 (\theta')^2} \, dz = 0,$$

where $\rho' = d\rho/dz$ and $\theta' = d\theta/dz$. Note that the integrand $F \equiv n\Lambda$ is independent of z and θ and therefore

$$N \equiv F - \rho' \frac{\partial F}{\partial \rho'} - \theta' \frac{\partial F}{\partial \theta'}, \tag{1.79}$$

$$R \equiv \frac{\partial F}{\partial \theta'}, \tag{1.80}$$

are constants along the light path (i.e., $dN/dz = 0 = dR/dz$).

(a) Using the conservation law (1.80), show that, by solving for θ' as a function of ρ and ρ', we obtain

$$\Lambda(\rho, \rho') \equiv n(\rho) \rho \sqrt{\frac{1 + (\rho')^2}{n^2(\rho) \, \rho^2 - R^2}}. \tag{1.81}$$

(b) Using the conservation law (1.79), with Eq. (1.81), obtain the integral solution

$$z(\rho) \equiv z_a + \int_a^\rho \frac{N \, \sigma \, d\sigma}{\sqrt{\sigma^2 [n^2(\sigma) - N^2] - R^2}},$$

which can then be inverted to obtain $\rho(z; N, R)$.

Chapter 2

Lagrangian Mechanics

Newtonian mechanics discusses the dynamics of particles in terms of (vector) forces acting on them. Within the context of Newtonian mechanics, we distinguish between two classes of forces, depending on whether a force is able to do work or not. In the first class, an *active* force $\mathbf{F}_w$ is involved in performing infinitesimal work $dW = \mathbf{F}_w \cdot d\mathbf{x}$ evaluated along the infinitesimal displacement $d\mathbf{x}$; the class of active forces includes conservative (e.g., gravity) and nonconservative (e.g., friction) forces. In the second class, a *passive* force $\mathbf{F}_0$ is defined as a force not involved in performing work, which includes constraint forces such as normal and tension forces. Here, the infinitesimal work performed by a passive force is $\mathbf{F}_0 \cdot d\mathbf{x} = 0$ because the infinitesimal displacement $d\mathbf{x}$ is required to satisfy the constraints. In contrast, the investigation of particle dynamics within Lagrangian mechanics uses the concepts of kinetic and potential energies, which are both scalar quantities. The difference may seem academic until we realize that it is the Lagrangian method which generalizes to physical theories that lie well beyond the classical dynamics of particles.

In this Chapter, we present four principles by which single-particle dynamics may be derived. The reader is referred to Refs. [4, 12, 21] for comments regarding the history of the Principles of Least Action of Maupertuis, Jacobi, and Hamilton as well as Refs. [9–11] for some additional comments concerning recent developments. The primary focus of this Chapter will be applications of Hamilton's Principle on which Lagrangian Mechanics is based.

2.1   Maupertuis-Jacobi Principle of Least Action

The publication of Fermat's Principle of Least Time in 1657 generated an intense controversy between Fermat and disciples of René Descartes (1596-1650) regarding whether light travels slower (Fermat) or faster (Descartes) in a dense medium as compared to free space (air).

In 1744, Pierre Louis Moreau de Maupertuis (1698-1759) stated (without proof) that, in analogy with Fermat's Principle of Least Time for light, a particle of mass m under the influence of an active force $\mathbf{F} = -\nabla U$ moves along a path that satisfies the Principle of *Least Action*: $\delta S = 0$, where the action integral is defined as

$$S[\mathbf{x}] = \int \mathbf{p} \cdot d\mathbf{x} = \int mv \, ds. \tag{2.1}$$

Here, $v = ds/dt \equiv \mathbf{v} \cdot d\mathbf{x}/ds$ denotes the particle's speed, which can also be expressed as

$$v(s) = \sqrt{(2/m)\,[E - U(s)]}, \tag{2.2}$$

with the particle's kinetic energy $K = mv^2/2 = E - U$ written in terms of its total energy E and its potential energy $U(s)$.

2.1.1   *Maupertuis' Principle*

In 1744, Euler proved Maupertuis' Principle of Least Action $\delta S = 0$ for particle motion in the (x, y)-plane as follows [21]. For this purpose, we use the Frenet-Serret curvature formula for the planar path $y(x)$, where we define the tangent unit vector $\widehat{t}$ and the principal normal unit vector $\widehat{n}$ as

$$\widehat{t} = \frac{d\mathbf{x}}{ds} = \frac{\widehat{x} + y'\,\widehat{y}}{\sqrt{1 + (y')^2}} \quad \text{and} \quad \widehat{n} = \frac{\widehat{y} - y'\,\widehat{x}}{\sqrt{1 + (y')^2}} \equiv \widehat{z} \times \widehat{t}, \tag{2.3}$$

where $y' = dy/dx$ and $ds = dx\,\sqrt{1 + (y')^2}$. The Frenet-Serret formula for the curvature of a two-dimensional curve (see App. A) is

$$\frac{d\widehat{t}}{ds} = \frac{y''\,\widehat{n}}{[1 + (y')^2]^{3/2}} \equiv \kappa\,\widehat{n}.$$

Note that the binormal unit vector $\widehat{b} \equiv \widehat{t} \times \widehat{n}$ is defined in Eq. (2.3) as $\widehat{b} = \widehat{z}$, which is a constant vector and, therefore, the Frenet-Serret torsion for Newtonian planar motion is zero.

Next, we introduce Newton's Second Law of Motion $m\,d\mathbf{v}/dt = \mathbf{F}$, where

$$\frac{d\mathbf{v}}{dt} = v\,\frac{d(v\widehat{t})}{ds} = \mathbf{v}\,\frac{dv}{ds} + v^2\,\frac{d\widehat{t}}{ds}.$$

By using the energy conservation law ($\nabla K = -\nabla U \equiv \mathbf{F}$), Newton's Second Law becomes

$$mv \left(\frac{dv}{ds} \widehat{\mathbf{t}} + v \frac{d\widehat{\mathbf{t}}}{ds} \right) = \widehat{\mathbf{t}} \left(\widehat{\mathbf{t}} \cdot \nabla K \right) + mv^2 \, \kappa \, \widehat{\mathbf{n}}$$

$$= \nabla K = m v \, \nabla v \tag{2.4}$$

between the unit vectors $\widehat{\mathbf{t}}$ and $\widehat{\mathbf{n}}$ associated with the path, the Frenet-Serret curvature κ, and the kinetic energy $K = \frac{1}{2} mv^2(x, y)$ of the particle. Note that Eq. (2.4) can be re-written as

$$\frac{d\widehat{\mathbf{t}}}{ds} = \widehat{\mathbf{t}} \times \left(\nabla \ln v \times \widehat{\mathbf{t}} \right), \tag{2.5}$$

which highlights a deep connection with the light-ray Frenet-Serret curvature equation (1.53) derived from Fermat's Principle of Least Time, where the index of refraction n is now replaced by the speed (2.2).

Lastly, we now show that Eq. (2.5) can be derived from the Maupertuis action functional (2.1), which is expressed as

$$\mathcal{S} = \int m \, v(x, y) \, \sqrt{1 + (y')^2} \, dx \equiv \int F(y, y'; x) \, dx. \tag{2.6}$$

We now construct Euler's first equation for Eq. (2.1), where

$$\frac{\partial F}{\partial y'} = \frac{mv \, y'}{\sqrt{1 + (y')^2}} \quad \text{and} \quad \frac{\partial F}{\partial y} = m \sqrt{1 + (y')^2} \, \frac{\partial v}{\partial y},$$

so that we obtain the Euler-Maupertuis equation

$$\frac{m \, v \, y''}{[1 + (y')^2]^{3/2}} = \frac{m}{\sqrt{1 + (y')^2}} \frac{\partial v}{\partial y} - \frac{m \, y'}{\sqrt{1 + (y')^2}} \frac{\partial v}{\partial x}$$

$$= \frac{m}{\sqrt{1 + (y')^2}} \left(\widehat{\mathbf{y}} - y'\widehat{\mathbf{x}} \right) \cdot \nabla v \equiv m \widehat{\mathbf{n}} \cdot \nabla v, \tag{2.7}$$

which can also be expressed as $mv \, \kappa = m \widehat{\mathbf{n}} \cdot \nabla v$. Using the relation $\mathbf{F} = \nabla K$ and the Frenet-Serret formulas (2.3), the Maupertuis-Euler equation (2.7) becomes $mv^2 \, \kappa = \mathbf{F} \cdot \widehat{\mathbf{n}}$, from which we recover Newton's Second Law (2.4).

2.1.2 *Jacobi's Principle*

Jacobi emphasized the connection between Fermat's Principle of Least Time (1.36) and Maupertuis' Principle of Least Action (2.1) by introducing a different form of the Principle of Least Action $\delta \mathcal{S} = 0$, where Jacobi's action integral is

$$\mathcal{S}[\mathbf{x}] = \int \sqrt{2m \, (E - U)} \, ds = 2 \int K \, dt, \tag{2.8}$$

where the particle momentum is written as $p = \sqrt{2m\,(E - U)}$. To obtain the second expression of the action integral (2.8), Jacobi made use of the fact that, by introducing a path parameter τ such that $v = ds/dt = s'/t'$ (where a prime denotes a τ-derivative), we find

$$K = \frac{m\,(s')^2}{2\,(t')^2} = E - U,$$

so that $2\,K\,t' = s'\,p$, and the second form of Jacobi's action integral results. Next, Jacobi used the Principle of Least Action (2.8) to establish the geometric foundations of particle mechanics. Here, the Euler-Jacobi equation resulting from Jacobi's Principle of Least Action is expressed as

$$\frac{d}{ds}\left(\sqrt{E - U}\,\frac{d\mathbf{x}}{ds}\right) = \nabla\sqrt{E - U}\,, \tag{2.9}$$

which is identical to the Euler-Fermat equation (1.52), with the index of refraction n substituted with $\sqrt{E - U}$.

Note that the connection between Fermat's Principle of Least Time and Maupertuis-Jacobi's Principle of Least Action yields the relation

$$|\mathbf{p}| = \alpha\,n, \tag{2.10}$$

where α is a constant (see Table 3.1). This connection was later used by Prince Louis Victor Pierre Raymond de Broglie (1892-1987) to establish the relation $|\mathbf{p}| = \hbar|\mathbf{k}| = n\,(\hbar\omega/c)$ between the momentum of a particle and its wavenumber $|\mathbf{k}| = 2\pi/\lambda = n\omega/c$; in Eq. (2.10), the constant α is $\alpha = (\hbar\omega/c)$. Using de Broglie's relation $\mathbf{p} = \hbar\mathbf{k}$ between the particle momentum $\mathbf{p}$ and the wave vector $\mathbf{k}$, we note that

$$\frac{|\mathbf{p}|^2}{2m} = \frac{\hbar^2|\mathbf{k}|^2}{2m} \equiv \frac{\hbar^2}{2m}\,|\nabla\Theta|^2 = E - U, \tag{2.11}$$

where Θ is the dimensionless (eikonal) phase. The time-independent Hamilton-Jacobi equation

$$E = \frac{|\nabla\mathcal{S}_E|^2}{2m} + U \tag{2.12}$$

is obtained from Eq. (2.11) by using the relation $\mathcal{S}_E \equiv \hbar\,\Theta$ between Hamilton's principal function $\mathcal{S}_E$ (at constant energy E) and the eikonal phase Θ (see Chap. 3 for additional details). Further historical comments concerning the variational derivation of Schroedinger's equation is discussed by Dugas [4], Lanczos [12], and Yourgrau and Mandelstam [21], as well as Refs. [10, 11].

2.2   d'Alembert's Principle

So far, the Maupertuis-Jacobi principles (2.1) and (2.8) make use of the length variable s as the orbit parameter to describe particle motion. We now turn our attention to two principles that will provide a clear path toward the ultimate action principle called Hamilton's Principle, from which equations of motion are derived in terms of generalized spatial coordinates in *configuration* space.

2.2.1   *Principle of Virtual Work*

The Principle of Virtual Work is one of the oldest principles in Physics, which may find its origin in the work of Aristotle (384-322 B.C.) on the static equilibrium of levers [4]. The Principle of Virtual Work was finally written in its current form in 1717 by Jean (Johann) Bernoulli and it states that a system composed of N particles is in static equilibrium if the virtual work

$$\delta W = \sum_{i=1}^{N} \mathbf{F}_i \cdot \delta \mathbf{x}^i = 0 \tag{2.13}$$

vanishes for all virtual displacements $(\delta \mathbf{x}^1, ..., \delta \mathbf{x}^N)$ that satisfy physical constraints.

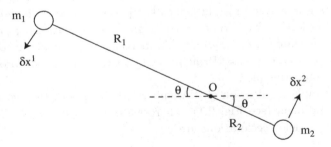

Fig. 2.1   Static equilibrium of a lever.

As an application of the Principle of Virtual Work (2.13), we consider the static equilibrium of a lever (see Fig. 2.1) composed of two masses m_1 and m_2 placed on a massless rod at distances R_1 and R_2, respectively, from the fulcrum point O. Here, the only active forces acting on the masses are due to gravity: $\mathbf{F}_i = -m_i g \, \widehat{\mathbf{y}}$, and the position vectors of m_1 and m_2 are

$$\mathbf{r}_1 = R_1 \left(-\cos\theta \, \widehat{\mathbf{x}} + \sin\theta \, \widehat{\mathbf{y}} \right) \quad \text{and} \quad \mathbf{r}_2 = R_2 \left(\cos\theta \, \widehat{\mathbf{x}} - \sin\theta \, \widehat{\mathbf{y}} \right),$$

respectively (see Fig. 2.1). By using the virtual displacements

$$\delta\mathbf{x}^i = \epsilon\widehat{\mathbf{z}} \times \mathbf{r}_i \tag{2.14}$$

(where ϵ is an infinitesimal angular displacement and the axis of rotation is directed along the z-axis, i.e., out of the page), the Principle of Virtual Work (2.13) yields the following condition for static equilibrium:

$$0 = \epsilon\cos\theta\,(m_1 g\, R_1 - m_2 g\, R_2) \quad\rightarrow\quad m_1\, R_1 = m_2\, R_2.$$

Note that, although the static equilibrium of the lever is based on the concept of *torque* (moment of force) equilibrium, the Principle of Virtual Work shows that all static equilibria are encompassed by the Principle.

2.2.2 Lagrange's Equations from d'Alembert's Principle

It was Jean Le Rond d'Alembert (1717-1783) who generalized the Principle of Virtual Work (in 1742) by including the *accelerating* force $-m_i\,\ddot{\mathbf{x}}^i$ in the Principle of Virtual Work (2.13):

$$\sum_{i=1}^{N}\left(\mathbf{F}_i - m_i\,\frac{d^2\mathbf{x}^i}{dt^2}\right)\cdot\delta\mathbf{x}^i = 0, \tag{2.15}$$

so that the equations of dynamics could be obtained from Eq. (2.15). Hence, d'Alembert's Principle, in effect, states that the work done by all active forces acting in a system is algebraically equal to the work done by all the accelerating forces. Note that, in contrast to the variational principles of Classical Mechanics (e.g., Fermat, Maupertuis, and Jacobi), d'Alembert's Principle (2.15) and Gauss' Principle of Least Constraint (see problem 10) are *constraint* principles.

As a simple application of d'Alembert's Principle (2.15), we return to the lever problem (see Fig. 2.1), where we now assume $m_2\, R_2 > m_1\, R_1$. Here, the particle accelerations are

$$\ddot{\mathbf{x}}^i = -(\dot{\theta})^2\,\mathbf{r}_i - \ddot{\theta}\,\widehat{\mathbf{z}} \times \mathbf{r}_i,$$

so that, with Eq. (2.14), d'Alembert's Principle (2.15) yields

$$0 = \epsilon\left[(m_1 g\, R_1 - m_2 g\, R_2)\,\cos\theta + \left(m_1\, R_1^2 + m_2\, R_2^2\right)\ddot{\theta}\right].$$

Hence, according to d'Alembert's Principle, the angular acceleration $\ddot{\theta}$ of the unbalanced lever is

$$\ddot{\theta} = \frac{g\cos\theta}{I}\,(m_2\, R_2 - m_1\, R_1),$$

where $I = m_1 R_1^2 + m_2 R_2^2$ denotes the moment of inertia of the lever as it rotates about the fulcrum point O. Thus, we see that rotational dynamics associated with unbalanced torques can be described in terms of d'Alembert's Principle. (See Chap. 7 for additional details concerning rotational dynamics.)

The most historically significant application of d'Alembert's Principle (2.15), however, came from Lagrange who transformed it as follows. Consider, for simplicity, the following infinitesimal-work identity

$$0 = \left(\mathbf{F} - m \frac{d^2\mathbf{x}}{dt^2} \right) \cdot \delta\mathbf{x}$$

$$= \delta W - \frac{d}{dt} \left(m \frac{d\mathbf{x}}{dt} \cdot \delta\mathbf{x} \right) + m \frac{d\mathbf{x}}{dt} \cdot \frac{d\,\delta\mathbf{x}}{dt}, \qquad (2.16)$$

where $\mathbf{F}$ represents an active force applied to a particle of mass m so that $\delta W = \mathbf{F} \cdot \delta\mathbf{x}$ denotes the virtual work calculated along the virtual displacement $\delta\mathbf{x}$. We note that if the position vector $\mathbf{x}(q_1, ..., q_k; t)$ is a time-dependent function of k *generalized* coordinates, then we find

$$\delta\mathbf{x} = \sum_{i=1}^{k} \frac{\partial\mathbf{x}}{\partial q_i} \delta q_i,$$

and

$$\mathbf{v} = \frac{d\mathbf{x}}{dt} = \frac{\partial\mathbf{x}}{\partial t} + \sum_{i=1}^{k} \frac{\partial\mathbf{x}}{\partial q_i} \dot{q}_i. \qquad (2.17)$$

Next, we introduce the variation of the kinetic energy $K = mv^2/2$:

$$\delta K = m \frac{d\mathbf{x}}{dt} \cdot \frac{d\,\delta\mathbf{x}}{dt} = \sum_i \delta q_i \frac{\partial K}{\partial q_i},$$

since the virtual variation operator δ (introduced by Lagrange) commutes with the time derivative d/dt, and we introduce the generalized force

$$Q^i \equiv \mathbf{F} \cdot \frac{\partial\mathbf{x}}{\partial q_i},$$

so that $\delta W = \sum_i Q^i \delta q_i$. We shall also use the identity

$$m \frac{d^2\mathbf{x}}{dt^2} \cdot \frac{\partial\mathbf{x}}{\partial q_i} = \frac{d}{dt} \left(m\mathbf{v} \cdot \frac{\partial\mathbf{x}}{\partial q_i} \right) - m\mathbf{v} \cdot \frac{d}{dt} \left(\frac{\partial\mathbf{x}}{\partial q_i} \right),$$

with

$$\frac{d}{dt} \left(\frac{\partial\mathbf{x}}{\partial q_i} \right) \equiv \frac{\partial^2\mathbf{x}}{\partial t\,\partial q_i} + \dot{q}_j \frac{\partial^2\mathbf{x}}{\partial q_j\,\partial q_i} \equiv \frac{\partial\mathbf{v}}{\partial q_i},$$

and

$$\frac{\partial \mathbf{v}}{\partial \dot{q}_i} = \frac{\partial \mathbf{x}}{\partial q_i},$$

which both follow from Eq. (2.17). Our first result derived from d'Alembert's Principle (2.16) is now expressed in terms of the generalized coordinates $(q_1, ..., q_k)$ as

$$0 = \sum_i \delta q_i \left[\frac{d}{dt} \left(\frac{\partial K}{\partial \dot{q}_i} \right) - \frac{\partial K}{\partial q_i} - Q^i \right].$$

Since this relation must hold for any variation δq_i $(i = 1, ..., k)$, we obtain the d'Alembert-Lagrange equation

$$\frac{d}{dt} \left(\frac{\partial K}{\partial \dot{q}_i} \right) - \frac{\partial K}{\partial q_i} = Q^i, \tag{2.18}$$

where the generalized force Q^i is associated with any active (conservative or nonconservative) force $\mathbf{F}$. Hence, for a *conservative* active force derivable from a single potential energy U (i.e., $\mathbf{F} = -\nabla U$), the ith-component of the generalized force is $Q^i = -\partial U / \partial q_i$, and the d'Alembert-Lagrange equation (2.18) becomes

$$\frac{d}{dt} \left(\frac{\partial K}{\partial \dot{q}_i} \right) - \frac{\partial K}{\partial q_i} = -\frac{\partial U}{\partial q_i}. \tag{2.19}$$

We shall return to this important equation [see Eq. (2.31)].

Our second result based on d'Alembert's Principle (2.16), now expressed as

$$\delta K + \delta W = \frac{d}{dt} \left(m \frac{d\mathbf{x}}{dt} \cdot \delta \mathbf{x} \right), \tag{2.20}$$

is obtained as follows. For a *conservative* active force derivable from a single potential energy U (i.e., $\mathbf{F} = -\nabla U$), the virtual work is $\delta W = -\delta U$, so that time integration of Eq. (2.20) yields an important principle known as Hamilton's Principle

$$\int_{t_1}^{t_2} \left(\delta K - \delta U \right) dt \equiv \delta \int_{t_1}^{t_2} L \, dt = 0, \tag{2.21}$$

where $\delta \mathbf{x}$ vanishes at $t = t_1$ and t_2 and the function $L = K - U$, obtained by subtracting the potential energy U from the kinetic energy K, is known as the Lagrangian function of the system. Note that the Maupertuis-Jacobi Principle (2.8) leads to Hamilton's Principle (2.21) if we use the identity $2K \equiv (K - U) + E$ and use the variation operator δ_E at constant total energy E.

2.3 Hamilton's Principle

2.3.1 *Constraint Forces*

To illustrate Hamilton's Principle (2.21), we consider a pendulum composed of an object of mass m and a massless string of constant length ℓ in a constant gravitational field with acceleration g. For this problem, Newton's Second Law of Motion $m\ddot{\mathbf{x}} = \mathbf{F}$ is expressed in terms of the net force $\mathbf{F} = \mathbf{T} + m\,\mathbf{g}$, where the weight force $m\,\mathbf{g} = -mg\,\widehat{\mathbf{y}} = -mg\,(\sin\theta\,\widehat{\theta} - \cos\theta\,\widehat{\mathbf{r}})$ and the radially-inward tension force $\mathbf{T} = -T\widehat{\mathbf{r}}$ are expressed in terms of the polar coordinates (r, θ) defined in Fig. 2.2. Using these polar coordinates, the accelerating force is $m\ddot{\mathbf{x}} = m\ell\,(\ddot{\theta}\,\widehat{\theta} - \dot{\theta}^2\,\widehat{\mathbf{r}})$, and Newton's Second Law in the radial direction states that, since the pendulum length ℓ is constant, the total radial force must vanish, which yields the tension

$$T = mg\cos\theta + m\ell\,\dot{\theta}^2, \tag{2.22}$$

expressed as the sum of the radial component of the weight force and the centrifugal force. Newton's Second Law in the polar direction, on the other hand, yields the pendulum equation

$$m\ell\,\ddot{\theta} = -mg\sin\theta, \tag{2.23}$$

where the polar component of the weight force of the pendulum acts as a restoring force. We note, here, that the tension force is passive (since it does no work) while the weight force is active.

We now investigate the motion of the pendulum as a dynamical problem in two dimensions with a single constraint (i.e., constant length) and later reduce this problem to a single dimension by carefully choosing a single generalized coordinate. Using Cartesian coordinates (x, y) for the pendulum mass shown in Fig. 2.2, the kinetic energy is $K = \frac{1}{2}m(\dot{x}^2 + \dot{y}^2)$ and the gravitational potential energy is $U = mg\,y$, where the length of the pendulum string ℓ is *constrained* to be constant:

$$\ell = \sqrt{x^2 + y^2}. \tag{2.24}$$

Hence, we consider the constrained action integral defined as

$$\mathcal{A}_\lambda[\mathbf{x}] = \int \left[\frac{1}{2}m\left(\dot{x}^2 + \dot{y}^2\right) - mg\,y + \lambda\left(\ell - \sqrt{x^2 + y^2}\right) \right] dt$$

$$\equiv \int F(\mathbf{x}, \dot{\mathbf{x}}; \lambda)\,dt,$$

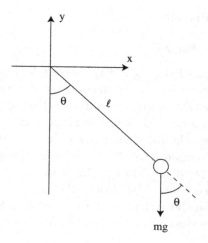

Fig. 2.2   The two-dimensional pendulum problem.

where λ represents a Lagrange multiplier used to enforce the constant-length constraint (see Sec. 1.2.1), i.e., by definition, $\partial F/\partial \lambda = 0$ yields the constraint (2.24) for all $\mathbf{x}$. Euler's equations for x and y, respectively, are

$$m\ddot{x} = -\lambda \frac{x}{\ell} \quad \text{and} \quad m\ddot{y} = -mg - \lambda \frac{y}{\ell}. \tag{2.25}$$

The Lagrange multiplier λ is constructed from Eqs. (2.25) as

$$\lambda \equiv -\frac{m}{\ell}\left[\, gy + (x\ddot{x} + y\ddot{y})\,\right]. \tag{2.26}$$

Next, using the second time derivative of the constant-length constraint (2.24), we obtain

$$x\ddot{x} + y\ddot{y} = -\left(\dot{x}^2 + \dot{y}^2\right),$$

so that Eq. (2.26) becomes

$$\lambda = \frac{m}{\ell}\left(\dot{x}^2 + \dot{y}^2\right) - mg\frac{y}{\ell} \equiv m\ell\,\dot{\theta}^2 + mg\cos\theta.$$

The Lagrange multiplier λ is, thus, interpreted as the (passive) tension force (2.22) in the pendulum string: the constrained displacement for the pendulum is expressed as $d\mathbf{x} = \ell\, d\theta\, \widehat{\theta}$, so that the tension force $\mathbf{T} \equiv -T\widehat{r}$ is indeed passive since $\mathbf{T} \cdot d\mathbf{x} = 0$.

It turns out that a (passive) constraint force in a dynamical system can most often be represented in terms of a constraint involving spatial coordinates. We shall now see that each constraint force can be eliminated from the dynamical problem by making use of new spatial coordinates that enforce the constraint. For example, in the case of the pendulum problem discussed above, we note that the constant-length constraint can be enforced by expressing the Cartesian coordinates $x = \ell \sin\theta$ and $y = -\ell \cos\theta$ in terms of the angle θ (see Fig. 2.2). We shall return to the pendulum problem in Sec. 2.4.1.

2.3.2 Generalized Coordinates in Configuration Space

The *configuration* space of a mechanical system with $n-k$ constraints evolving in n-dimensional space, with spatial coordinates $\mathbf{x} = (x^1, x^2, ..., x^n)$, can sometimes be described in terms of *generalized* coordinates $\mathbf{q} = (q^1, q^2, ..., q^k)$ in a k-dimensional *configuration* space, with $k \leq n$. Each generalized coordinate q^i is said to describe motion along a *degree of freedom* of the mechanical system.

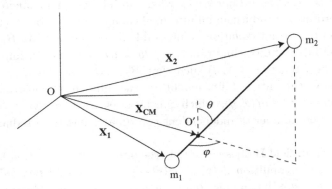

Fig. 2.3   Configuration space.

For example, consider a mechanical system composed of two particles (see Fig. 2.3), with masses (m_1, m_2) and three-dimensional coordinate positions $(\mathbf{x}_1, \mathbf{x}_2)$, tied together with a massless rigid rod (so that the distance $|\mathbf{x}_1 - \mathbf{x}_2|$ is constant). The configuration of this two-particle system (in six-dimensional space) can be described in terms of the five-dimensional coordinates $(\mathbf{X}_{\mathrm{CM}}; \theta, \varphi)$, where the position of the center-of-mass in the laboratory frame (O) is

$$\mathbf{X}_{\mathrm{CM}} \equiv \frac{\sum_i m_i \mathbf{x}_i}{\sum_i m_i} = \frac{m_1 \mathbf{x}_1 + m_2 \mathbf{x}_2}{m_1 + m_2}, \tag{2.27}$$

and the orientation of the rod in the CM frame (O') is expressed in terms of the two angles (θ, φ). Hence, as a result of the existence of a single constraint $(\ell = |\mathbf{x}_1 - \mathbf{x}_2|)$, we have reduced the number of coordinates needed to describe the state of the system from six to five. Each generalized coordinate is said to describe dynamics along a *degree of freedom* of the mechanical system; for example, in the case of the two-particle system discussed above, the generalized coordinates $\mathbf{x}_{\mathrm{CM}}$ describe an arbitrary translation of the center-of-mass while the generalized coordinates (θ, φ) describe an arbitrary rotation about the center-of-mass.

Constraints are found to be of two different types referred to as *holonomic* and *nonholonomic* constraints [12]. For example, the differential (kinematic) constraint equation $dq(\mathbf{r}) = \mathbf{B}(\mathbf{r}) \cdot d\mathbf{r}$ is said to be *holonomic* (or integrable) if the vector field $\mathbf{B}$ satisfies the integrability condition $\nabla \times \mathbf{B} = 0$. If this condition is satisfied, the function $q(\mathbf{r})$ can be explicitly constructed and, thus, the number of independent coordinates can be reduced by one. For example, consider the differential constraint equation $dz = B_x(x, y)\, dx + B_y(x, y)\, dy$, where an infinitesimal change in the x and y coordinates produce an infinitesimal change in the z coordinate. This differential constraint equation is integrable if the components B_x and B_y satisfy the integrability condition $\partial B_x / \partial y = \partial B_y / \partial x$, which implies that there exists a function $f(x, y)$ such that $B_x = \partial f / \partial x$ and $B_y = \partial f / \partial y$. Hence, under this integrability condition, the differential constraint equation becomes $dz = df(x, y)$, which can be integrated to give $z = f(x, y)$ and, thus, the number of independent coordinates has been reduced from 3 to 2.

If the vector field $\mathbf{B}$ does not satisfy the integrability condition $\nabla \times \mathbf{B} = 0$, however, the condition $dq(\mathbf{r}) = \mathbf{B}(\mathbf{r}) \cdot d\mathbf{r}$ is said to be *non-holonomic*. An example of non-holonomic condition is the case of the rolling of a solid body on a surface. Moreover, we note that a kinematic condition is called *rheonomic* if it is time-dependent, otherwise it is called *scleronomic*.

In summary, the presence of holonomic constraints can always be treated by the introduction of generalized coordinates. The treatment of nonholonomic constraints, on the other hand, requires the addition of constraint forces on the right side of Lagrange's equation (2.18), which falls outside the scope of this course.

2.3.3   *Constrained Motion on a Surface*

As an example of motion under an holonomic constraint, we consider the general problem associated with the motion of a particle constrained to move on a surface described by the relation $F(x, y, z) = 0$. First, since the velocity $d\mathbf{x}/dt$ of the particle along its trajectory must be perpendicular to the gradient ∇F (i.e., tangent to the surface $F = 0$), the displacement $d\mathbf{x}$ is required to satisfy the constraint condition

$$d\mathbf{x} \cdot \nabla F = 0. \tag{2.28}$$

Next, any point $\mathbf{x}$ on the surface $F(x, y, z) = 0$ may be parameterized by two *surface* coordinates (u, v) such that

$$\frac{\partial \mathbf{x}}{\partial u}(u, v) \cdot \nabla F = 0 = \frac{\partial \mathbf{x}}{\partial v}(u, v) \cdot \nabla F.$$

Hence, we may write an expression for $d\mathbf{x}$ that satisfies Eq. (2.28) as

$$d\mathbf{x} = \frac{\partial \mathbf{x}}{\partial u} \, du + \frac{\partial \mathbf{x}}{\partial v} \, dv \quad \text{and} \quad \frac{\partial \mathbf{x}}{\partial u} \times \frac{\partial \mathbf{x}}{\partial v} = \mathcal{J} \, \nabla F,$$

where the function $\mathcal{J}$ depends on the surface coordinates (u, v). It is thus quite clear that the surface coordinates (u, v) are the generalized coordinates for this constrained motion.

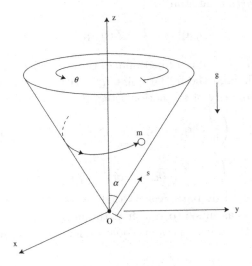

Fig. 2.4   Motion on the surface of a cone.

For example, we consider the motion of a particle constrained to move on the surface of a cone of apex angle α (see Fig. 2.4). Here, the constraint is expressed as $F(x, y, z) = \sqrt{x^2 + y^2} - z \tan \alpha = 0$ with $\nabla F = \widehat{\rho} - \tan \alpha \, \widehat{\mathbf{z}}$, where $\rho^2 = x^2 + y^2$ and $\widehat{\rho} = (x/\rho)\,\widehat{x} + (y/\rho)\,\widehat{y}$. The surface coordinates can be chosen to be the polar angle θ and the function

$$s(x, y, z) = \sqrt{x^2 + y^2 + z^2} \equiv \sqrt{\rho^2 + z^2},$$

which measures the distance from the apex of the cone (defining the origin), with $\rho = s\,\sin\alpha$ and $z = s\,\cos\alpha$. Hence, using $\mathbf{x} = \rho\,(\cos\theta\,\widehat{x} + \sin\theta\,\widehat{y}) + z\,\widehat{z}$, we find

$$\frac{\partial\mathbf{x}}{\partial\theta} = \rho\,\widehat{\theta} = \rho\,\widehat{z}\times\widehat{\rho} \quad\text{and}\quad \frac{\partial\mathbf{x}}{\partial s} = \sin\alpha\,\widehat{\rho} + \cos\alpha\,\widehat{z} = \widehat{s},$$

with

$$\frac{\partial\mathbf{x}}{\partial\theta}\times\frac{\partial\mathbf{x}}{\partial s} = \rho\,\cos\alpha\,\nabla F$$

and thus $\mathcal{J} = \rho\,\cos\alpha$. We shall return to this example in Sec. 2.5.4.

2.3.4   *Euler-Lagrange Equations*

Hamilton's principle (sometimes called THE Principle of Least Action) is expressed in terms of a function $L(\mathbf{q}, \dot{\mathbf{q}}; t)$ known as the *Lagrangian*, which appears in the *action* integral

$$\mathcal{S}[\mathbf{q}] = \int_{t_i}^{t_f} L(\mathbf{q}, \dot{\mathbf{q}}; t)\,dt, \tag{2.29}$$

where the action integral is a functional of the generalized coordinates $\mathbf{q}(t)$, providing a path from the initial point $\mathbf{q}_i = \mathbf{q}(t_i)$ to the final point $\mathbf{q}_f = \mathbf{q}(t_f)$. The stationarity of the action integral

$$0 = \delta\mathcal{S}[\mathbf{q}; \delta\mathbf{q}] = \left(\frac{d}{d\epsilon}\mathcal{S}[\mathbf{q} + \epsilon\,\delta\mathbf{q}]\right)_{\epsilon=0} = \int_{t_i}^{t_f}\left(\delta\mathbf{q}\cdot\frac{\partial L}{\partial\mathbf{q}} + \delta\dot{\mathbf{q}}\cdot\frac{\partial L}{\partial\dot{\mathbf{q}}}\right) dt$$

$$= \int_{t_i}^{t_f}\delta\mathbf{q}\cdot\left[\frac{\partial L}{\partial\mathbf{q}} - \frac{d}{dt}\left(\frac{\partial L}{\partial\dot{\mathbf{q}}}\right)\right] dt, \tag{2.30}$$

where an integration by parts was carried out on the term $\delta\dot{\mathbf{q}}\cdot\partial L/\partial\dot{\mathbf{q}}$ and the variation $\delta\mathbf{q}$ is assumed to vanish at the integration boundaries ($\delta\mathbf{q}_i = 0 = \delta\mathbf{q}_f$), yields the *Euler-Lagrange* equation for the generalized coordinate q^j ($j = 1, ..., k$):

$$\frac{d}{dt}\left(\frac{\partial L}{\partial\dot{q}^j}\right) = \frac{\partial L}{\partial q^j}. \tag{2.31}$$

The Lagrangian also satisfies the Euler's Second Equation:

$$\frac{d}{dt}\left(L - \dot{\mathbf{q}}\cdot\frac{\partial L}{\partial\dot{\mathbf{q}}}\right) = \frac{\partial L}{\partial t}, \tag{2.32}$$

and thus, for time-independent Lagrangian systems ($\partial L/\partial t = 0$), we find that $L - \dot{\mathbf{q}} \cdot \partial L/\partial \dot{\mathbf{q}}$ is a conserved quantity whose interpretation will be discussed shortly.

Note that, according to d'Alembert's Principle (2.19), the form of the Lagrangian function $L(\mathbf{r}, \dot{\mathbf{r}}; t)$ is dictated by our requirement that Newton's Second Law $m\ddot{\mathbf{r}} = -\nabla U(\mathbf{r}, t)$, which describes the motion of a particle of mass m in a nonuniform (possibly time-dependent) potential $U(\mathbf{r}, t)$, be written in the Euler-Lagrange form (2.31). One easily obtains the form

$$L(\mathbf{r}, \dot{\mathbf{r}}; t) = \frac{m}{2} |\dot{\mathbf{r}}|^2 - U(\mathbf{r}, t), \tag{2.33}$$

for the Lagrangian of a particle of mass m, which is simply the kinetic energy of the particle **minus** its potential energy. The minus sign in Eq. (2.33) is important; not only does this form give us the correct equations of motion but, without the minus sign, energy would not be conserved. In fact, we note that Jacobi's action integral (2.8) can also be written as $A = \int [(K-U)+E]\, dt$, using the energy conservation law $E = K+U$; hence, energy conservation is the important connection between the Principles of Least Action of Maupertuis-Jacobi and Euler-Lagrange.

2.3.5 *Four-step Lagrangian Method*

For a simple mechanical system, the Lagrangian function is obtained by computing the kinetic energy of the system and its potential energy and then constructing Eq. (2.33). The construction of a Lagrangian function for a system of N particles, therefore, proceeds in four steps as follows.

• **Step I.** Define k generalized coordinates $\mathbf{q}(t) = (q^1(t), ..., q^k(t))$ that represent the instantaneous *configuration* of the mechanical system of N particles at time t. Hence, for each particle (labeled $a = 1, ..., N$), the Cartesian-coordinate position vector

$$\mathbf{x}_a \equiv \mathbf{x}_a(\mathbf{q}; t) \tag{2.34}$$

is expressed as an explicit function of the generalized coordinates.

• **Step II.** For each particle, use the position vector (2.34) to construct the velocity

$$\mathbf{v}_a(\mathbf{q}, \dot{\mathbf{q}}; t) = \frac{\partial \mathbf{x}_a}{\partial t} + \sum_{j=1}^{k} \dot{q}^j \frac{\partial \mathbf{x}_a}{\partial q^j}. \tag{2.35}$$

• **Step III.** From the position (2.34) and velocity (2.35) of each particle, construct the kinetic energy

$$K(\mathbf{q}, \dot{\mathbf{q}}; t) = \sum_a \frac{m_a}{2} |\mathbf{v}_a(\mathbf{q}, \dot{\mathbf{q}}; t)|^2$$

and the potential energy

$$U(\mathbf{q};t) = \sum_a U(\mathbf{x}_a(\mathbf{q};t), t)$$

for the system and combine them to obtain the Lagrangian

$$L(\mathbf{q}, \dot{\mathbf{q}};t) = K(\mathbf{q}, \dot{\mathbf{q}};t) - U(\mathbf{q};t). \tag{2.36}$$

• **Step IV.** From the Lagrangian (2.36), the Euler-Lagrange equations (2.31) are derived for each generalized coordinate q^j:

$$\sum_a \frac{d}{dt}\left(\frac{\partial \mathbf{x}_a}{\partial q^j} \cdot m_a \mathbf{v}_a\right) = \sum_a \left(m_a \mathbf{v}_a \cdot \frac{\partial \mathbf{v}_a}{\partial q^j} - \frac{\partial \mathbf{x}_a}{\partial q^j} \cdot \nabla_a U\right), \tag{2.37}$$

where we have used the identity $\partial \mathbf{v}_a/\partial \dot{q}^j = \partial \mathbf{x}_a/\partial q^j$.

2.3.6 *Lagrangian Mechanics in Curvilinear Coordinates**

Jacobi was the first to investigate the relation between particle dynamics and Riemannian geometry. The Euler-Lagrange equation (2.37) can be framed within the context of Riemannian geometry as follows. The kinetic energy of a single particle of mass m, with generalized coordinates $\mathbf{q} = (q^1, ..., q^k)$, is expressed as

$$K = \frac{m}{2}|\mathbf{v}|^2 = \frac{m}{2}\frac{\partial \mathbf{r}}{\partial q^i} \cdot \frac{\partial \mathbf{r}}{\partial q^j}\dot{q}^i \dot{q}^j \equiv \frac{m}{2}g_{ij}\dot{q}^i \dot{q}^j,$$

where g_{ij} denotes the metric tensor on configuration space (i.e., $ds^2 = g_{ij}dq^i dq^j$). When the particle moves in a potential $U(\mathbf{q})$, the Euler-Lagrange equation (2.37) becomes

$$\frac{d}{dt}\left(m\,g_{ij}\dot{q}^j\right) = m\,g_{ij}\ddot{q}^j + \frac{m}{2}\left(\frac{\partial g_{ij}}{\partial q^k} + \frac{\partial g_{ik}}{\partial q^j}\right)\dot{q}^j \dot{q}^k$$

$$= \frac{m}{2}\frac{\partial g_{jk}}{\partial q^i}\dot{q}^j \dot{q}^k - \frac{\partial U}{\partial q^i},$$

or

$$m\,g_{ij}\left(\frac{d^2 q^j}{dt^2} + \Gamma^j{}_{k\ell}\frac{dq^k}{dt}\frac{dq^\ell}{dt}\right) = -\frac{\partial U}{\partial q^i}, \tag{2.38}$$

where the Christoffel symbol (1.19) is defined as

$$\Gamma^j{}_{k\ell} \equiv \frac{g^{ij}}{2}\left(\frac{\partial g_{ik}}{\partial q^\ell} + \frac{\partial g_{i\ell}}{\partial q^k} - \frac{\partial g_{k\ell}}{\partial q^i}\right).$$

Thus, the concepts associated with Riemannian geometry that appear extensively in the theory of General Relativity have natural antecedents in classical Lagrangian mechanics.

2.4 Lagrangian Mechanics in Configuration Space

In this Section, we explore the Lagrangian formulation of several mechanical systems, which are listed here in order of increasing complexity. As we proceed with our examples, we should note how the Lagrangian formulation maintains its relative simplicity compared to the application of the more familiar Newton's method (Isaac Newton, 1643-1727) associated with the vectorial decomposition of forces. Here, all constraint forces are eliminated in terms of generalized coordinates and all active conservative forces are expressed in terms of gradients of suitable potential-energy functions.

2.4.1 *Example I: Pendulum*

As a first example, we reconsider the pendulum (see Sec. 2.3.1) composed of an object of mass m and a massless string of constant length ℓ in a constant gravitational field with acceleration g. Although the motion of the pendulum is two-dimensional, a single generalized coordinate is needed to describe the configuration of the pendulum: the angle θ measured from the negative y-axis (see Fig. 2.2). Here, the position of the object is given as (Step I)

$$x(\theta) = \ell \sin \theta \quad \text{and} \quad y(\theta) = -\ell \cos \theta,$$

with associated velocity components (Step II)

$$\dot{x}(\theta, \dot{\theta}) = \ell \dot{\theta} \cos \theta \quad \text{and} \quad \dot{y}(\theta, \dot{\theta}) = \ell \dot{\theta} \sin \theta.$$

Hence, the kinetic energy of the pendulum is (Step III)

$$K = \frac{m}{2} \left(\dot{x}^2 + \dot{y}^2 \right) = \frac{m}{2} \ell^2 \dot{\theta}^2,$$

and choosing the zero potential energy point when $\theta = 0$, the gravitational potential energy is (Step III)

$$U = mg\ell \left(1 - \cos \theta \right).$$

The Lagrangian $L = K - U$ is, therefore, written as (Step III)

$$L(\theta, \dot{\theta}) = \frac{m}{2} \ell^2 \dot{\theta}^2 - mg\ell \left(1 - \cos \theta \right),$$

and the Euler-Lagrange equation for θ is (Step IV)

$$\frac{\partial L}{\partial \dot{\theta}} = m\ell^2 \dot{\theta} \quad \rightarrow \quad \frac{d}{dt} \left(\frac{\partial L}{\partial \dot{\theta}} \right) = m\ell^2 \ddot{\theta} = \frac{\partial L}{\partial \theta} = -mg\ell \sin \theta$$

or

$$\ddot{\theta} + \frac{g}{\ell}\sin\theta = 0. \tag{2.39}$$

The pendulum equation (2.39), which is identical to Eq. (2.23) derived by the Newtonian method, is solved in the next Chapter through the use of the Energy method. Note that, whereas the tension force **T** in the pendulum string must be considered explicitly in the Newtonian method, **T** is replaced by the constraint $d\mathbf{x} = \ell\, d\theta\, \widehat{\theta}$ in the Lagrangian method.

2.4.2   *Example II: Bead on a Rotating Hoop*

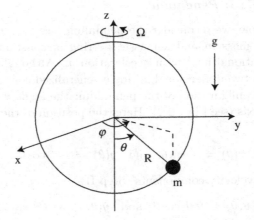

Fig. 2.5   Generalized coordinates for the bead-on-a-rotating-hoop problem.

As a second example, we consider a bead of mass m sliding freely on a hoop of radius R rotating with angular velocity Ω in a constant gravitational field with acceleration g (see Fig. 2.5). Here, since the bead on the rotating hoop effectively moves on the surface of a sphere of radius R, we use the generalized coordinates given by the two angles θ (measured from the negative z-axis) and φ (measured from the positive x-axis), where $\dot{\varphi} = \Omega$ is used as an additional constraint (i.e., expressed as $d\varphi = \Omega\, dt$). The position of the bead is given in terms of Cartesian coordinates as

$$x(\theta,t) = R\,\sin\theta\,\cos(\varphi_0 + \Omega t),$$
$$y(\theta,t) = R\,\sin\theta\,\sin(\varphi_0 + \Omega t),$$
$$z(\theta,t) = -\,R\,\cos\theta,$$

where $\varphi(t) = \varphi_0 + \Omega t$, and its associated Cartesian velocity components are

$$\dot{x}(\theta, \dot{\theta}; t) = R\left(\dot{\theta}\,\cos\theta\,\cos\varphi \;-\; \Omega\,\sin\theta\,\sin\varphi\right),$$

$$\dot{y}(\theta, \dot{\theta}; t) = R\left(\dot{\theta}\,\cos\theta\,\sin\varphi \;+\; \Omega\,\sin\theta\,\cos\varphi\right),$$

$$\dot{z}(\theta, \dot{\theta}; t) = R\,\dot{\theta}\,\sin\theta,$$

so that the kinetic energy of the bead is

$$K(\theta, \dot{\theta}) \;=\; \frac{m}{2}\,|\mathbf{v}|^2 \;=\; \frac{m\,R^2}{2}\left(\dot{\theta}^2 \;+\; \Omega^2\,\sin^2\theta\right).$$

The gravitational potential energy is

$$U(\theta) \;=\; mgR\,(1 \;-\; \cos\theta),$$

where the zero-potential energy point is chosen at the bottom of the hoop ($\theta = 0$ in Fig. 2.5).

The Lagrangian $L = K - U$ is, therefore, written as

$$L(\theta, \dot{\theta}) \;=\; \frac{m\,R^2}{2}\left(\dot{\theta}^2 \;+\; \Omega^2\,\sin^2\theta\right) \;-\; mgR\,(1 - \cos\theta),$$

and the Euler-Lagrange equation for θ is

$$\frac{\partial L}{\partial \dot{\theta}} \;=\; mR^2\dot{\theta} \;\;\rightarrow\;\; \frac{d}{dt}\left(\frac{\partial L}{\partial \dot{\theta}}\right) = mR^2\,\ddot{\theta}$$

$$\frac{\partial L}{\partial \theta} = -\,mgR\,\sin\theta$$
$$+\; mR^2\Omega^2\,\cos\theta\,\sin\theta$$

or

$$\ddot{\theta} \;+\; \sin\theta\left(\frac{g}{R} \;-\; \Omega^2\,\cos\theta\right) \;=\; 0. \tag{2.40}$$

Note that the support (constraint) force provided by the hoop (necessary in the Newtonian method) is now replaced by the constraint $R = $ constant in the Lagrangian method. Furthermore, although the motion intrinsically takes place on the surface of a sphere of radius R, the azimuthal motion is constrained ($d\varphi(t) = \Omega\,dt$) and, thus, the motion of the bead takes place in a one-dimensional configuration space (with coordinate θ).

Lastly, we note that Eq. (2.40) displays *bifurcation* behavior, which is investigated in Chap. 8. For $\Omega^2 < g/R$, the equilibrium point $\theta = 0$ is stable while, for $\Omega^2 > g/R$, the equilibrium point $\theta = 0$ is now unstable and the new equilibrium point $\theta = \arccos(g/\Omega^2 R)$ is stable.

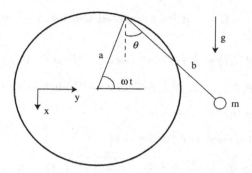

Fig. 2.6　Generalized coordinates for the rotating-pendulum problem.

2.4.3　*Example III: Rotating Pendulum*

As a third example, we consider a pendulum of mass m and length b attached to the edge of a disk of radius a rotating at angular velocity ω in a constant gravitational field with acceleration g. Placing the origin at the center of the disk, the coordinates of the pendulum mass are

$$(x,\, y) = a\,(-\sin\omega t,\, \cos\omega t) + b\,(\cos\theta,\, \sin\theta),$$

so that the velocity components are

$$(\dot{x},\, \dot{y}) = -\,a\omega\,(\cos\omega t,\, \sin\omega t) + b\dot{\theta}\,(-\sin\theta,\, \cos\theta),$$

and the squared velocity is

$$v^2 = a^2\omega^2 + b^2\dot{\theta}^2 + 2\,ab\,\omega\,\dot{\theta}\,\sin(\theta - \omega t).$$

Setting the zero potential energy at $x = 0$, the gravitational potential energy is

$$U = -\,mg\,x = mga\,\sin\omega t - mgb\,\cos\theta.$$

The Lagrangian $L = K - U$ is, therefore, written as

$$L(\theta, \dot{\theta}; t) = \frac{m}{2}\left[\,a^2\omega^2 + b^2\dot{\theta}^2 + 2\,ab\,\omega\,\dot{\theta}\,\sin(\theta - \omega t)\,\right]$$
$$-\,mga\,\sin\omega t + mgb\,\cos\theta, \tag{2.41}$$

and the Euler-Lagrange equation for θ is

$$\frac{\partial L}{\partial \dot{\theta}} = mb^2\,\dot{\theta} + m\,ab\,\omega\,\sin(\theta - \omega t) \;\rightarrow$$
$$\frac{d}{dt}\left(\frac{\partial L}{\partial \dot{\theta}}\right) = mb^2\,\ddot{\theta} + m\,ab\,\omega\,(\dot{\theta} - \omega)\,\cos(\theta - \omega t)$$

and

$$\frac{\partial L}{\partial \theta} = m\,ab\,\omega\,\dot{\theta}\,\cos(\theta - \omega t) - mg\,b\,\sin\theta$$

or

$$\ddot{\theta} + \frac{g}{b}\,\sin\theta - \frac{a}{b}\,\omega^2\,\cos(\theta - \omega t) = 0$$

We recover the standard equation of motion for the pendulum when a or ω vanish.

Note that the terms

$$\frac{m}{2}\,a^2\,\omega^2 - mga\,\sin\omega t$$

in the Lagrangian (2.41) play no role in determining the dynamics of the system. In fact, as can easily be shown (see Sec. 2.5), a Lagrangian L is always defined up to an exact time derivative, i.e., the Lagrangians L and $L' = L - df/dt$, where $f(\mathbf{q}, t)$ is an arbitrary function, lead to the same Euler-Lagrange equations. In the present case,

$$f(t) = [(m/2)\,a^2\omega^2]\,t + (mga/\omega)\,\cos\omega t$$

and thus this term can be omitted from the Lagrangian (2.41) without changing the equations of motion.

2.4.4 *Example IV: Compound Atwood Machine*

As a fourth example, we consider a compound Atwood machine (see Fig. 2.7) composed three masses (labeled m_1, m_2, and m_3) attached by two massless ropes through two massless pulleys in a constant gravitational field with acceleration g.

The two generalized coordinates for this system are the distance x of mass m_1 from the top of the first pulley and the distance y of mass m_2 from the top of the second pulley; here, the lengths ℓ_a and ℓ_b are constants. The coordinates and velocities of the three masses m_1, m_2, and m_3 are

$$x_1 = x \rightarrow v_1 = \dot{x},$$
$$x_2 = \ell_a - x + y \rightarrow v_2 = \dot{y} - \dot{x},$$
$$x_3 = \ell_a - x + \ell_b - y \rightarrow v_3 = -\dot{x} - \dot{y},$$

respectively, so that the total kinetic energy is

$$K = \frac{m_1}{2}\,\dot{x}^2 + \frac{m_2}{2}\,(\dot{y} - \dot{x})^2 + \frac{m_3}{2}\,(\dot{x} + \dot{y})^2.$$

Fig. 2.7   Generalized coordinates for the compound-Atwood problem.

Placing the zero potential energy at the top of the first pulley, the total gravitational potential energy, on the other hand, can be written as

$$U = -g\,x\,(m_1 - m_2 - m_3) - g\,y\,(m_2 - m_3),$$

where constant terms were omitted. The Lagrangian $L = K - U$ is, therefore, written as

$$L(x, \dot{x},\, y, \dot{y}) = \frac{m_1}{2}\,\dot{x}^2 + \frac{m_2}{2}\,(\dot{x} - \dot{y})^2 + \frac{m_3}{2}\,(\dot{x} + \dot{y})^2$$
$$+ g\,x\,(m_1 - m_2 - m_3) + g\,y\,(m_2 - m_3).$$

The Euler-Lagrange equation for x is

$$\frac{\partial L}{\partial \dot{x}} = (m_1 + m_2 + m_3)\,\dot{x} + (m_3 - m_2)\,\dot{y} \;\;\rightarrow$$
$$\frac{d}{dt}\left(\frac{\partial L}{\partial \dot{x}}\right) = (m_1 + m_2 + m_3)\,\ddot{x} + (m_3 - m_2)\,\ddot{y}$$

$$\frac{\partial L}{\partial x} = g\,(m_1 - m_2 - m_3)$$

while the Euler-Lagrange equation for y is

$$\frac{\partial L}{\partial \dot{y}} = (m_3 - m_2)\,\dot{x} + (m_2 + m_3)\,\dot{y} \;\;\rightarrow$$
$$\frac{d}{dt}\left(\frac{\partial L}{\partial \dot{y}}\right) = (m_3 - m_2)\,\ddot{x} + (m_2 + m_3)\,\ddot{y}$$

$$\frac{\partial L}{\partial y} = g\,(m_2 - m_3).$$

We combine these two Euler-Lagrange equations

$$(m_1 + m_2 + m_3)\ddot{x} + (m_3 - m_2)\ddot{y} = g\,(m_1 - m_2 - m_3), \quad (2.42)$$

$$(m_3 - m_2)\ddot{x} + (m_2 + m_3)\ddot{y} = g\,(m_2 - m_3), \quad (2.43)$$

to describe the dynamical evolution of the compound Atwood machine. This set of equations can, in fact, be solved explicitly as

$$\ddot{x} = g\left(\frac{m_1\,m_+ - (m_+^2 - m_-^2)}{m_1\,m_+ + (m_+^2 - m_-^2)}\right)$$

and

$$\ddot{y} = g\left(\frac{2\,m_1\,m_-}{m_1\,m_+ + (m_+^2 - m_-^2)}\right),$$

where $m_\pm = m_2 \pm m_3$. Note also that, by using the energy conservation law $E = K + U$, it can be shown that the position z of the center of mass of the mechanical system (as measured from the top of the first pulley) satisfies the relation

$$Mg\,(z - z_0) = \frac{m_1}{2}\,\dot{x}^2 + \frac{m_2}{2}\,(\dot{y} - \dot{x})^2 + \frac{m_3}{2}\,(\dot{x} + \dot{y})^2 > 0, \quad (2.44)$$

where $M = (m_1 + m_2 + m_3)$ denotes the total mass of the system and we have assumed that the system starts from rest (with its center of mass located at z_0). This important relation tells us that, as the masses start to move ($\dot{x} \neq 0$ and $\dot{y} \neq 0$), the center of mass must fall: $z > z_0$.

Before proceeding to our last example, we introduce a convenient technique (henceforth known as *Frozen Degrees of Freedom*) for checking on the physical accuracy of any set of coupled Euler-Lagrange equations. Hence, for the Euler-Lagrange equation (2.42), we may freeze the degree of freedom associated with the y-coordinate (i.e., we set $\dot{y} = 0 = \ddot{y}$ or $m_- = 0$) to obtain $\ddot{x} = g\,(m_1 - m_+)/(m_1 + m_+)$, in agreement with the analysis of a *simple* Atwood machine composed of a mass m_1 on one side and a mass $m_+ = m_2 + m_3$ on the other side. Likewise, for the Euler-Lagrange equation (2.43), we may freeze the degree of freedom associated with the x-coordinate (i.e., we set $\dot{x} = 0 = \ddot{x}$ or $m_1 m_+ = m_+^2 - m_-^2$) to obtain $\ddot{y} = g\,(m_-/m_+)$, again in agreement with the analysis of a *simple* Atwood machine.

2.4.5 *Example V: Pendulum with Oscillating Fulcrum*

As our final example, we consider the case of a pendulum of mass m and length ℓ attached to a massless block which is attached to a fixed wall by

Fig. 2.8 Generalized coordinates for the oscillating-pendulum problem.

a massless spring of constant k. Here, we assume that the massless block moves without friction on a set of rails (see problem 5). We use the two generalized coordinates x and θ shown in Fig. 2.8 and write the Cartesian coordinates (y, z) of the pendulum mass as $y = x + \ell \sin \theta$ and $z = -\ell \cos \theta$, with its associated velocity components $\dot{y} = \dot{x} + \ell \dot{\theta} \cos \theta$ and $\dot{z} = \ell \dot{\theta} \sin \theta$. The kinetic energy of the pendulum is thus

$$K = \frac{m}{2} \left(\dot{y}^2 + \dot{z}^2 \right) = \frac{m}{2} \left(\dot{x}^2 + \ell^2 \dot{\theta}^2 + 2\ell \cos \theta \, \dot{x}\dot{\theta} \right).$$

The potential energy $U = U_k + U_g$ has two terms: one term $U_k = \frac{1}{2} kx^2$ associated with displacement of the spring away from its equilibrium position and one term $U_g = mgz$ associated with gravity. Hence, the Lagrangian for this system is

$$L(x, \theta, \dot{x}, \dot{\theta}) = \frac{m}{2} \left(\dot{x}^2 + \ell^2 \dot{\theta}^2 + 2\ell \cos \theta \, \dot{x}\dot{\theta} \right) - \frac{k}{2} x^2 + mg\ell \cos \theta.$$

The Euler-Lagrange equation for x is

$$\frac{\partial L}{\partial \dot{x}} = m \left(\dot{x} + \ell \cos \theta \, \dot{\theta} \right) \rightarrow$$

$$\frac{d}{dt} \left(\frac{\partial L}{\partial \dot{x}} \right) = m \, \ddot{x} + m\ell \left(\ddot{\theta} \cos \theta - \dot{\theta}^2 \sin \theta \right)$$

$$\frac{\partial L}{\partial x} = -k x$$

while the Euler-Lagrange equation for θ is

$$\frac{\partial L}{\partial \dot{\theta}} = m\ell \left(\ell \dot{\theta} + \dot{x} \cos\theta \right) \ \rightarrow$$

$$\frac{d}{dt} \left(\frac{\partial L}{\partial \dot{\theta}} \right) = m\ell^2 \ddot{\theta} + m\ell \left(\ddot{x} \cos\theta - \dot{x}\dot{\theta} \sin\theta \right)$$

$$\frac{\partial L}{\partial \theta} = - m\ell \dot{x}\dot{\theta} \sin\theta - mg\ell \sin\theta$$

or[1]

$$m\ddot{x} + kx = m\ell \left(\dot{\theta}^2 \sin\theta - \ddot{\theta} \cos\theta \right), \tag{2.45}$$

$$\ddot{\theta} + (g/\ell) \sin\theta = - (\ddot{x}/\ell) \cos\theta. \tag{2.46}$$

Here, we recover the dynamical equation for a block-and-spring harmonic oscillator from Eq. (2.45) by freezing the degree of freedom associated with the θ-coordinate (i.e., by setting $\dot{\theta} = 0 = \ddot{\theta}$) and the dynamical equation for the pendulum from Eq. (2.46) by freezing the degree of freedom associated with the x-coordinate (i.e., by setting $\dot{x} = 0 = \ddot{x}$). It is easy to see from this last example how powerful and yet simple the Lagrangian method is compared to the Newtonian method.

2.5 Symmetries and Conservation Laws

We are sometimes faced with a Lagrangian function that is either independent of time, independent of a linear spatial coordinate, or independent of an angular coordinate. The Noether theorem (Amalie Emmy Noether, 1882-1935) states that *for each symmetry of the Lagrangian there corresponds a conservation law (and vice versa).* When the Lagrangian L is invariant under a time translation, a space translation, or a spatial rotation, the conservation law involves energy, linear momentum, or angular momentum, respectively.

We begin our discussion with a general expression for the variation δL of the Lagrangian $L(\mathbf{q}, \dot{\mathbf{q}}, t)$:

$$\delta L = \delta\mathbf{q} \cdot \left[\frac{\partial L}{\partial \mathbf{q}} - \frac{d}{dt} \left(\frac{\partial L}{\partial \dot{\mathbf{q}}} \right) \right] + \frac{d}{dt} \left(\delta\mathbf{q} \cdot \frac{\partial L}{\partial \dot{\mathbf{q}}} \right),$$

obtained after re-arranging the term $\delta\dot{\mathbf{q}} \cdot \partial L / \partial \dot{\mathbf{q}}$ in Eq. (2.30). Next, we make use of the Euler-Lagrange equations for $\mathbf{q}$ (which enables us to drop

---

[1]Do not attempt to integrate these equations of motion since, without taking into account the mass of the block, these equations are singular. (See problem 5.)

the term $\delta\mathbf{q} \cdot [\cdots]$) and we find

$$\delta L = \frac{d}{dt}\left(\delta\mathbf{q} \cdot \frac{\partial L}{\partial\dot{\mathbf{q}}}\right). \tag{2.47}$$

Lastly, the variation δL can only be generated by a time translation δt:

$$\delta L \equiv \frac{d}{d\epsilon}\left[L\left(\mathbf{q}(t + \epsilon\,\delta t),\ \dot{\mathbf{q}}(t + \epsilon\,\delta t), t\right)\right]_{\epsilon=0} = \delta t\left(\frac{dL}{dt} - \frac{\partial L}{\partial t}\right).$$

By combining this expression with Eq. (2.47), we find

$$\frac{d}{dt}\left(\delta\mathbf{q} \cdot \frac{\partial L}{\partial\dot{\mathbf{q}}} - \delta t\, L\right) \equiv -\,\delta t\left(\frac{\partial L}{\partial t}\right)_{\mathbf{q},\dot{\mathbf{q}}}, \tag{2.48}$$

where $(\partial L/\partial t)_{\mathbf{q},\dot{\mathbf{q}}}$ denotes the partial time derivative of $L(\mathbf{q},\dot{\mathbf{q}},t)$ at constant $(\mathbf{q},\dot{\mathbf{q}})$. Equation (2.48) is, henceforth, refered to as the Noether equation for finite-dimensional mechanical systems; see Chap. 9, Eq. (9.14), for the infinite-dimensional case.

We now apply Noether's Theorem, based on the Noether equation (2.48), to investigate the connection between symmetries of the Lagrangian with conservation laws.

2.5.1 *Energy Conservation Law*

First, we consider time translations, $t \to t + \delta t$ (with $\delta\mathbf{q} = \dot{\mathbf{q}}\,\delta t$), so that the Noether equation (2.48) becomes Euler's Second Equation

$$-\frac{\partial L}{\partial t} = \frac{d}{dt}\left(\dot{\mathbf{q}} \cdot \frac{\partial L}{\partial\dot{\mathbf{q}}} - L\right).$$

Noether's Theorem states that if the Lagrangian is invariant under time translations (i.e., $\partial L/\partial t = 0$), then energy is conserved, $dE/dt = 0$, where

$$E \equiv \dot{\mathbf{q}} \cdot \frac{\partial L}{\partial\dot{\mathbf{q}}} - L \tag{2.49}$$

defines the energy invariant.

2.5.2 *Momentum Conservation Laws*

Next, we consider invariance under spatial translations, $\mathbf{q} \to \mathbf{q} + \boldsymbol{\epsilon}$ (where $\delta\mathbf{q} = \boldsymbol{\epsilon}$ denotes a constant infinitesimal displacement in an arbitrary direction and $\delta t = 0$), so that the Noether equation (2.48) yields the *linear* momentum conservation law

$$0 = \frac{d}{dt}\left(\frac{\partial L}{\partial\dot{\mathbf{q}}}\right) = \frac{d\mathbf{P}}{dt},$$

where

$$\mathbf{P} \equiv \frac{\partial L}{\partial \dot{\mathbf{q}}} \qquad (2.50)$$

denotes the total linear momentum of the mechanical system.

On the other hand, when the Lagrangian is invariant under spatial rotations, $\mathbf{q} \to \mathbf{q} + \delta\theta\,\hat{\mathbf{z}} \times \mathbf{q}$ (where a constant infinitesimal rotation $\delta\theta$ is carried out about an arbitrary symmetry z-axis), the Noether equation (2.48) yields the *angular* momentum conservation law

$$0 = \frac{d}{dt}\left(\mathbf{q} \times \frac{\partial L}{\partial \dot{\mathbf{q}}}\right) = \frac{d\mathbf{L}}{dt},$$

where $\mathbf{L} = \mathbf{q} \times \mathbf{P}$ denotes the total angular momentum of the mechanical system.

2.5.3 *Invariance Properties of a Lagrangian*

Lastly, an important invariance property of the Lagrangian is related to the fact that the Euler-Lagrange equations themselves are invariant under the *gauge* transformation

$$L \to L + \frac{dF}{dt} \qquad (2.51)$$

on the Lagrangian itself, where $F(\mathbf{q}, t)$ is an arbitrary time-dependent function so that

$$\frac{dF(\mathbf{q}, t)}{dt} = \frac{\partial F}{\partial t} + \sum_j \dot{q}^j\,\frac{\partial F}{\partial q^j}.$$

To investigate the invariance property (2.51), we call $L' = L + dF/dt$ the new Lagrangian and L the old Lagrangian, and consider the new Euler-Lagrange equations

$$\frac{d}{dt}\left(\frac{\partial L'}{\partial \dot{q}^i}\right) = \frac{\partial L'}{\partial q^i}.$$

We now express each term in terms of the old Lagrangian L and the function F. Let us begin with

$$\frac{\partial L'}{\partial \dot{q}^i} = \frac{\partial}{\partial \dot{q}^i}\left(L + \frac{\partial F}{\partial t} + \sum_j \dot{q}^j\,\frac{\partial F}{\partial q^j}\right) = \frac{\partial L}{\partial \dot{q}^i} + \frac{\partial F}{\partial q^i},$$

so that

$$\frac{d}{dt}\left(\frac{\partial L'}{\partial \dot{q}^i}\right) = \frac{d}{dt}\left(\frac{\partial L}{\partial \dot{q}^i}\right) + \frac{\partial^2 F}{\partial t \partial q^i} + \sum_k \dot{q}^k\,\frac{\partial^2 F}{\partial q^k \partial q^i}.$$

Next, we find

$$\frac{\partial L'}{\partial q^i} = \frac{\partial}{\partial q^i}\left(L + \frac{\partial F}{\partial t} + \sum_k \dot{q}^k \frac{\partial F}{\partial q^k}\right)$$

$$= \frac{\partial L}{\partial q^i} + \frac{\partial^2 F}{\partial q^i \partial t} + \sum_k \dot{q}^k \frac{\partial^2 F}{\partial q^i \partial q^k}.$$

Using the symmetry properties

$$\dot{q}^j \frac{\partial^2 F}{\partial q^i \partial q^j} = \dot{q}^j \frac{\partial^2 F}{\partial q^j \partial q^i} \quad \text{and} \quad \frac{\partial^2 F}{\partial t \partial q^i} = \frac{\partial^2 F}{\partial q^i \partial t},$$

we easily verify that

$$\frac{d}{dt}\left(\frac{\partial L'}{\partial \dot{q}^i}\right) - \frac{\partial L'}{\partial q^i} = \frac{d}{dt}\left(\frac{\partial L}{\partial \dot{q}^i}\right) - \frac{\partial L}{\partial q^i} = 0.$$

Hence, since L and $L' = L + dF/dt$ lead to the same Euler-Lagrange equations, they are said to be equivalent.

Using this invariance property, for example, we note that the Lagrangian is also invariant under the Galilean velocity transformation $\mathbf{v} \to \mathbf{v} + \boldsymbol{\alpha}$, so that the Lagrangian variation

$$\delta L = \boldsymbol{\alpha} \cdot \left(\mathbf{v} \frac{\partial L}{\partial v^2}\right) \equiv \boldsymbol{\alpha} \cdot \frac{d\mathbf{x}}{dt} \frac{\partial L}{\partial v^2},$$

using the kinetic identity $\partial L/\partial v^2 = m/2$, can be written as an exact time derivative

$$\delta L = \frac{d}{dt}\left(\boldsymbol{\alpha} \cdot \frac{m}{2}\mathbf{x}\right) \equiv \frac{d\delta F}{dt}.$$

Hence, because Lagrangian mechanics is invariant under the *gauge* transformation (2.51), the Lagrangian L is said to be Galilean invariant.

2.5.4 *Lagrangian Mechanics with Symmetries*

As an example of Lagrangian mechanics with symmetries, we return to the motion of a particle of mass m constrained to move on the surface of a cone of apex angle α (such that $\sqrt{x^2 + y^2} = z \tan\alpha$) in the presence of a gravitational field (see Fig. 2.4 and Sec. 2.3.3).

The Lagrangian for this constrained mechanical system is expressed in terms of the generalized coordinates (s, θ), where s denotes the distance from the cone's apex (labeled O in Fig. 2.4) and θ is the standard polar angle in the (x, y)-plane. Hence, by combining the kinetic energy $K = $

$\frac{1}{2} m(\dot{s}^2 + s^2 \dot{\theta}^2 \sin^2 \alpha)$ with the potential energy $U = mgz = mg\,s\,\cos\alpha$, we construct the Lagrangian

$$L(s, \theta; \dot{s}, \dot{\theta}) = \frac{1}{2} m \left(\dot{s}^2 + s^2 \dot{\theta}^2 \sin^2 \alpha \right) - mg\,s\,\cos\alpha. \qquad (2.52)$$

Since the Lagrangian is independent of the polar angle θ, the canonical angular momentum

$$p_\theta = \frac{\partial L}{\partial \dot{\theta}} = ms^2 \dot{\theta} \sin^2 \alpha \qquad (2.53)$$

is a constant of the motion (as predicted by Noether's Theorem). The Euler-Lagrange equation for s, on the other hand, is expressed as

$$\ddot{s} + g \cos \alpha = s \dot{\theta}^2 \sin^2 \alpha = \frac{p_\theta^2}{m^2 s^3 \sin^2 \alpha}, \qquad (2.54)$$

where $g \cos \alpha$ denotes the component of the gravitational acceleration parallel to the surface of the cone. The right side of Eq. (2.54), which represents the effect of the centrifugal force, becomes a function of s only after using $\dot{\theta} = p_\theta/(m\,s^2 \sin^2 \alpha)$, which follows from the conservation of angular momentum.

2.5.5 *Routh's Procedure*

Edward John Routh (1831-1907) introduced a simple procedure for eliminating ignorable degrees of freedom while introducing their corresponding conserved momenta within the context of Lagrangian Mechanics.

Consider, for example, two-dimensional motion on the (x, y)-plane represented by the Lagrangian $L(r; \dot{r}, \dot{\theta})$, where r and θ are the polar coordinates. Since the Lagrangian under consideration is independent of the angle θ, the canonical momentum $p_\theta = \partial L/\partial \dot{\theta}$ is conserved. Routh's procedure involves the construction of the *Routh-Lagrange* function (or Routhian)

$$R(r, \dot{r}; p_\theta) \equiv L(r; \dot{r}, \dot{\theta}) - p_\theta \dot{\theta}(r, p_\theta), \qquad (2.55)$$

where $\dot{\theta}(r, p_\theta)$ is expressed as a function of r and p_θ. Note that the sign convention used in Eq. (2.55), which is different from Landau's convention [13], implies that the Routhian R can be treated as a *reduced* Lagrangian.

Returning to the case of the constrained motion of a particle on the surface of a cone in the presence of gravity, the Lagrangian (2.52) can be reduced to the Routhian:

$$R(s, \dot{s}; p_\theta) \equiv L \left(s, \dot{s}; \dot{\theta}(s, p_\theta) \right) - p_\theta \dot{\theta}(s, p_\theta)$$

$$= \frac{1}{2} m \dot{s}^2 - \left(mg\,s\,\cos\alpha + \frac{p_\theta^2}{2\,ms^2 \sin^2 \alpha} \right), \qquad (2.56)$$

where the function $\dot{\theta}(s, p_\theta)$ is obtained from Eq. (2.53). The equation of motion (2.54) can thus be expressed in Euler-Lagrange form

$$\frac{d}{ds}\left(\frac{\partial R}{\partial \dot{s}}\right) = \frac{\partial R}{\partial s} \quad \rightarrow \quad m\,\ddot{s} = -V'(s),$$

in terms of the effective potential

$$V(s) = mg\,s\,\cos\alpha + \frac{p_\theta^2}{2\,ms^2\,\sin^2\alpha}.$$

Here, the effective potential $V(s)$ has a single minimum at $s = s_0$, where

$$s_0 = \left(\frac{p_\theta^2}{m^2g\,\sin^2\alpha\,\cos\alpha}\right)^{\frac{1}{3}}$$

and $V_0 \equiv V(s_0) = \frac{3}{2}\,mg\,s_0\cos\alpha$.

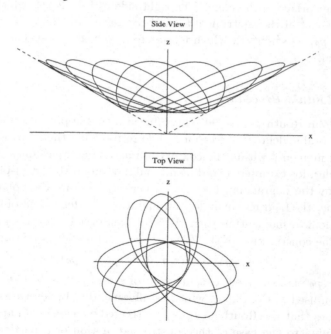

Fig. 2.9   Particle orbits on the surface of a cone.

Figure 2.9 shows the results of the numerical integration of the dimensionless Euler-Lagrange equations for $\theta(\tau)$ and $\sigma(\tau) \equiv s(\tau)/s_0$, where $\tau \equiv t\,\sqrt{(g/s_0)}\,\cos\alpha$; see Appendix A.5 for some advice concerning the numerical solution of coupled ordinary differential equations. The top figure in Fig. 2.9 shows a projection of the path of the particle on the (x, z)-plane (side view), which clearly shows that the motion is periodic as the σ-coordinate oscillates between two finite values of σ. The bottom figure in Fig. 2.9 shows a projection of the path of the particle on the (x, y)-plane (top view), which shows the slow precession motion in the θ-coordinate.

In the next Chapter, we will show that the doubly-periodic motion of the particle moving on the surface of the inverted cone is a result of the conservation law of angular momentum and energy (since the Lagrangian system is also independent of time).

2.6 Lagrangian Mechanics in the CM Frame

An important frame of reference associated with the dynamical description of the motion of interacting particles and rigid bodies is provided by the center-of-mass (CM) frame. The following discussion focuses on the Lagrangian for an isolated two-particle system expressed as

$$L = \frac{m_1}{2}|\dot{\mathbf{r}}_1|^2 + \frac{m_2}{2}|\dot{\mathbf{r}}_2|^2 - U(\mathbf{r}_1 - \mathbf{r}_2),$$

where $\mathbf{r}_1$ and $\mathbf{r}_2$ represent the positions of the particles of mass m_1 and m_2, respectively, and $U(\mathbf{r}_1, \mathbf{r}_2) = U(\mathbf{r}_1 - \mathbf{r}_2)$ is the potential energy for an isolated two-particle system (see Fig. 2.10).

Fig. 2.10   Center-of-Mass frame.

Let us now define the position $\mathbf{R}$ of the center of mass

$$\mathbf{R} = \frac{m_1\,\mathbf{r}_1 + m_2\,\mathbf{r}_2}{m_1 + m_2},$$

and define the relative inter-particle position vector $\mathbf{r} = \mathbf{r}_1 - \mathbf{r}_2$, so that the particle positions can be expressed as

$$\mathbf{r}_1 = \mathbf{R} + \frac{m_2}{M}\mathbf{r} \quad \text{and} \quad \mathbf{r}_2 = \mathbf{R} - \frac{m_1}{M}\mathbf{r},$$

where $M = m_1 + m_2$ is the total mass of the two-particle system (see Fig. 2.10). The Lagrangian of the isolated two-particle system, thus, becomes

$$L = \frac{M}{2}|\dot{\mathbf{R}}|^2 + \frac{\mu}{2}|\dot{\mathbf{r}}|^2 - U(\mathbf{r}),$$

where

$$\mu = \frac{m_1 m_2}{m_1 + m_2} = \left(\frac{1}{m_1} + \frac{1}{m_2}\right)^{-1}$$

denotes the *reduced* mass of the two-particle system. We note that the angular momentum of the two-particle system is expressed as

$$\mathbf{L} = \sum_a \mathbf{r}_a \times \mathbf{p}_a = \mathbf{R} \times \mathbf{P} + \mathbf{r} \times \mathbf{p}, \tag{2.57}$$

where the canonical momentum of the center-of-mass $\mathbf{P}$ and the canonical momentum $\mathbf{p}$ of the two-particle system in the CM frame are defined, respectively, as

$$\mathbf{P} = \frac{\partial L}{\partial \dot{\mathbf{R}}} = M\dot{\mathbf{R}} \quad \text{and} \quad \mathbf{p} = \frac{\partial L}{\partial \dot{\mathbf{r}}} = \mu\dot{\mathbf{r}}.$$

For an isolated system ($\partial L/\partial \mathbf{R} = 0$), the canonical momentum $\mathbf{P}$ of the center-of-mass is a constant of the motion. The CM reference frame can be defined by the condition $\mathbf{R} = 0$, i.e., we move the origin of our coordinate system to the CM position.

In the CM frame, the Lagrangian for an isolated two-particle system is

$$L(\mathbf{r}, \dot{\mathbf{r}}) = \frac{\mu}{2}|\dot{\mathbf{r}}|^2 - U(\mathbf{r}), \tag{2.58}$$

which describes the motion of a *fictitious* particle of mass μ at position $\mathbf{r}$, where the positions of the two *real* particles of masses m_1 and m_2 are

$$\mathbf{r}_1 = \frac{m_2}{M}\mathbf{r} \quad \text{and} \quad \mathbf{r}_2 = -\frac{m_1}{M}\mathbf{r}. \tag{2.59}$$

Hence, once the Euler-Lagrange equation

$$\frac{d}{dt}\left(\frac{\partial L}{\partial \dot{\mathbf{r}}}\right) = \frac{\partial L}{\partial \mathbf{r}} \quad \rightarrow \quad \mu\ddot{\mathbf{r}} = -\nabla U(\mathbf{r})$$

is solved for $\mathbf{r}(t)$, the motion of the two particles in the CM frame is determined through Eqs. (2.59).

The angular momentum $\mathbf{L} = \mu\mathbf{r} \times \dot{\mathbf{r}}$ in the CM frame satisfies the evolution equation

$$\frac{d\mathbf{L}}{dt} = \mathbf{r} \times \mu\ddot{\mathbf{r}} = -\mathbf{r} \times \nabla U(\mathbf{r}). \tag{2.60}$$

Table 2.1   Summary of Chapter 2: Lagrangian Mechanics.

Topic	Equation
Maupertuis' Principle	(2.1)
Jacobi's Principle	(2.8)
Euler-Jacobi Equation	(2.9)
Principle of Virtual Work	(2.13)
d'Alembert's Principle	(2.15)
Hamilton's Principle	(2.21)
Lagrangian Method	(2.34)-(2.37)
Noether Equation	(2.48)
Energy-Momentum Conservation Laws	(2.49)-(2.50)
Routh-Lagrange Function	(2.55)
Lagrangian Function in CM Frame	(2.58)

Here, using spherical coordinates (r, θ, φ), we find

$$\frac{d\mathbf{L}}{dt} = -\widehat{\varphi} \, \frac{\partial U}{\partial \theta} + \frac{\widehat{\theta}}{\sin\theta} \, \frac{\partial U}{\partial \varphi}.$$

If motion is originally taking place on the (x, y)-plane (i.e., at $\theta = \pi/2$) and the potential $U(r, \varphi)$ is independent of the polar angle θ, then the angular momentum vector is $\mathbf{L} = \ell\,\widehat{\mathbf{z}}$ and its magnitude ℓ satisfies the evolution equation

$$\frac{d\ell}{dt} = -\frac{\partial U}{\partial \varphi}.$$

Hence, for motion in a potential $U(r)$ that depends only on the radial position r, the angular momentum remains along the z-axis, and $\mathbf{L} = \ell\,\widehat{\mathbf{z}}$ represents an additional constant of motion. Motion in such central-force potentials will be studied in Chap. 4.

2.7   Summary

Chapter 2 presented various variational principles used to describe particle dynamics in force fields that are derived from potential-energy functions. The four-step Lagrangian method was introduced as powerful way of deriving equations of motion in configuration space and several examples were given. Table 2.1 presents a summary of the important topics of Chapter 2.

2.8   Problems

1. Consider a physical system composed of two blocks of mass m_1 and m_2 resting on incline planes placed at angles θ_1 and θ_2, respectively, as measured from the horizontal (see Fig. 2.11). The only active force acting on the blocks is due to gravity ($\mathbf{g} = -g\widehat{\mathbf{y}}$): $\mathbf{F}_i = -m_i g\widehat{\mathbf{y}}$ and, thus, the Principle of Virtual Work (2.13) implies that the system is in static equilibrium if

$$0 = m_1 g\widehat{\mathbf{y}} \cdot \delta\mathbf{x}^1 + m_2 g\widehat{\mathbf{y}} \cdot \delta\mathbf{x}^2.$$

Find the virtual displacements $\delta\mathbf{x}^1$ and $\delta\mathbf{x}^2$ needed to show that, according to the Principle of Virtual Work, the condition for static equilibrium is $m_1 \sin\theta_1 = m_2 \sin\theta_2$.

Fig. 2.11   Problem 1.

2. A particle of mass m is constrained to slide down a curve $y = V(x)$ under the action of gravity without friction. Show that the Euler-Lagrange equation for this system yields the equation

$$\ddot{x} = -V' \left(g + \ddot{V} \right),$$

where $\dot{V} = \dot{x}\,V'$ and $\ddot{V} = \ddot{x}\,V' + \dot{x}^2\,V''$.

3. Derive Eq. (2.44) for the compound Atwood machine.

4. A bead (of mass m) slides without friction on a wire in the shape of a cycloid: $x(\theta) = a\,(\theta - \sin\theta)$ and $y(\theta) = a\,(1 + \cos\theta)$.

(a) Show that the Lagrangian for this problem is

$$L(\theta, \dot{\theta}) = m\,a^2\,(1 - \cos\theta)\,\dot{\theta}^2 - mg\,a\,(1 + \cos\theta)$$

and derive the Euler-Lagrange equation for the angle θ.

(b) Show that the equation of motion for $u = \cos(\theta/2)$ is $\ddot{u} + \Omega^2\,u = 0$ and find an expression for Ω.

5. A cart of mass M is placed on rails and attached to a wall with the help of a massless spring with constant k (Fig. 2.12); the spring is in its equilibrium state when the cart is at a distance x_0 from the wall. A pendulum of mass m and length ℓ is attached to the cart (as shown).

Fig. 2.12   Problem 5.

(a) Show that the Lagrangian for the cart-pendulum system is

$$L(x, \dot{x}, \theta, \dot{\theta}) = \frac{1}{2}(m + M)\dot{x}^2 + \frac{1}{2}m\ell^2\dot{\theta}^2 + m\ell\dot{x}\dot{\theta}\cos\theta$$

$$- \frac{1}{2}kx^2 + mg\ell\cos\theta, \tag{2.61}$$

where x denotes the position of the cart (as measured from a suitable origin) and θ denotes the angular position of the pendulum.

(b) From the Lagrangian (2.61), write the Euler-Lagrange equations for the generalized coordinates x and θ.

(c) Write the normalized equations for $\xi \equiv x/\ell$ and θ in terms of the normalized time $\tau \equiv t\sqrt{g/\ell}$ and the two dimensionless parameters $\mu \equiv m/(m + M) < 1$ and $\Omega^2 \equiv k\ell/[(m + M)g]$.

6. An Atwood machine is composed of two masses m and M attached by means of a massless rope into which a massless spring (with constant k) is inserted (as shown in Fig. 2.13). When the spring is in a relaxed state, the spring-rope length is ℓ.

(a) Find suitable generalized coordinates to describe the motion of the two

Fig. 2.13   Problem 6.

masses (allowing for elongation or compression of the spring).

(b) Using these generalized coordinates, construct the Lagrangian and derive the appropriate Euler-Lagrange equations.

7. An Atwood machine is composed of two masses m and M attached by means of a massless rope. The massless pulley is attached to a massless spring with constant k (as shown in Fig. 2.14).

(a) Find suitable generalized coordinates to describe the motion of the two masses (allowing for elongation or compression of the spring).

(b) Using these generalized coordinates, construct the Lagrangian and derive the appropriate Euler-Lagrange equations.

8. A pendulum of length ℓ and mass m is attached to a point of mass M that is constrained to only move horizontally.

(a) Derive the coupled Euler-Lagrange equations for the horizontal displacement x and the angular displacement θ.

(b) Show that this system possesses a symmetry related to translations along the x-axis. Using the corresponding conservation law, show that the

Fig. 2.14 Problem 7.

coupled equations derived in Part (a) can be expressed as

$$\left(1 - \mu \, \cos^2 \theta\right) \ddot\theta + \omega_g^2 \, \sin \theta + \frac{\mu}{2} \, \dot\theta^2 \, \sin 2\theta = 0,$$

where $\mu \equiv m/(m+M)$ is the mass ratio and $\omega_g \equiv \sqrt{g/\ell}$ is the pendulum angular frequency.

9. Dissipative effects can be included within the Lagrangian formalism through the Rayleigh dissipation function $\mathcal{R}(\dot x)$, such that

$$\frac{d}{dt}\left(\frac{\partial L}{\partial \dot x}\right) - \frac{\partial L}{\partial x} = -\frac{\partial \mathcal{R}}{\partial \dot x}.$$

Find the Rayleigh dissipation and Lagrangian functions for the equation of motion $m \ddot x + \lambda \dot x + k x = 0$.

10*. The equations of motion for a particle of mass m moving under the influence of a potential $U(x, y, z)$ and the constraint $z = f(x, y)$ are

$$m \ddot x - F_x = (F_z - m \ddot z) \frac{\partial f}{\partial x} \quad \text{and} \quad m \ddot y - F_y = (F_z - m \ddot z) \frac{\partial f}{\partial y},$$

where $\mathbf{F} \equiv - \nabla U$ is the force derived from the potential U.

(a) Show that these equations are derived from the constrained Lagrangian $L = \frac{1}{2} m|\dot{\mathbf{x}}|^2 - U - \lambda [z - f(x, y)]$.

(b) Show that these equations follow from Gauss' Least Constraint Principle

$$\delta \left| m \, \ddot{\mathbf{x}} - \mathbf{F} \right|^2 \equiv (m \, \ddot{\mathbf{x}} - \mathbf{F}) \cdot m \, \delta \ddot{\mathbf{x}} = 0,$$

where the variation is applied with respect to the acceleration $\ddot{\mathbf{x}}$ only.

11*. Rocket propulsion is described in terms of the equations of motion $\ddot{\mathbf{x}} + \nabla \Phi = (\dot{m}/m) \, \mathbf{c}$, where $\mathbf{c}$ is the exhaust velocity relative to the rocket and $\Phi(\mathbf{x})$ is the potential energy per unit mass. By definition, the mass loss rate is given as $\dot{m}/m = - \frac{1}{c} |\ddot{\mathbf{x}} + \nabla \Phi| \equiv - a/c$, which implies that, if the magnitude $c \equiv |\mathbf{c}|$ of the exhaust velocity is constant, the ratio m_f/m_i of the final mass to the initial mass of the rocket is expressed as

$$\frac{m_f}{m_i} = \exp \left(- \frac{1}{c} \int_{t_i}^{t_f} a(t) \, dt \right).$$

The mass ratio is therefore maximum (i.e., the rocket uses the least amount of fuel) if the integral

$$\int_{t_i}^{t_f} a(\mathbf{x}, \ddot{\mathbf{x}}) \, dt \equiv \int_{t_i}^{t_f} |\ddot{\mathbf{x}} + \nabla \Phi(\mathbf{x})| \, dt$$

is minimum.

(a) Show that the Euler-Lagrange equation for this problem is

$$\frac{d^2}{dt^2} \left(\frac{\partial a}{\partial \ddot{\mathbf{x}}} \right) + \frac{\partial a}{\partial \mathbf{x}} = 0,$$

and derive that equation in terms of the unit vector $\hat{\mathbf{c}} \equiv \mathbf{c}/c$.

(b) By using the fact that $\hat{\mathbf{c}}$ is a unit vector, show that the Euler-Lagrange equation derived in Part (a) can be written as $|d\hat{\mathbf{c}}/dt|^2 = \hat{\mathbf{c}} \cdot \nabla \nabla \Phi \cdot \hat{\mathbf{c}}$, so that a minimum solution exists provided the condition $\hat{\mathbf{c}} \cdot \nabla \nabla \Phi \cdot \hat{\mathbf{c}} \geq 0$ is satisfied (where the equality applies to the case $d\hat{\mathbf{c}}/dt \equiv 0$).

(c) Show that the minimum condition $\hat{\mathbf{c}} \cdot \nabla \nabla \Phi \cdot \hat{\mathbf{c}} \geq 0$ yields $\cos \psi \equiv \hat{\mathbf{c}} \cdot \hat{\mathbf{r}} \leq 1/\sqrt{3}$ for the case of the attractive gravitational potential $\Phi(\mathbf{x}) = - G M/r$, where M denotes the mass of the object to which the rocket is attracted.

12. An oscillating pendulum consists of a bob of mass m attached to a spring of constant k and relaxed length ℓ (see Fig. 2.15). The generalized

Fig. 2.15   Problem 13.

coordinates for this system are the angle θ and the displacement r away from the spring's equilibrium. Find the Lagrangian $L(r, \theta; \dot{r}, \dot{\theta})$ and derive the Euler-Lagrange equations for r and θ.

Chapter 3

Hamiltonian Mechanics

In the previous Chapter, the Lagrangian method was introduced as a powerful alternative to the Newtonian method for deriving equations of motion for multi-particle mechanical systems. In the present Chapter, a complementary approach to the Lagrangian method, known as the Hamiltonian method, is presented.

Although much of the Hamiltonian method is outside the scope of this course (e.g., the canonical and noncanonical Hamiltonian formulations of Classical Mechanics and the Hamiltonian formulation of Quantum Mechanics), a simplified version (the Energy method) is presented here as a practical method for *solving* the Euler-Lagrange equations by quadrature. See Appendix C for a brief introduction to the modern formulation of Hamiltonian Mechanics.

3.1 Hamilton's Canonical Equations

The Euler-Lagrange equations on the k-dimensional configuration space $\mathbf{q}$ are k second-order differential equations:

$$\frac{d}{dt}\left(\frac{\partial L}{\partial \dot{q}^j}\right) = \frac{\partial L}{\partial q^j}. \tag{3.1}$$

This set of second-order differential equations can be written as $2k$ first-order differential equations on a $2k$-dimensional *phase* space with coordinates $\mathbf{z} = (q^1, ..., q^k; p_1, ..., p_k)$, where

$$p_j(\mathbf{q}, \dot{\mathbf{q}}; t) = \frac{\partial L}{\partial \dot{q}^j}(\mathbf{q}, \dot{\mathbf{q}}; t) \tag{3.2}$$

defines the jth-component of the *canonical* momentum. In terms of these new coordinates, the Euler-Lagrange equations (3.1) are transformed into

Hamilton's canonical equations (William Rowan Hamilton, 1805-1865)

$$\frac{dq^j}{dt} = \frac{\partial H}{\partial p_j} \quad \text{and} \quad \frac{dp_j}{dt} = -\frac{\partial H}{\partial q^j}, \tag{3.3}$$

where the Hamiltonian function $H(\mathbf{q}, \mathbf{p}; t)$ is defined from the Lagrangian function $L(\mathbf{q}, \dot{\mathbf{q}}; t)$ by the Legendre transformation (Adrien-Marie Legendre, 1752-1833)

$$H(\mathbf{q}, \mathbf{p}; t) = \mathbf{p} \cdot \dot{\mathbf{q}}(\mathbf{q}, \mathbf{p}, t) - L[\mathbf{q}, \dot{\mathbf{q}}(\mathbf{q}, \mathbf{p}, t), t]. \tag{3.4}$$

Hamilton's *canonical* equations of motion (3.3) are completely equivalent to the Lagrangian formulation.

We note that Hamilton's equations (3.3) can also be derived from a variational principle in $2k$-dimensional phase space $\mathbf{z} = (\mathbf{q}, \mathbf{p})$ as follows. First, we use the inverse of the Legendre transformation

$$L(\mathbf{z}, \dot{\mathbf{z}}; t) = \mathbf{p} \cdot \dot{\mathbf{q}} - H(\mathbf{z}; t) \tag{3.5}$$

to obtain an expression for the Lagrangian function in phase space. Next, we calculate the first-variation of the action integral

$$\delta \int L(\mathbf{q}, \mathbf{p}; t)\, dt = \int \left[\delta\mathbf{p} \cdot \left(\dot{\mathbf{q}} - \frac{\partial H}{\partial \mathbf{p}} \right) + \left(\mathbf{p} \cdot \delta\dot{\mathbf{q}} - \delta\mathbf{q} \cdot \frac{\partial H}{\partial \mathbf{q}} \right) \right] dt,$$

where the variations δq^i and δp_i are now considered independent (and they are both assumed to vanish at the end points). By integrating by parts the term $\mathbf{p} \cdot \delta\dot{\mathbf{q}}$, we find

$$\delta \int L(\mathbf{q}, \mathbf{p}; t)\, dt = \int \left[\delta\mathbf{p} \cdot \left(\dot{\mathbf{q}} - \frac{\partial H}{\partial \mathbf{p}} \right) - \delta\mathbf{q} \cdot \left(\dot{\mathbf{p}} + \frac{\partial H}{\partial \mathbf{q}} \right) \right] dt,$$

so that the Principle of Least Action $\int \delta L\, dt = 0$ now yields Hamilton's equations (3.3) for arbitrary variations $(\delta\mathbf{q}, \delta\mathbf{p})$ in $2k$-dimensional phase space.

Lastly, an important equation associated with Hamilton's principal function $\mathcal{S}$ can be derived from the infinitesimal action

$$d\mathcal{S}(\mathbf{q}, t) = \mathbf{p} \cdot d\mathbf{q} - H\, dt, \tag{3.6}$$

from which we obtain the relations

$$\left. \begin{array}{l} H = -\partial\mathcal{S}/\partial t \\[2mm] \mathbf{p} = \partial\mathcal{S}/\partial\mathbf{q} \end{array} \right\}. \tag{3.7}$$

These relations can be used to obtain the Hamilton-Jacobi equation for particle dynamics [7]

$$\frac{\partial\mathcal{S}}{\partial t} + H\left(\mathbf{q}, \frac{\partial\mathcal{S}}{\partial\mathbf{q}}; t \right) = 0. \tag{3.8}$$

The solution to this equation is said to generate a canonical transformation that annihilates the Hamiltonian, i.e., the function S generates a time-dependent canonical transformation $\mathbf{z} = (\mathbf{q}, \mathbf{p}) \rightarrow \mathbf{Z} = (\mathbf{Q}, \mathbf{P})$ such that the new Hamiltonian $K(\mathbf{Z}; t) \equiv H(\mathbf{z}(\mathbf{Z}, t); t) + \partial_t S(\mathbf{q}(\mathbf{Z}, t), t)$ vanishes. Applications of the Hamilton-Jacobi equation fall outside the scope of the present course [7, 13]. We simply mention here that the Hamilton-Jacobi equation (3.8) figures prominently in the historical connection between particle dynamics and wave mechanics (as discussed in Sec. 3.3), as well as the connection between Classical Mechanics and Quantum Mechanics (as discussed in Sec. 9.3 and problem 1 in Chap. 9).

3.2 Legendre Transformation*

Before proceeding with the Hamiltonian formulation of particle dynamics, we investigate the condition under which the Legendre transformation (3.4) is possible. It turns out that this condition is associated with the condition under which the inversion of the relation $\mathbf{p}(\mathbf{r}, \dot{\mathbf{r}}, t) \rightarrow \dot{\mathbf{r}}(\mathbf{r}, \mathbf{p}, t)$ is possible. To simplify our discussion, we focus on motion in two dimensions (x, y).

The general expression of the kinetic energy term of a Lagrangian with two degrees of freedom $L(x, \dot{x}, y, \dot{y}) = K(x, \dot{x}, y, \dot{y}) - U(x, y)$ is

$$K(x, \dot{x}, y, \dot{y}) = \frac{\alpha}{2}\,\dot{x}^2 + \beta\,\dot{x}\dot{y} + \frac{\gamma}{2}\,\dot{y}^2 = \frac{1}{2}\,\dot{\mathbf{r}}^\mathsf{T} \cdot \mathsf{M} \cdot \dot{\mathbf{r}}, \qquad (3.9)$$

where $\dot{\mathbf{r}}^\mathsf{T} = (\dot{x}, \dot{y})$ denotes the transpose of $\dot{\mathbf{r}}$ (see Appendix A for additional details concerning linear algebra) and the *mass* matrix M is

$$\mathsf{M} = \begin{pmatrix} \alpha & \beta \\ \beta & \gamma \end{pmatrix}.$$

Here, the coefficients α, β, and γ may be functions of x and y. The canonical momentum vector (3.2) is thus defined as

$$\mathbf{p} = \frac{\partial L}{\partial \dot{\mathbf{r}}} = \mathsf{M} \cdot \dot{\mathbf{r}} \rightarrow \begin{pmatrix} p_x \\ p_y \end{pmatrix} = \begin{pmatrix} \alpha & \beta \\ \beta & \gamma \end{pmatrix} \cdot \begin{pmatrix} \dot{x} \\ \dot{y} \end{pmatrix}$$

or

$$\left. \begin{array}{l} p_x = \alpha\,\dot{x} + \beta\,\dot{y} \\[2mm] p_y = \beta\,\dot{x} + \gamma\,\dot{y} \end{array} \right\}. \qquad (3.10)$$

The Lagrangian is said to be *regular* if the mass matrix M is invertible, i.e., if its determinant

$$\Delta = \det(M) = \alpha\gamma - \beta^2 \neq 0.$$

In the case of a regular Lagrangian, we readily invert (3.10) to obtain

$$\dot{\mathbf{r}}(\mathbf{r}, \mathbf{p}, t) = M^{-1} \cdot \mathbf{p} \;\rightarrow\; \begin{pmatrix} \dot{x} \\ \dot{y} \end{pmatrix} = \frac{1}{\Delta} \begin{pmatrix} \gamma & -\beta \\ -\beta & \alpha \end{pmatrix} \cdot \begin{pmatrix} p_x \\ p_y \end{pmatrix}$$

or

$$\left. \begin{aligned} \dot{x} &= (\gamma\, p_x - \beta\, p_y)/\Delta \\ \dot{y} &= (\alpha\, p_y - \beta\, p_x)/\Delta \end{aligned} \right\}, \tag{3.11}$$

and the kinetic energy term becomes

$$K(x, p_x,\, y, p_y) = \frac{1}{2}\, \mathbf{p}^\top \cdot M^{-1} \cdot \mathbf{p}.$$

Lastly, under the Legendre transformation (3.4), we find

$$H = \mathbf{p}^\top \cdot (M^{-1} \cdot \mathbf{p}) - \left(\frac{1}{2}\, \mathbf{p}^\top \cdot M^{-1} \cdot \mathbf{p} - U \right)$$

$$= \frac{1}{2}\, \mathbf{p}^\top \cdot M^{-1} \cdot \mathbf{p} + U \equiv K + U.$$

Hence, we clearly see that the Legendre transformation is applicable only if the mass matrix M in the kinetic energy (3.9) is invertible. Lastly, we note that the Legendre transformation is also used in other areas in physics such as Thermodynamics.

3.3   Hamiltonian Optics and Wave-Particle Duality*

Historically, the Hamiltonian method was first introduced as a formulation of the dynamics of light rays [4, 21]. Consider the following *phase* integral

$$\Theta[\mathbf{z}] \equiv \int_{t_1}^{t_2} \theta(\mathbf{x}, \mathbf{k}; t)\, dt = \int_{t_1}^{t_2} \Big[\mathbf{k} \cdot \dot{\mathbf{x}} - \omega(\mathbf{x}, \mathbf{k}; t) \Big]\, dt, \tag{3.12}$$

where $\Theta[\mathbf{z}]$ is a functional of the light-path $\mathbf{z}(t) = (\mathbf{x}(t), \mathbf{k}(t))$ in *ray phase space*, expressed in terms of the instantaneous position $\mathbf{x}(t)$ of a light ray and its associated instantaneous wave vector $\mathbf{k}(t)$; here, the dispersion relation $\omega(\mathbf{x}, \mathbf{k}; t)$ is obtained as a root of the dispersion equation $\det D(\mathbf{x}, t; \mathbf{k}, \omega) = 0$, and a dot denotes a total time derivative: $\dot{\mathbf{x}} = d\mathbf{x}/dt$.

Assuming that the phase integral $\Theta[\mathbf{z}]$ acquires a *stationary* value for a *physical* ray orbit $\mathbf{z}(t)$, henceforth called the Principle of *Stationary Phase* $\delta\Theta = 0$, we can show that Euler's First Equation leads to Hamilton's (canonical) ray equations:

$$\frac{d\mathbf{x}}{dt} = \frac{\partial\omega}{\partial\mathbf{k}} \quad \text{and} \quad \frac{d\mathbf{k}}{dt} = -\nabla\omega. \tag{3.13}$$

The first ray equation states that a ray travels at the *group* velocity while the second ray equation states that the wave vector $\mathbf{k}$ is refracted as the ray propagates in a non-uniform medium (see Chap. 1). Hence, the frequency function $\omega(\mathbf{x}, \mathbf{k}; t)$ is the Hamiltonian of ray dynamics in a nonuniform medium. The Hamiltonian theory of wave dynamics in ray phase space is covered extensively by Tracy, Brizard, Richardson, and Kaufman [19].

The Hamilton-Jacobi equation for ray optics is obtained from the infinitesimal phase

$$d\Theta(\mathbf{x}, t) = \mathbf{k}\cdot d\mathbf{x} - \omega\,dt, \tag{3.14}$$

from which we obtain the *eikonal* relations

$$\left.\begin{array}{l} \omega = -\,\partial\Theta/\partial t \\[2mm] \mathbf{k} = \nabla\Theta \end{array}\right\}. \tag{3.15}$$

These relations can then be used to obtain the Hamilton-Jacobi equation

$$\frac{\partial\Theta}{\partial t} + \omega(\mathbf{x}, \nabla\Theta; t) = 0, \tag{3.16}$$

where $\omega(\mathbf{x}, \mathbf{k}; t)$ is the Hamiltonian for the ray equations (3.13). The analogy between the Hamilton-Jacobi equation (3.8) for particle dynamics and the Hamilton-Jacobi equation (3.16) for ray optics leads us to recognize the deep connections between Classical Mechanics and Wave Mechanics (see Table 3.1 for a detailed correspondence).

It was de Broglie who noted (as a graduate student well versed in Classical Mechanics) the similarities between Hamilton's equations (3.3) and (3.13), on the one hand, and the Maupertuis-Jacobi (2.1) and Euler-Lagrange (2.29) Principles of Least Action and Fermat's Principle of Least Time (1.36) and Principle of Stationary Phase (3.12), on the other hand (see Table 3.1). By using the *quantum of action* $\hbar = h/2\pi$ defined in terms of Planck's constant h and Planck's energy hypothesis $E = \hbar\omega$, de Broglie suggested that a particle's momentum $\mathbf{p}$ be related to its wavevector $\mathbf{k}$ according to de Broglie's formula $\mathbf{p} = \hbar\mathbf{k}$ and introduced the wave-particle synthesis based on the identity

$$\mathcal{S}[\mathbf{z}] = \hbar\,\Theta[\mathbf{z}] \tag{3.17}$$

Table 3.1 Correspondence between Particle and Wave Mechanics.

	Particle	Wave
Phase Space	$\mathbf{z} = (\mathbf{q}, \mathbf{p})$	$\mathbf{z} = (\mathbf{x}, \mathbf{k})$
Hamiltonian	$H(\mathbf{z}; t)$	$\omega(\mathbf{z}; t)$
Variational Principle I	Maupertuis-Jacobi	Fermat
$\int (\cdots) \, ds$	$\sqrt{2m \, (E - U(\mathbf{q}))}$	$n(\mathbf{x})$
Variational Principle II	Hamilton	Stationary Phase
$\int (\cdots) \, dt$	$L = \mathbf{p} \cdot \dot{\mathbf{q}} - H$	$\theta = \mathbf{k} \cdot \dot{\mathbf{x}} - \omega$
Hamilton-Jacobi	$\partial_t S + H(\mathbf{q}, \partial_\mathbf{q} S, t) = 0$	$\partial_t \Theta + \omega(\mathbf{x}, \nabla\Theta, t) = 0$
Hamilton's equations	$(\dot{\mathbf{q}}, \dot{\mathbf{p}}) = (\partial_\mathbf{p} H, -\partial_\mathbf{q} H)$	$(\dot{\mathbf{x}}, \dot{\mathbf{k}}) = (\partial_\mathbf{k}\omega, -\nabla\omega)$

involving the action integral $S[\mathbf{z}]$ and the phase integral $\Theta[\mathbf{z}]$.

The final synthesis between Classical and Quantum Mechanics came from Richard Phillips Feynman (1918-1988) who provided an explicit derivation of Schroedinger's equation (Erwin Rudolf Josef Alexander Schroedinger, 1887-1961) by associating the probability that a particle follow a particular path $\mathbf{z}(t; \mathbf{z}_0)$ with the expression $\exp(i \, \hbar^{-1} S[\mathbf{z}])$, where $S[\mathbf{z}]$ denotes the action integral for the path [21].

3.4 Motion in an Electromagnetic Field

Although the problem of the motion of a charged particle in an electromagnetic field can be considered outside the scope of the present course, it represents a important paradigm that beautifully illustrates the connection between Lagrangian and Hamiltonian mechanics and it is well worth studying.

3.4.1 *Euler-Lagrange Equations*

The equations of motion for a charged particle of mass m and charge e moving in an electromagnetic field represented by the electric field $\mathbf{E}$ and magnetic field $\mathbf{B}$ are

$$\frac{d\mathbf{x}}{dt} = \mathbf{v} \tag{3.18}$$

$$\frac{d\mathbf{v}}{dt} = \frac{e}{m} \left(\mathbf{E} + \frac{d\mathbf{x}}{dt} \times \frac{\mathbf{B}}{c} \right), \tag{3.19}$$

where $\mathbf{x}$ denotes the position of the particle and $\mathbf{v}$ its velocity (Note: Gaussian units are used whenever electromagnetic fields are involved).

By treating the coordinates $(\mathbf{x}, \mathbf{v})$ as generalized coordinates (i.e., $\delta\mathbf{v}$ is treated independently from $\delta\mathbf{x}$), we can show that the equations of motion (3.18) and (3.19) can be obtained as Euler-Lagrange equations from the Lagrangian (3.5):

$$L(\mathbf{x}, \dot{\mathbf{x}}, \mathbf{v}, \dot{\mathbf{v}}; t) = \left(m\,\mathbf{v} + \frac{e}{c}\,\mathbf{A}(\mathbf{x}, t) \right) \cdot \dot{\mathbf{x}} - \left(e\,\Phi(\mathbf{x}, t) + \frac{m}{2}\,|\mathbf{v}|^2 \right), \quad (3.20)$$

where Φ and $\mathbf{A}$ are the electromagnetic potentials in terms of which electric and magnetic fields are defined

$$\mathbf{E} = -\nabla\Phi - \frac{1}{c}\frac{\partial\mathbf{A}}{\partial t} \text{ and } \mathbf{B} = \nabla \times \mathbf{A}. \quad (3.21)$$

Note that these expressions for $\mathbf{E}$ and $\mathbf{B}$ satisfy Faraday's law $\nabla \times \mathbf{E} = -c^{-1}\,\partial\mathbf{B}/\partial t$ and Gauss's Law $\nabla \cdot \mathbf{B} = 0$.

First, we look at the Euler-Lagrange equation for $\mathbf{x}$:

$$\frac{\partial L}{\partial\dot{\mathbf{x}}} = m\,\mathbf{v} + \frac{e}{c}\,\mathbf{A} \;\rightarrow\; \frac{d}{dt}\left(\frac{\partial L}{\partial\dot{\mathbf{x}}} \right) = m\,\dot{\mathbf{v}} + \frac{e}{c}\left(\frac{\partial\mathbf{A}}{\partial t} + \dot{\mathbf{x}} \cdot \nabla\mathbf{A} \right)$$

$$\frac{\partial L}{\partial\mathbf{x}} = \frac{e}{c}\,\nabla\mathbf{A} \cdot \dot{\mathbf{x}} - e\,\nabla\Phi,$$

which yields the Lorentz force equation (3.19), since

$$m\,\dot{\mathbf{v}} = -e\left(\nabla\Phi + \frac{1}{c}\frac{\partial\mathbf{A}}{\partial t} \right) + \frac{e}{c}\,\dot{\mathbf{x}} \times \nabla \times \mathbf{A} = e\,\mathbf{E} + \frac{e}{c}\,\dot{\mathbf{x}} \times \mathbf{B}, \quad (3.22)$$

where the definitions (3.21) were used.

Next, we look at the Euler-Lagrange equation for $\mathbf{v}$:

$$\frac{\partial L}{\partial\dot{\mathbf{v}}} = 0 \;\rightarrow\; \frac{d}{dt}\left(\frac{\partial L}{\partial\dot{\mathbf{v}}} \right) = 0 = \frac{\partial L}{\partial\mathbf{v}} = m\,(\dot{\mathbf{x}} - \mathbf{v}),$$

which yields Eq. (3.18). Because $\partial L/\partial\dot{\mathbf{v}} = 0$, we note that we could use Eq. (3.18) as a constraint, which could be imposed *a priori* on the Lagrangian (3.20), to give

$$L(\mathbf{x}, \dot{\mathbf{x}}; t) = \frac{m}{2}\,|\dot{\mathbf{x}}|^2 + \frac{e}{c}\,\mathbf{A}(\mathbf{x}, t) \cdot \dot{\mathbf{x}} - e\,\Phi(\mathbf{x}, t). \quad (3.23)$$

The Euler-Lagrange equation for $\mathbf{x}$ in this case is identical to Eq. (3.22) with $\dot{\mathbf{v}} = \ddot{\mathbf{x}}$.

Lastly, Euler's second equation yields

$$\frac{d}{dt}\left(L - \dot{\mathbf{x}} \cdot \frac{\partial L}{\partial\dot{\mathbf{x}}} \right) - \frac{\partial L}{\partial t} = -\frac{d}{dt}\left(\frac{m}{2}\,|\mathbf{v}|^2 + e\,\Phi \right) + e\frac{\partial\Phi}{\partial t} - \frac{e}{c}\frac{\partial\mathbf{A}}{\partial t} \cdot \dot{\mathbf{x}}$$

$$= -e\,\dot{\mathbf{x}} \cdot \mathbf{E} - e\,\dot{\mathbf{x}} \cdot \left(\nabla\Phi + \frac{1}{c}\frac{\partial\mathbf{A}}{\partial t} \right) = 0,$$

which follows from the definitions (3.21).

3.4.2   Gauge Invariance

The electric and magnetic fields defined in (3.21) are invariant under the gauge transformation

$$\Phi \rightarrow \Phi - \frac{1}{c} \frac{\partial \chi}{\partial t} \text{ and } \mathbf{A} \rightarrow \mathbf{A} + \nabla \chi, \qquad (3.24)$$

where $\chi(\mathbf{x}, t)$ is an arbitrary scalar field. Although the equations of motion (3.18) and (3.19) are *manifestly* gauge invariant, the Lagrangian (3.23) is not manifestly gauge invariant since the electromagnetic potentials Φ and $\mathbf{A}$ appear explicitly. Under a gauge transformation (3.24), however, we find

$$L \rightarrow L + \frac{e}{c} \dot{\mathbf{x}} \cdot \nabla \chi - e \left(- \frac{1}{c} \frac{\partial \chi}{\partial t} \right) = L + \frac{d}{dt} \left(\frac{e}{c} \chi \right).$$

Since Lagrangian Mechanics is invariant under the transformation (2.51), the Lagrangian (3.23) is invariant under the gauge transformation (3.24).

3.4.3   Canonical Hamilton's Equations

The canonical momentum $\mathbf{p}$ for a particle of mass m and charge e in an electromagnetic field is defined as

$$\mathbf{p}(\mathbf{x}, \mathbf{v}, t) = \frac{\partial L}{\partial \dot{\mathbf{x}}} = m \mathbf{v} + \frac{e}{c} \mathbf{A}(\mathbf{x}, t), \qquad (3.25)$$

which is the sum of the kinetic momentum ($m \mathbf{v}$) and a magnetic contribution (represented by the vector potential $\mathbf{A}$). The canonical Hamiltonian function $H(\mathbf{x}, \mathbf{p}, t)$ is now constructed through the Legendre transformation

$$H(\mathbf{x}, \mathbf{p}, t) = \mathbf{p} \cdot \dot{\mathbf{x}}(\mathbf{x}, \mathbf{p}, t) - L[\mathbf{x}, \dot{\mathbf{x}}(\mathbf{x}, \mathbf{p}, t), t] \qquad (3.26)$$

$$= \frac{1}{2m} \left| \mathbf{p} - \frac{e}{c} \mathbf{A}(\mathbf{x}, t) \right|^2 + e \Phi(\mathbf{x}, t), \qquad (3.27)$$

where $\mathbf{v}(\mathbf{x}, \mathbf{p}, t)$ was obtained by inverting $\mathbf{p}(\mathbf{x}, \mathbf{v}, t)$ from Eq. (3.25). Using the canonical Hamiltonian function (3.27), we immediately find

$$\dot{\mathbf{x}} = \frac{\partial H}{\partial \mathbf{p}} = \frac{1}{m} \left(\mathbf{p} - \frac{e}{c} \mathbf{A} \right), \qquad (3.28)$$

$$\dot{\mathbf{p}} = - \frac{\partial H}{\partial \mathbf{x}} = - e \nabla \Phi - \frac{e}{c} \nabla \mathbf{A} \cdot \dot{\mathbf{x}}, \qquad (3.29)$$

from which we recover the equations of motion (3.18) and (3.19) once we use the definition (3.25) for the canonical momentum. We also note that the total energy is not invariant in time-dependent electromagnetic fields:

$$\frac{\partial H}{\partial t} = e\frac{\partial \Phi}{\partial t} - \frac{e}{c}\frac{\partial \mathbf{A}}{\partial t}\cdot\dot{\mathbf{x}} = -\frac{\partial L}{\partial t} + \left(\mathbf{p} - \frac{\partial L}{\partial \dot{\mathbf{x}}}\right)\cdot\frac{\partial \dot{\mathbf{x}}}{\partial t} \equiv -\frac{\partial L}{\partial t},$$

(3.30)

where we used the definitions (3.25)-(3.26), and $\partial L/\partial t$ is calculated in configuration space from Eq. (3.23) at constant $(\mathbf{x}, \dot{\mathbf{x}})$.

We should warn the reader that the simplicity of the canonical Hamiltonian formalism comes at a price: the canonical momentum $\mathbf{p} \equiv \partial L/\partial\dot{\mathbf{q}}$ and the Hamiltonian are not physical quantities. Indeed, under the gauge transformation (3.24), the canonical momentum and Hamiltonian are transformed as

$$\mathbf{p} \to \mathbf{p} + \frac{e}{c}\nabla\chi \quad \text{and} \quad H \to H - \frac{e}{c}\frac{\partial\chi}{\partial t}.$$

These transformations, however, leave Hamilton's canonical equations (3.28)-(3.29) invariant.

3.4.4 *Maupertuis' Principle for Particle-Beam Optics*

Because of the close connection between Fermat's Principle of Least Time and Maupertuis' Principle of Least Action (see Table 3.1), it is instructive to derive a variational principle suitable for applications in particle-beam *optics*, where the path of a charged-particle beam is guided by electric and magnetic lenses.

We begin with the classical action integral

$$\int L\,dt = \int\left(L - \frac{\partial L}{\partial\dot{\mathbf{r}}}\cdot\dot{\mathbf{r}}\right)dt + \int\frac{\partial L}{\partial\dot{\mathbf{r}}}\cdot d\mathbf{r}$$

$$= -\int H\,dt + \int \mathbf{p}\cdot d\mathbf{r},$$

where we used the Legendre transformation (3.4) in the first integral and used the definition (3.2) for the canonical momentum in the second integral. If we consider time-independent guiding magnetic fields, the electron energy is constant $(H = E)$ and Maupertuis' Principle of Least Action $\delta\mathcal{S}_E = 0$ is expressed in terms of the action functional (at constant energy E) [3]

$$\mathcal{S}_E = \int \mathbf{p}\cdot d\mathbf{r} = \int\left(m\mathbf{v} + \frac{q}{c}\mathbf{A}\right)\cdot d\mathbf{r}$$

$$= \int\left(mv + \frac{q}{c}\mathbf{A}\cdot\hat{\mathbf{s}}\right)ds,$$

(3.31)

where electrons have mass m and charge q, the unit vector $\hat{\mathbf{s}}$ is defined as $\hat{\mathbf{s}} \equiv d\mathbf{r}/ds$, and $mv \equiv \sqrt{2m(E - q\Phi)}$ is defined in terms of the total energy E and the electric scalar potential Φ.

3.5 One-degree-of-freedom Hamiltonian Dynamics

In this Section, we investigate Hamiltonian dynamics with one degree of freedom in a time-independent potential. In particular, we show that, by using the Energy Method, such systems are always integrable (i.e., they can always be solved by quadrature).

After introducing the Energy Method (Sec. 3.5.1), we consider four examples of integrable motion in one-dimensional Hamiltonian dynamics: the simple harmonic oscillator (Sec. 3.5.2); the motion of a particle in the Morse potential (Sec. 3.5.3); the pendulum (Sec. 3.5.4); and the constrained motion of a particle on the surface of a cone (Sec. 3.5.5). While the second example (the motion of a particle in the Morse potential) appears overly complicated, it is still solvable in terms of trigonometric functions.

3.5.1 *Energy Method*

The one degree-of-freedom Hamiltonian dynamics of a particle of mass m is based on the Hamiltonian

$$H(x,p) = \frac{p^2}{2m} + U(x), \qquad (3.32)$$

where $p = m\dot{x}$ is the particle's momentum and $U(x)$ is the time-independent potential energy. The Hamilton's equations (3.3) for this Hamiltonian are

$$\frac{dx}{dt} = \frac{p}{m} \text{ and } \frac{dp}{dt} = -\frac{dU(x)}{dx}. \qquad (3.33)$$

Since the Hamiltonian (and Lagrangian) is time independent, the energy conservation law states that $H(x,p) = E$. In turn, this conservation law implies that the particle's velocity $\dot{x}$ can be expressed as

$$\dot{x}(x,E) = \pm\sqrt{\frac{2}{m}[E - U(x)]}, \qquad (3.34)$$

where the sign of $\dot{x}$ is determined from the initial conditions.

It is immediately clear from Eq. (3.34) that physical motion is possible only if $E \geq U(x)$; points where $E = U(x)$ are known as *turning* points since the particle velocity $\dot{x}$ vanishes at these points. In Fig. 3.1, which represents the dimensionless potential $U(x) = x - x^3/3$, each horizontal line corresponds to a constant energy value (called an energy *level*). For the top energy level, only one turning point (labeled a in Fig. 3.1) exists and a particle coming from the right will be *reflected* at point a and return to large (positive) values of x; the motion in this case is said to be along an

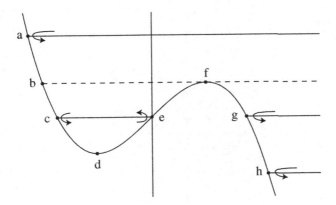

Fig. 3.1   Bounded and unbounded energy levels in a cubic potential $U(x) = x - x^3/3$.

unbounded orbit (see orbits I in Fig. 3.2). As the energy value is lowered, two turning points (labeled b and f) appear and motion can either be *bounded* (between points b and f) or unbounded (if the initial position is to the right of point f); this energy level is known as a *separatrix* level since bounded and unbounded motions share one turning point (see orbits II and III in Fig. 3.2). As energy is lowered below the separatrix level, three turning points (labeled c, e, and g) appear and, once again, motion can either be along a *bounded* orbit (with turning points c and e) or an unbounded orbit if the initial position is to the right of point g (see orbits IV and V in Fig. 3.2).[1] Lastly, we note that point d in Fig. 3.1 is actually an equilibrium point (as is point f), where $\dot{x}$ and $\ddot{x}$ both vanish; only unbounded motion is allowed as energy is lowered below point d (e.g., point h) and the corresponding unbounded orbits are analogous to orbit V in Fig. 3.2.

The dynamical solution $x(t; E)$ of the Hamilton's equations (3.33) is first expressed an integration by quadrature using Eq. (3.34) as

$$t(x; E) = \sqrt{\frac{m}{2}} \int_{x_0}^{x} \frac{ds}{\sqrt{E - U(s)}}, \qquad (3.35)$$

where the particle's initial position x_0 is between the turning points $x_1 < x_2$ (allowing $x_2 \to \infty$) and we assume that $\dot{x}(0) > 0$. Next, inversion of the relation (3.35) yields the solution $x(t; E)$.

Lastly, for bounded motion in one dimension, the particle bounces back and forth between the two turning points x_1 and $x_2 > x_1$, and the period

---

[1]Note: Quantum tunneling establishes a connection between the bounded and unbounded solutions separated by unphysical regions (where $E < U$).

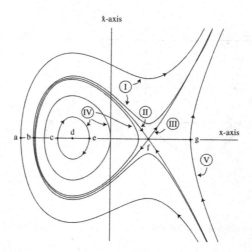

Fig. 3.2 Bounded and unbounded orbits in the cubic potential shown in Fig. 3.1: orbits I correspond to the energy level with turning point a; the bounded orbit II and unbounded orbit III correspond to the separatrix energy level with turning points b and f; bounded orbits IV correspond to the energy level with turning points c and e; and the unbounded orbit V corresponds to the energy level with turning point g. (These orbits are explicitly solved in Appendix B in terms of the Weierstrass elliptic function.)

of oscillation $T(E)$ is a function of energy alone

$$T(E) = 2 \int_{x_1}^{x_2} \frac{dx}{|\dot{x}(x, E)|} = \sqrt{2m} \int_{x_1}^{x_2} \frac{dx}{\sqrt{E - U(x)}}. \tag{3.36}$$

Thus, Eqs. (3.35) and (3.36) describe applications of the Energy Method in one dimension. We now look at a series of one-dimensional problems solvable by the Energy Method.

3.5.2 *Simple Harmonic Oscillator*

As a first example, we consider the case of a particle of mass m attached to a spring of constant k, for which the potential energy is $U(x) = \frac{1}{2} k x^2$. The motion of a particle with total energy E is always bounded, with turning points

$$x_{1,2}(E) = \pm \sqrt{2E/k} = \pm a$$

determined from the turning-point equation $E = \frac{1}{2} k x^2$.

We start with the solution (3.35) for $t(x; E)$ for the case of $x(0; E) = +a$, so that $\dot{x}(t; E) < 0$ for $t > 0$, and

$$t(x; E) = -\sqrt{\frac{m}{k}} \int_a^x \frac{ds}{\sqrt{a^2 - s^2}} = \sqrt{\frac{m}{k}} \int_x^a \frac{ds}{\sqrt{a^2 - s^2}}$$

$$= \sqrt{\frac{m}{k}} \arccos\left(\frac{x}{a}\right). \tag{3.37}$$

Inversion of this relation yields the well-known solution

$$x(t; E) = a \cos(\omega_0 t), \tag{3.38}$$

where $\omega_0 = \sqrt{k/m}$. Next, we compute the particle velocity $v(t) = dx/dt$:

$$v(t) = -a\omega_0 \sin(\omega_0 t), \tag{3.39}$$

which shows that the graph of Eqs. (3.38)-(3.39) in the phase-space portrait (x, v) is an ellipse: $x^2/a^2 + v^2/(a\omega_0)^2 = 1$.

Lastly, using Eq. (3.36), we find the period of oscillation

$$T(E) = \frac{4}{\omega_0} \int_0^a \frac{dx}{\sqrt{a^2 - x^2}} = \frac{2\pi}{\omega_0}, \tag{3.40}$$

which turns out to be independent of energy E. Hence, as the energy E is raised, the distance $2a = \sqrt{8E/k}$ between the two turning points increases, but then so does the average speed $\bar{v} \equiv 2a/T$, so that the period T is a constant.

3.5.3 *Morse Potential*

While the solution (3.35) for the Hamilton equation (3.34) appears straightforward, there are few potentials $U(x)$ for which Eq. (3.35) can actually be solved explicitly. One such potential is the Morse potential

$$U(x) = U_0 \left(e^{-2\alpha x} - 2e^{-\alpha x}\right), \tag{3.41}$$

which has the minimum $-U_0$ at $x = 0$ (see Fig. 3.3). The constant α determines where the potential changes sign: $U(x) > 0$ for $x < x_0 \equiv -(1/\alpha) \ln(2)$; and $U(x) < 0$ for $x > x_0$; the Morse potential also vanishes as $x \to \infty$. For motion near the vicinity of the minimum at $x = 0$, we note that the Morse potential $U(x) \simeq -U_0 + U_0 \alpha^2 x^2$ is approximated as a shifted simple harmonic oscillator.

The motion of a particle (of mass m) in the Morse potential (3.41) is bounded when $-U_0 < E < 0$, while the motion is unbounded when $E \geq 0$. This can be seen by calculating the turning points defined as the solution of the turning-point equation $E = U(x)$. By defining the dimensionless

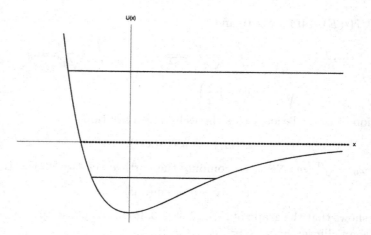

Fig. 3.3   Graph of $U(x)$ versus x for the Morse potential (3.41), with three energy levels shown: $E = -U_0/2$ (lowest), $E = 0$ (middle), and $E = U_0$ (highest).

energy parameter $\epsilon \equiv E/U_0$ and $r \equiv \exp(-\alpha x) > 0$, the turning-point equation becomes $r^2 - 2r - \epsilon = 0$, whose solutions are $r_\pm \equiv 1 \pm \sqrt{1 + \epsilon}$ or

$$x_\pm = -\alpha^{-1} \ln\left(1 \pm \sqrt{1 + \epsilon}\right) = \alpha^{-1} \ln\left(\pm \epsilon^{-1} \sqrt{1 + \epsilon} - \epsilon^{-1}\right). \quad (3.42)$$

It is easy to see that for the case $-1 < \epsilon < 0$ (i.e., $-U_0 < E < 0$), the two roots $r_\pm = 1 \pm \sqrt{1 - |\epsilon|} > 0$ exist and the motion is bounded (i.e., it is periodic between $x_+ < x < x_-$); see lowest energy level in Fig. 3.3. For the case $\epsilon > 0$ (i.e., $E > 0$), the root $r_- = 1 - \sqrt{1 + \epsilon} < 0$ is not allowed (since r cannot be negative by definition), and the motion is unbounded with a single turning point at $x = x_+ = -(1/\alpha) \ln(1 + \sqrt{1 + \epsilon})$; see highest energy level in Fig. 3.3. The dotted line in Fig. 3.3 corresponds to the energy level $E = 0$.

The solution (3.35) for the Morse potential can be expressed as

$$\begin{aligned}
t(x) &= \sqrt{\frac{m}{2E}} \int_{x_+}^{x} \frac{dy}{\sqrt{1 - \epsilon^{-1}(e^{-2\alpha y} - 2e^{-\alpha y})}} \\
&= \sqrt{\frac{m}{2E}} \int_{x_+}^{x} \frac{e^{\alpha y}\, dy}{\sqrt{e^{2\alpha y} + 2\epsilon^{-1} e^{\alpha y} - \epsilon^{-1}}} \\
&= \sqrt{\frac{m}{2\alpha^2 E}} \int_{\exp(\alpha x_+)}^{\exp(\alpha x)} \frac{ds}{\sqrt{s^2 + 2\epsilon^{-1} s - \epsilon^{-1}}},
\end{aligned} \quad (3.43)$$

where the initial condition $x(0) = x_+$ is chosen as the leftmost turning point (see Fig. 3.3), where $\dot{x}(0) = 0$, and the last integral is obtained through the substitution $s(y) = \exp(\alpha y)$. We now solve this integral explicitly for the three cases corresponding to $-1 < \epsilon < 0$ (bounded motion), $\epsilon = 0$ (unbounded motion), and $\epsilon > 0$ (unbounded).

3.5.3.1 *Bounded Motion*

For the case $-1 < \epsilon < 0$, the integral solution (3.43) becomes

$$
t(x) = \sqrt{\frac{m}{2\,\alpha^2|E|}} \int_{\exp(\alpha x_+)}^{\exp(\alpha x)} \frac{ds}{\sqrt{-s^2 + 2|\epsilon|^{-1}s - |\epsilon|^{-1}}}
$$

$$
= \sqrt{\frac{m}{2\,\alpha^2|E|}} \int_{\exp(\alpha x_+)}^{\exp(\alpha x)} \frac{ds}{\sqrt{|\epsilon|^{-2}(1 - |\epsilon|) - (s - |\epsilon|^{-1})^2}},
$$

where we completed the square $-s^2 + 2a\,s - a \equiv (a^2 - a) - (s - a)^2$, with $a = |\epsilon|^{-1}$, to obtain the last integral. By using the trigonometric substitution $s(\theta) = |\epsilon|^{-1}(1 - \sqrt{1 - |\epsilon|}\,\cos\theta)$, so that $\exp(\alpha x_+) \equiv s(\theta = 0)$, we easily find

$$
t(x) = \sqrt{\frac{m}{2\,\alpha^2|E|}}\,\arccos\left(\frac{1 - |\epsilon|\,\exp(\alpha x)}{\sqrt{1 - |\epsilon|}}\right).
$$

This solution can be easily inverted to yield the equation of motion for the bounded case:

$$
x(t) = \frac{1}{\alpha}\ln\left(\frac{1}{|\epsilon|} - \frac{\sqrt{1 - |\epsilon|}}{|\epsilon|}\,\cos(\omega t)\right), \tag{3.44}
$$

where $\omega(E) \equiv \sqrt{2\alpha^2|E|/m}$, and the particle velocity $v(t) = \dot{x}(t)$ is obtained from Eq. (3.44)

$$
v(t) = \sqrt{\frac{2\,|E|}{m}}\left(\frac{\sqrt{1 - |\epsilon|}\,\sin(\omega t)}{1 - \sqrt{1 - |\epsilon|}\,\cos(\omega t)}\right), \tag{3.45}
$$

The period of the bounded motion is

$$
T(E) = \sqrt{\frac{2m}{\alpha^2\,|E|}} \int_{\exp(\alpha x_+)}^{\exp(\alpha x_-)} \frac{ds}{\sqrt{|\epsilon|^{-2}(1 - |\epsilon|) - (s - |\epsilon|^{-1})^2}}
$$

$$
= \sqrt{\frac{2m}{\alpha^2\,|E|}} \int_0^\pi d\theta = \frac{\pi}{\alpha}\sqrt{\frac{2m}{|E|}}. \tag{3.46}
$$

One peculiar aspect of the period (3.46) is that, in the limit $|E| \to 0$, the period becomes infinite. This is easily understood from the fact that the second turning point $x_- = -(1/\alpha)\ln(1 - \sqrt{1 - |\epsilon|}) \to \infty$ as $|\epsilon| \to 0$.

3.5.3.2 Unbounded Motion

For the case $\epsilon > 0$, the integral solution (3.43) may be expressed as

$$t(x) = \sqrt{\frac{m}{2\,\alpha^2 E}} \int_{\exp(\alpha x_+)}^{\exp(\alpha x)} \frac{ds}{\sqrt{(s + \epsilon^{-1})^2 - \epsilon^{-2}(1 + \epsilon)}},$$

after completing the square. By using the hyperbolic-trigonometric substitution $s(z) = \epsilon^{-1}(-1 + \sqrt{1 + \epsilon}\,\cosh z)$, so that $\exp(\alpha x_+) \equiv s(z = 0)$, we easily find

$$t(x) = \sqrt{\frac{m}{2\,\alpha^2 E}}\,\operatorname{arccosh}\left(\frac{1 + \epsilon\,\exp(\alpha x)}{\sqrt{1 + \epsilon}}\right).$$

This solution can be easily inverted to yield the equation of motion for the unbounded case ($\epsilon > 0$):

$$x(t) = \frac{1}{\alpha}\ln\left(\frac{\sqrt{1 + \epsilon}}{\epsilon}\cosh(\alpha\,v_\infty\,t) - \frac{1}{\epsilon}\right), \tag{3.47}$$

where $v_\infty \equiv \sqrt{2E/m}$ is the asymptotic velocity at $x = \infty$. The particle velocity $v(t) = \dot{x}(t)$, on the other hand, is obtained from Eq. (3.47) as

$$v(t) = v_\infty\left(\frac{\sqrt{1 + \epsilon}\,\sinh(\alpha\,v_\infty\,t)}{\sqrt{1 + \epsilon}\,\cosh(\alpha\,v_\infty\,t) - 1}\right), \tag{3.48}$$

which reaches the asymptotic value v_∞ as $t \to \infty$.

Lastly, the special unbounded case $\epsilon = 0$ is solved as

$$t(x) = \sqrt{\frac{m}{2\,\alpha^2 U_0}} \int_{1/2}^{\exp(\alpha x)} \frac{ds}{\sqrt{2\,s - 1}},$$

which yields the equation of motion

$$x(t) = \frac{1}{\alpha}\ln\left(\frac{1}{2} + \frac{1}{2}\,(\alpha\,v_0\,t)^2\right), \tag{3.49}$$

where $v_0 \equiv \sqrt{2U_0/m}$ is the velocity at $x = 0$ (for $E = 0$) and the particle velocity is

$$v(t) = v_0\left(\frac{2\,\alpha\,v_0\,t}{1 + (\alpha\,v_0\,t)^2}\right), \tag{3.50}$$

which vanishes as $t \to \infty$. Note that this solution can either be obtained from the bounded solution (3.44) in the limit $|\epsilon| \to 0$ or from the unbounded solution (3.47) in the limit $\epsilon \to 0$.

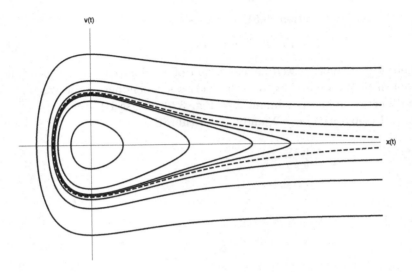

Fig. 3.4 Phase-space portrait for the motion in the Morse potential. The inner curves, which are labeled by $\epsilon = (-0.8, -0.3, -0.1, -0.05)$, describe the bounded motion (3.44)-(3.45). The outer curves, which are labeled by $\epsilon = (0.1, 0.5, 2)$, describe the unbounded motion (3.47)-(3.48). The dashed curve (labeled by $\epsilon = 0$) describes the separatrix solution (3.49)-(3.50), which separates the bounded and unbounded solutions.

3.5.3.3 *Phase-space Portrait*

The phase-space portrait for the motion in the Morse potential is shown in Fig. 3.4. The dashed curve corresponds to the special unbounded case given by Eqs. (3.49)-(3.50). Because this special curve separates the inner (bounded) curves (3.44)-(3.45) from the outer (unbounded) curves (3.47)-(3.48), it is referred to as the *separatrix* curve. The asymptotic values $\pm v_\infty$ for the unbounded velocities correspond to the case of a free particle.

3.5.4 *Pendulum*

Our third example involves the case of the pendulum problem discussed in Sec. 2.4.1. The energy equation in this case is

$$E = \frac{1}{2} m\ell^2 \dot{\theta}^2 + mg\ell (1 - \cos\theta), \qquad (3.51)$$

where the potential energy term is $mg\ell (1 - \cos\theta) \leq 2\,mg\ell$. It is convenient to rescale the pendulum dynamics by introducing a dimensionless time $\tau = \omega_0 t$, where $\omega_0 \equiv \sqrt{g/\ell}$, and a dimensionless energy $\epsilon \equiv E/(mg\ell)$, so that the energy equation (3.51) becomes

$$\epsilon = \frac{1}{2}(\theta')^2 + (1 - \cos\theta), \qquad (3.52)$$

where $\theta(\tau)$ is now viewed as a function of τ. Solutions of the pendulum problem (3.52) are divided into three classes depending on the value of the total energy of the pendulum (see Fig. 3.5): Class III (rotation) $\epsilon > 2$, Class II (separatrix) $\epsilon = 2$, and Class I (libration) $\epsilon < 2$.

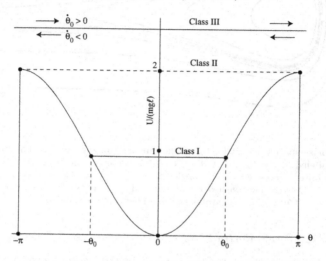

Fig. 3.5   Normalized pendulum potential $U(\theta)/(mg\ell) = 1 - \cos\theta$.

In the rotation class ($\epsilon > 2$), the kinetic energy can never vanish and the pendulum keeps rotating either clockwise or counter-clockwise depending on the sign of θ_0'. In the libration class ($\epsilon < 2$), on the other hand, the kinetic energy vanishes at turning points easily determined by initial conditions if the pendulum starts from rest ($\theta_0' = 0$) – in this case, the turning points are $\pm\theta_0 \equiv \pm\arccos(1 - \epsilon)$. In the separatrix class ($\epsilon = 2$), the turning points are $\pm\arccos(-1) = \pm\pi$.

3.5.4.1 *Libration Class* $(E < 2mg\ell)$

We now look at an explicit solution for pendulum librations (class I), where the (dimensionless) angular velocity θ' is

$$\theta'(\theta; E) = \pm\sqrt{2(\cos\theta - \cos\theta_0)} = \pm 2\sqrt{\sin^2(\theta_0/2) - \sin^2(\theta/2)},$$

$$(3.53)$$

where $\theta_0 = \arccos(1 - \epsilon)$. By making the substitution $\sin \theta/2 = k \sin \varphi$, where

$$k(\epsilon) = \sin(\theta_0/2) = \sqrt{\epsilon/2} < 1 \tag{3.54}$$

and $\varphi = \pm \pi/2$ when $\theta = \pm \theta_0$, Eq. (3.53) becomes

$$\varphi' = \pm \sqrt{1 - k^2 \sin^2 \varphi}. \tag{3.55}$$

The libration solution of the pendulum problem is thus

$$\tau(\theta) = \int_{\Theta(\theta)}^{\pi/2} \frac{d\varphi}{\sqrt{1 - k^2 \sin^2 \varphi}}, \tag{3.56}$$

where $\Theta(\theta) = \arcsin(k^{-1} \sin \theta/2)$. The inversion of this relation yields the libration equation of motion

$$\theta(\tau) = 2 \arcsin\left[k \, \mathrm{sn}(\tau|k^2)\right] \tag{3.57}$$

expressed in terms of the Jacobi elliptic function $\mathrm{sn}(\tau|k^2)$ (see Appendix B.1 for more details). From Eq. (3.57), we also calculate the angular velocity

$$\theta'(\tau) = \frac{2k \, \mathrm{cn}(\tau|k^2) \, \mathrm{dn}(\tau|k^2)}{\sqrt{1 - k^2 \, \mathrm{sn}^2(\tau|k^2)}} = 2k \, \mathrm{cn}(\tau|k^2), \tag{3.58}$$

where we used the Jacobi-function identity $\mathrm{dn}^2(\tau|k^2) \equiv 1 - k^2 \, \mathrm{sn}^2(\tau|k^2)$. Equations (3.57)-(3.58) can be used to recover the energy equation (3.52):

$$\frac{1}{2}(\theta')^2 + (1 - \cos\theta) = 2k^2 \left(\mathrm{cn}^2 + \mathrm{sn}^2\right) = 2k^2 = \epsilon,$$

which makes use of the Jacobi-function identity $\mathrm{cn}^2(\tau|k^2) + \mathrm{sn}^2(\tau|k^2) = 1$. We note that in the limit $k \ll 1$, we find $\mathrm{cn}(\tau|k^2) \simeq \mathrm{cn}(\tau|0) = \cos(\tau)$ and $\mathrm{sn}(\tau|k^2) \simeq \mathrm{sn}(\tau|0) = \sin(\tau)$, and Eqs. (3.57)-(3.58) become $\theta(\tau) \simeq \theta_0 \sin(\tau)$ and $\theta' = \theta_0 \cos(\tau)$.

Lastly, the period of oscillation is obtained from Eq. (3.56) and is defined as

$$\begin{aligned}
T(E) &= \frac{4}{\omega_0} \int_0^{\pi/2} \frac{d\varphi}{\sqrt{1 - k^2 \sin^2 \varphi}} \\
&= \frac{4}{\omega_0} \int_0^{\pi/2} d\varphi \left(1 + \frac{k^2}{2} \sin^2 \varphi + \cdots\right), \\
&= \frac{2\pi}{\omega_0} \left(1 + \frac{k^2}{4} + \cdots\right) = 4 K(k^2)/\omega_0, \tag{3.59}
\end{aligned}$$

where $K(k^2)$ denotes the complete elliptic integral of the first kind (see Fig. 3.6 and Appendix B.1). We note here that if $k \ll 1$ (or $\theta_0 \ll 1$) the libration period of a pendulum is nearly independent of energy, $T \simeq 2\pi/\omega_0$. However, we also note that as $\epsilon \to 2$ (i.e., $k \to 1$ or $\theta_0 \to \pi$), the libration period of the pendulum becomes infinitely large, i.e., $T \to \infty$ in Eq. (3.59).

3.5.4.2   Separatrix Class ($E = 2mg\ell$)

In the separatrix case ($\theta_0 = \pi$), the pendulum equation (3.53) yields the *separatrix* equation $\varphi' = \cos\varphi$, where $\varphi = \theta/2$. The integral solution

$$\tau(\varphi) = \int_0^\varphi \sec\psi \, d\psi = \ln(\sec\varphi + \tan\varphi)$$

can be inverted to yield the separatrix solution

$$\theta(\tau) = 2 \arcsec(\cosh(\tau)) = 2 \arccos(\operatorname{sech}(\tau)), \qquad (3.60)$$

where $\theta(0) = 0$ was chosen as the initial condition. The angular velocity associated with the separatrix solution is

$$\theta' = 2 \operatorname{sech}(\tau). \qquad (3.61)$$

Equations (3.60)-(3.61) can be used to recover the energy equation (3.52):

$$\frac{1}{2}(\theta')^2 + (1 - \cos\theta) = 2\operatorname{sech}^2(\tau) + 2\left(1 - \operatorname{sech}^2(\tau)\right) = 2 = \epsilon.$$

We again note that $\theta \to \pi$ only as $\tau \to \infty$.

3.5.4.3   Rotation Class ($E > 2mg\ell$)

The solution for rotations (class III) associated with the initial conditions $\theta_0 = 0$ and the energy equation (3.52) becomes

$$\frac{1}{2}(\theta_0')^2 = \epsilon = \frac{1}{2}(\theta')^2 - \left(1 - \cos\theta\right),$$

or $\theta' = \pm 2\sqrt{\epsilon/2 - \sin^2(\theta/2)}$, which shows that θ' does not vanish for rotations (since $\epsilon > 2$). We now define $\varphi \equiv \theta/2$ to obtain

$$\varphi' = \pm k\sqrt{1 - k^{-2}\sin^2\varphi}, \qquad (3.62)$$

where $k \equiv \sqrt{\epsilon/2} > 1$. Hence, the integral solution for rotations

$$\tau(\theta) = \frac{1}{k}\int_0^{\theta/2} \frac{d\varphi}{\sqrt{1 - k^{-2}\sin^2\varphi}} \qquad (3.63)$$

can be inverted and expressed in terms the Jacobi elliptic function $\operatorname{sn}(k\,\tau|k^{-2})$ (see Appendix B.1 for more details) as

$$\theta(\tau) = 2 \arcsin\left[\operatorname{sn}(k\,\tau|k^{-2})\right] \qquad (3.64)$$

From Eq. (3.64), we also calculate the angular velocity

$$\theta'(\tau) = \frac{2k\,\operatorname{cn}(k\tau|k^{-2})\,\operatorname{dn}(k\tau|k^{-2})}{\sqrt{1 - \operatorname{sn}^2(k\tau|k^{-2})}} = 2k\,\operatorname{dn}(k\tau|k^{-2}). \qquad (3.65)$$

Equations (3.64)-(3.65) can be used to recover the energy equation (3.52):

$$\frac{1}{2}(\theta')^2 + (1 - \cos\theta) = 2k^2 \, dn^2(k\tau|k^{-2}) + 2\sin^2(k\tau|k^{-2})$$

$$= 2k^2 \left(dn^2(k\tau|k^{-2}) + k^{-2} \sin^2(k\tau|k^{-2}) \right) = 2k^2 = \epsilon,$$

which makes use of the Jacobi-function identity $dn^2(z|m) + m \sin^2(z|m) = 1$. Lastly, the rotation period is obtained from Eq. (3.63) and is defined as

$$T(E) = \frac{4}{k\omega_0} \int_0^{\pi/2} \frac{d\varphi}{\sqrt{1 - k^{-2}\sin^2\varphi}} = \frac{4}{k\omega_0} K(k^{-2}). \qquad (3.66)$$

Figure 3.6 shows the plot of the normalized pendulum period as a function of k^2 for the Libration Class I and Rotation Class III.

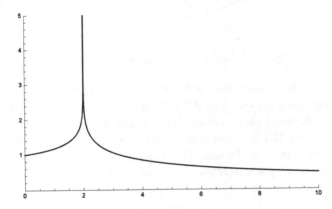

Fig. 3.6   Normalized pendulum period $\omega_0 T(E)/2\pi$ as a function of the normalized energy ϵ for Libration Class I ($\epsilon < 2$) and Rotation Class III ($\epsilon > 2$). The period is infinite for the Separatrix Class II ($\epsilon = 2$).

3.5.4.4   *Phase-space Portrait*

The phase-space portrait (θ, θ') of the pendulum is shown in Fig. 3.7. The dotted curves represent the separatrix solution (3.60)-(3.61), with $\epsilon = 2$ and the turning points at $\pm\pi$. The inner curves represent the libration solution (3.57)-(3.58), with $\epsilon = (0.08, 0.5, 1.28)$ and the turning points located at $\pm\arccos(1 - \epsilon)$. The outer curves represent the rotation solution (3.64)-(3.65), with $\epsilon = (3, 8)$.

The appearance of separatrices in periodic Hamiltonian systems is quite common whenever two different classes of motion are related by a single

Fig. 3.7   Phase-space portrait of the pendulum.

parameter. In the case of motion in the Morse potential (see Fig. 3.4) and the pendulum problem (see Fig. 3.7), the energy parameter ϵ is used to distinguish two different classes of motion: bounded and unbounded motions. In Secs. 7.2.3 and B.1.2, we shall study the case where two different classes of bounded motion exist. In general, a separatrix is associated with at least one turning point s_0 where $U'(s_0) = 0$ and $U''(s_0) < 0$.

3.5.5   *Constrained Motion on the Surface of a Cone*

For our last example, we return to the constrained motion of a particle of mass m on a cone in the presence of gravity. In Sec. 2.5.4, we showed that this motion is doubly periodic in the generalized coordinates s and θ. The fact that the Lagrangian (2.52) is independent of time leads to the conservation law of energy

$$E = \frac{m}{2}\dot{s}^2 + \left(\frac{\ell^2}{2m\sin^2\alpha\, s^2} + mg\cos\alpha\, s\right) = \frac{m}{2}\dot{s}^2 + V(s), \quad (3.67)$$

where the conservation law of angular momentum $\ell = ms^2\sin^2\alpha\,\dot{\theta}$ was used. The effective potential $V(s)$ in Eq. (3.67) has a single minimum $V_0 = \frac{3}{2}mgs_0\cos\alpha$ at

$$s_0 = \left(\frac{\ell^2}{m^2 g\,\sin^2\alpha\cos\alpha}\right)^{\frac{1}{3}},$$

and the only type of motion is bounded when $E > V_0$.

The turning points for this problem are solutions of the cubic equation

$$\frac{3}{2}\,\epsilon \;=\; \frac{1}{2\sigma^2} \,+\, \sigma, \qquad\qquad (3.68)$$

where $\epsilon = E/V_0 > 1$ and $\sigma = s/s_0$. The three roots of Eq. (3.68) are

$$\left.\begin{array}{l} \sigma_1 = \epsilon/2 \,+\, \epsilon\,\cos[\phi(\epsilon)/3] \\[1.5em] \sigma_2 = \epsilon/2 \,-\, \epsilon\,\cos[\pi/3 + \phi(\epsilon)/3] \\[1.5em] \sigma_3 = \epsilon/2 \,-\, \epsilon\,\cos[\pi/3 - \phi(\epsilon)/3] \end{array}\right\}, \qquad (3.69)$$

where $\phi(\epsilon) \equiv \arccos(1 - 2/\epsilon^3)$; see App. A for the calculation of the three roots of a general cubic polynomial. The root σ_3 remains negative for all normalized energies ϵ; this root is discarded as unphysical since s must be positive (by definition). The other two roots (σ_1, σ_2), which are complex for $\epsilon < 1$ (i.e., for energies below the minimum of the effective potential energy V_0), become real at $\epsilon = 1$ ($\phi = \pi$), where $\sigma_1 = \sigma_2 = 1$, and separate $(\sigma_1 > \sigma_2)$ for larger values of ϵ.

The dimensionless equations of motion for this problem are

$$\left.\begin{array}{l} \sigma'' = -1 \,+\, 1/\sigma^3 \\[1.5em] \sin\alpha\,\theta' = 1/\sigma^2 \end{array}\right\}, \qquad (3.70)$$

where we introduced a dimensionless time $\tau \equiv \omega_g t$, with $\omega_g \equiv \sqrt{(g/s_0)}\,\cos\alpha$. The top figure of Fig. 2.9 shows that the solution of the σ-equation is periodic between $\sigma_2 \le \sigma \le \sigma_1$, while the θ-dynamics is naturally periodic (bottom figure).

Lastly, the period of σ-oscillations is determined by the definite integral

$$\omega_g\, T(\epsilon) \;=\; 2\int_{\sigma_1}^{\sigma_2} \frac{\sigma\,d\sigma}{\sqrt{3\epsilon\,\sigma^2 - 1 - 2\,\sigma^3}}, \qquad (3.71)$$

whose solution is expressed in terms of Weierstrass elliptic function (see Appendix B). We note that in one period (see Fig. 2.9), the orbit precesses by an angular deviation

$$\Delta\theta(\epsilon) \;=\; \int_0^{\omega_g T(\epsilon)} \theta'\,d\tau \;=\; \frac{1}{\sin\alpha}\int_0^{\omega_g T(\epsilon)} \frac{d\tau}{\sigma^2(\tau)},$$

where we used the conservation law of angular momentum.

Table 3.2   Summary of Chapter 3: Hamiltonian Mechanics.

Topic	Equation
Hamilton's Equations	(3.3)
Legendre Transformation	(3.4)
Phase-space Lagrangian	(3.5)
Hamilton-Jacobi Equation	(3.8)
Hamiltonian for Charged-Particle Motion	(3.27)
Energy Method	(3.35)
Simple Harmonic Oscillator	(3.37)-(3.40)
Morse Potential	(3.41)-(3.50)
Pendulum	(3.51)-(3.66)
Constrained Motion on the Surface of a Cone	(3.67)-(3.71)

3.6   Summary

Chapter 3 presented a brief introduction of the Hamiltonian method, where the Hamiltonian function on phase space was derived from the Lagrangian function on configuration space by the Legendre transformation. The energy method was used to provide explicit solutions to bounded and unbounded motion in one-dimensional potential functions. Table 3.2 presents a summary of the important topics of Chapter 3.

3.7   Problems

1. A particle of mass m and total energy E moves periodically in a one-dimensional potential $U(x) = F\,|x|$, where F is a positive constant.

(a) Find the turning points for this potential $\pm x_0$.

(b) Show that the dynamical solution $x(t; E)$ for this potential is

$$
x(t) \;=\; \begin{cases}
x_0 - a\,t^2/2 & (0 \leq t \leq \tau) \\[2mm]
-x_0 + a\,(t - 2\,\tau)^2/2 & (\tau \leq t \leq 3\,\tau) \\[2mm]
x_0 - a\,(t - 4\,\tau)^2/2 & (3\,\tau \leq t \leq 4\,\tau)
\end{cases}
$$

where a and τ are suitable constants to be determined.

(c) Find the period $T(E) \equiv 4\,\tau$ for the motion.

2. Find the Hamiltonian $H(x, p) = p\,\dot{x}(x, p) - L(x, \dot{x}(x, p))$ for the following

Lagrangian
$$L(x, \dot{x}) = \frac{m}{2} \dot{x}^2 (1 + \alpha x^2) - \frac{m}{2} \omega^2 x^2,$$
where (m, ω, α) are constants and $p \equiv \partial L/\partial \dot{x}$ is inverted to obtain the relation $\dot{x}(x, p)$.

3. A block of mass m rests on the inclined plane (with angle θ) of a triangular block of mass M as shown in Fig. 3.8. Here, we consider the case where both blocks slide without friction (i.e., m slides on the inclined plane without friction and M slides without friction on the horizontal plane).

Fig. 3.8   Problem 2.

(a) Using the generalized coordinates (x, y) shown in Fig. 3.8, construct the Lagrangian $L(x, \dot{x}, y, \dot{y})$.

(b) Derive the Euler-Lagrange equations for x and y.

(c) Calculate the canonical momenta
$$p_x(x, \dot{x}, y, \dot{y}) = \frac{\partial L}{\partial \dot{x}} \quad \text{and} \quad p_y(x, \dot{x}, y, \dot{y}) = \frac{\partial L}{\partial \dot{y}},$$
and invert these expressions to find the functions $\dot{x}(x, p_x, y, p_y)$ and $\dot{y}(x, p_x, y, p_y)$.

(d) Calculate the Hamiltonian $H(x, p_x, y, p_y)$ for this system by using the Legendre transformation $H(x, p_x, y, p_y) = p_x \dot{x} + p_y \dot{y} - L(x, \dot{x}, y, \dot{y})$, where the functions $\dot{x}(x, p_x, y, p_y)$ and $\dot{y}(x, p_x, y, p_y)$ are used.

(e) Find which of the two momenta found in Part (c) is a constant of the motion and discuss why it is so. If the two blocks start from rest, what is the value of this constant of motion?

4. Consider all possible orbits of a unit-mass particle moving in the dimensionless potential $U(x) = 1 - x^2/2 + x^4/16$. Here, orbits are solutions of the equation of motion $\ddot{x} = -U'(x)$ and the dimensionless energy equation is $E = \dot{x}^2/2 + U(x)$.

(a) Draw the potential $U(x)$ and identify all possible unbounded and bounded orbits (with their respective energy ranges).

(b) For each orbit found in part (a), find the turning point(s) for each energy level.

(c) Sketch the phase portrait $(x, \dot{x})$ showing all orbits (including the separatrix orbit).

(d) Show that the separatrix orbit (with initial conditions $x_0 = \sqrt{8}$ and $\dot{x}_0 = 0$) is expressed as $x(t) = \sqrt{8}\,\text{sech}(t)$ by solving the integral

$$t(x) = \int_x^{\sqrt{8}} \frac{ds}{\sqrt{s^2\,(1 - s^2/8)}}.$$

(*Hint: use the hyperbolic trigonometric substitution* $s = \sqrt{8}\,\text{sech}\,\xi$.)

5. Write a numerical code to solve the second-order ordinary differential equation $\ddot{x} = x - x^3/3$ by choosing appropriate initial conditions needed to obtain all the possible (bounded and unbounded) orbits.

6. When a particle (of mass m) moving under the potential $U(x)$ is perturbed by the potential $\delta U(x)$, its period (3.36) is changed by a small amount defined as

$$\delta T = -\sqrt{2m}\,\frac{\partial}{\partial E}\left[\int_{x_1}^{x_2} \frac{\delta U(x)\,dx}{\sqrt{E - U(x)}} \right],$$

where $x_{1,2}(E)$ are the turning points of the unperturbed problem.

Show that the change in the period of a particle moving in the quadratic potential $U(x) = m\omega^2 x^2/2$ introduced by the perturbation potential $\delta U(x) = \epsilon\, x^4$ is

$$\delta T = -\epsilon\,\frac{6\pi\,E}{m^2\,\omega^5},$$

where the particle is trapped in the region $-a \le x \le a$ in the unperturbed potential, where $a = \sqrt{2\,E/(m\omega^2)}$.

7. Consider two simple-harmonic oscillators described by the coordinates x_1 and x_2 with respective frequencies ω_1 and $\omega_2 > \omega_1$ and subject to the (nonlinear) constraint $x_1^2 + x_2^2 = 1$.

(a) Derive the Euler-Lagrange equations for x_1 and x_2 by using the Lagrangian

$$L = \frac{1}{2}\left(\dot{x}_1^2 + \dot{x}_2^2\right) - \frac{1}{2}\left(\omega_1^2 x_1^2 + \omega_2^2 x_2^2\right) - \lambda\left(x_1^2 + x_2^2 - 1\right),$$

where λ is the Lagrange multiplier associated with the constraint, and verify that the energy $E = \frac{1}{2}\left(\dot{x}_1^2 + \dot{x}_2^2\right) + \frac{1}{2}\left(\omega_1^2 x_1^2 + \omega_2^2 x_2^2\right)$ is a constant of the motion (i.e., $dE/dt = 0$).

(b) Find an expression for the Lagrange multiplier λ in terms of x_1 and x_2.

(c) By substituting $x_1(t) = \cos\theta(t)$ and $x_2(t) = \sin\theta(t)$ into the Euler-Lagrange equations derived in Part (a), show that one obtains the nonlinear equation

$$\ddot{\theta} + \left(\omega_2^2 - \omega_1^2\right)\cos\theta\sin\theta = 0.$$

Note that this equation can also be written as $\varphi'' + (\Omega^2 - 1)\sin\varphi = 0$, where $\varphi(\tau) = 2\,\theta(\omega_1 t)$ and $\Omega \equiv \omega_2/\omega_1 > 1$.

8. Recreate the phase portrait shown in Fig. 3.7 for the pendulum by numerical integration of the normalized equation $\theta'' + \sin\theta = 0$ subject to the energy conservation law $\epsilon = \frac{1}{2}(\theta')^2 + (1 - \cos\theta)$, from which suitable initial conditions can be selected.

9. Show that Eq. (3.60) is the separatrix solution for the pendulum problem when $E = 2\,mg\,\ell$.

10. A particle of mass m is moving in the potential

$$U(x,y) = \begin{cases} U_1 & (x < 0) \\ \\ U_2 & (x > 0) \end{cases}$$

(a) Assuming that the particle is moving with velocity

$$\mathbf{v}_1 = v_1\left(\cos\theta_1\,\widehat{\mathbf{x}} + \sin\theta_1\,\widehat{\mathbf{y}}\right)$$

in the region $x < 0$, and velocity

$$\mathbf{v}_2 = v_2\left(\cos\theta_2\,\widehat{\mathbf{x}} + \sin\theta_2\,\widehat{\mathbf{y}}\right)$$

in the region $x > 0$ and along the direction $\theta_2 \neq \theta_1$, and using the conservation laws of energy and momentum in the y-direction, show that

$$\frac{\sin \theta_1}{\sin \theta_2} = \sqrt{1 + \frac{U_1 - U_2}{E - U_1}}.$$

(b) Discuss the cases $E > U_1 > U_2$ and $E > U_2 > U_1$.

11. Show that the period of oscillation for a particle of mass m moving in the potential $U(x) = U_0 \tan^2(kx)$ is given as

$$T(E) = 2\sqrt{2m} \int_0^a \frac{dx}{\sqrt{E - U_0 \tan^2(kx)}} = \frac{\pi}{k} \sqrt{\frac{2m}{E + U_0}},$$

where the turning point is $a = \arctan(\sqrt{E/U_0})$.

Hint: Use the substitution $\tan(kx) = \tan(ka) \sin \theta$ and the integral

$$\int_0^{\pi/2} \frac{d\theta}{1 + \alpha^2 \sin^2 \theta} = \frac{\pi/2}{\sqrt{1 + \alpha^2}}.$$

12. The relativistic Lagrangian for a particle of rest mass m moving in a potential $U(\mathbf{x})$ is

$$L(\mathbf{x}, \dot{\mathbf{x}}) = - mc^2 \sqrt{1 - |\dot{\mathbf{x}}|^2/c^2} - U(\mathbf{x}),$$

where c is the speed of light.

(a) Derive the equation of motion for $\mathbf{x}$.

(b) Using the Legendre transformation (3.4), derive the relativistic Hamiltonian.

Chapter 4

Motion in a Central-Force Field

The present Chapter introduces an important set of problems that are solvable by the Noether and Energy methods. Here, bounded and unbounded solutions are obtained for time-independent central-force planar problems in which energy and angular momentum are constants of the motion. The existence of two constants of motion for two-dimensional planar motion implies that exact solutions can be obtained if integral solutions (obtained by quadrature) can be inverted.

4.1 Motion in a Central-Force Field

A particle moves under the influence of a central-force field

$$\mathbf{F}(\mathbf{r}) \;=\; F(r)\,\widehat{r}(\theta,\varphi) \;\equiv\; -\,U'(r)\,\widehat{r}, \tag{4.1}$$

if the force $F(r) = -U'(r)$ on the particle is independent of the angular position (θ, φ) of the particle about the center of force and depends only on its distance r from the center of force. For a power-law central-force potential $U(r) \equiv (k/n)\, r^n$, the central force $\mathbf{F} = -\nabla U = -k\, r^{n-1}\widehat{r}$ is attractive (or repulsive) if the constant k is positive (or negative).

Note that, for a central-force potential $U(r)$, the angular momentum $\mathbf{L} = \ell\widehat{z}$ in the CM frame is a constant of the motion [see Eq. (2.60)] since $\mathbf{r} \times \nabla U(r) = 0$. When we consider the planar dynamics of two particles that interact through a time-independent central potential, the conservation of energy and angular momentum allows us to obtain an exact solution in terms of the polar coordinates $(r(t), \theta(t))$ associated with the relative position

$$\mathbf{r} \;\equiv\; \mathbf{r}_1 - \mathbf{r}_2 \;\equiv\; r\left(\cos\theta\,\widehat{x} + \sin\theta\,\widehat{y}\right) \;\equiv\; r\widehat{r} \tag{4.2}$$

between the two particles.

4.1.1 *Lagrangian Formalism*

The center-of-mass (CM) Lagrangian for two interacting particles (with masses m_1 and m_2) in a central-force potential $U(r)$ was presented in Eq. (2.58). Using the polar coordinates (4.2), the CM Lagrangian becomes

$$L(r, \dot{r}, \dot{\theta}) = \frac{\mu}{2} \left(\dot{r}^2 + r^2 \dot{\theta}^2 \right) - U(r), \tag{4.3}$$

where $\mu \equiv m_1 m_2 / (m_1 + m_2)$ denotes the reduced mass for the two-particle system.

Since the Lagrangian (4.3) is independent of θ, it follows from Noether's Theorem

$$\frac{d}{dt} \left(\frac{\partial L}{\partial \dot{\theta}} \right) = \frac{\partial L}{\partial \theta} \equiv 0$$

that the canonical angular momentum

$$p_\theta = \frac{\partial L}{\partial \dot{\theta}} = \mu r^2 \dot{\theta} \equiv \ell \tag{4.4}$$

is a constant of motion (labeled ℓ). The Routhian for this problem is, therefore, defined as

$$R(r, \dot{r}; p_\theta) \equiv L(r, \dot{r}, \dot{\theta}) - p_\theta \dot{\theta} = \frac{\mu}{2} \dot{r}^2 - V(r), \tag{4.5}$$

where the effective potential is

$$V(r) = \frac{\ell^2}{2\mu r^2} + U(r) \tag{4.6}$$

The Euler-Lagrange-Routh equation for r yields the radial force equation

$$\mu \ddot{r} = -V'(r) = \frac{\ell^2}{\mu r^3} - U'(r). \tag{4.7}$$

The radial solution $r(t)$ of the radial equation (4.7) is formally obtained from the energy equation

$$E = \frac{\mu \dot{r}^2}{2} + \frac{\ell^2}{2\mu r^2} + U(r) = \frac{\mu \dot{r}^2}{2} + V(r) \tag{4.8}$$

as

$$t(r; E, \ell) = \sqrt{\frac{\mu}{2}} \int_{r_0}^{r} \frac{d\rho}{\sqrt{E - V(\rho)}}, \tag{4.9}$$

where $r_0 = r(0)$ is the initial radial position.

If the integral solution (4.9) can be explicitly evaluated and inverted, the radial solution $r(t; E, \ell)$ is then found for a given pair of invariants

(E, ℓ). Once the radial solution $r(t)$ is found, the angular solution $\theta(t)$ is then obtained by integration

$$\theta(t) = \theta_0 + \frac{\ell}{\mu} \int_0^t \frac{d\tau}{r^2(\tau)}. \tag{4.10}$$

The planar dynamical orbit

$$\mathbf{r}(t) = r(t) \left(\cos\theta(t)\,\widehat{\mathbf{x}} + \sin\theta(t)\,\widehat{\mathbf{y}} \right) \tag{4.11}$$

is thus parameterized by time, for a given pair of invariants (E, ℓ).

The dynamical solution (4.11) appears simple enough to write down, however, it is often difficult to explicitly evaluate the integral radial solution (4.9). Another approach focuses on deriving the orbital solution $r(\theta)$, where θ now appears as the orbit parameter, so that the planar orbit

$$\mathbf{r}(\theta) \equiv r(\theta) \left(\cos\theta\,\widehat{\mathbf{x}} + \sin\theta\,\widehat{\mathbf{y}} \right) \tag{4.12}$$

can be found. While the planar orbit (4.12) is easier to find, we cannot determine the position of the particle as a function of time until the angular solution $\theta(t)$ is known.

4.1.1.1 *Radial Orbit Equation*

We now proceed with the derivation of the orbital solution $r(\theta)$ for a given potential $U(r)$. Since $\dot\theta = \ell/\mu r^2$ does not change sign along the orbit (as a result of the conservation of angular momentum $\ell \neq 0$), we may replace $\dot r$ and $\ddot r$ with $r'(\theta)$ and $r''(\theta)$ as follows. First, we begin with

$$\dot r = \dot\theta\, r' = \frac{\ell\, r'}{\mu\, r^2} = -\frac{\ell}{\mu} \left(\frac{1}{r} \right)' = -(\ell/\mu)\, s',$$

where we use the conservation of angular momentum and define the new dependent variable $s(\theta) = 1/r(\theta)$. Next, we write $\ddot r = -(\ell/\mu)\,\dot\theta\, s'' = -(\ell/\mu)^2\, s^2\, s''$, so that the radial force equation (4.7) becomes

$$s'' + s = -\frac{\mu}{\ell^2 s^2}\, F(1/s) \equiv -\frac{d\overline{U}(s)}{ds}, \tag{4.13}$$

where

$$\overline{U}(s) = \frac{\mu}{\ell^2}\, U(1/s) \tag{4.14}$$

denotes the normalized central potential expressed as a function of s.

4.1.1.2 *Inversion Problem*

Before proceeding with the solution of the orbit equation (4.13), we note that the form of the physical potential $U(r)$ can be calculated from the knowledge of the function $s(\theta) = 1/r(\theta)$ and the constants of motion (E, ℓ). This inverse-problem procedure requires: (a) expressing s'' as a function of s through $\theta(s)$; (b) solving for $\overline{U}(s)$ through Eq. (4.13):

$$\frac{d\overline{U}(s)}{ds} \equiv -s''(\theta(s)) - s; \tag{4.15}$$

and (c) using the definition (4.14): $U(r) \equiv (\ell^2/\mu)\,\overline{U}(1/r)$. The angular dynamics $\theta(t)$ can also be determined as

$$\dot{\theta} = \frac{\ell}{\mu r^2(\theta)} \quad \rightarrow \quad t(\theta) = \frac{\mu}{\ell} \int_0^\theta r^2(\phi)\, d\phi, \tag{4.16}$$

where we assumed that $t(\theta = 0) = 0$.

For example, consider the particle trajectory described in terms of the function $r(\theta) = r_0 \sec(\alpha\,\theta)$, where r_0 and α are constants.[1] The radial orbit equation (4.13) then becomes

$$s'' + s = -(\alpha^2 - 1)\, s = -\frac{d\overline{U}(s)}{ds},$$

where we used $s(\theta) = s_0 \cos(\alpha\theta)$ and $s'' = -\alpha^2 s_0 \cos(\alpha\theta) \equiv -\alpha^2 s$, with $s_0 = 1/r_0$. Hence, we readily find

$$\overline{U}(s) = \frac{1}{2}\,(\alpha^2 - 1)\, s^2 \quad \rightarrow \quad U(r) = \frac{\ell^2}{2\mu r^2}\,(\alpha^2 - 1).$$

As expected, the central potential is either repulsive for $\alpha > 1$ or attractive for $\alpha < 1$ (see Fig. 4.1).

We note that knowledge of the orbit $r(\theta)$ allows us to solve for the angular dynamics $\theta(t)$ from Eq. (4.16), which in turn gives us the radial dynamics $r(t)$. For our example, we find

$$t(\theta) = \frac{\mu r_0^2}{\alpha\ell} \int_0^{\alpha\theta} \sec^2\phi\, d\phi = \frac{\mu r_0^2}{\alpha\ell}\tan(\alpha\theta),$$

which can be inverted to yield the angular solution

$$\theta(t) = \alpha^{-1}\arctan(\alpha\,\omega\,t), \tag{4.17}$$

where $\omega \equiv \ell/(\mu r_0^2)$. Next, upon substituting Eq. (4.17) into $r(\theta) = r_0 \sec(\alpha\,\theta)$, we obtain the radial solution

$$r(t) = r_0\sqrt{1 + (\alpha\,\omega t)^2}. \tag{4.18}$$

---

[1]These constants could be determined by tracing the orbit $r(\theta)$ as a function of θ of a particle over an extended period of time.

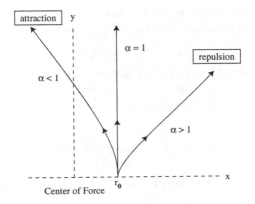

Fig. 4.1 Repulsive ($\alpha > 1$) and attractive ($\alpha < 1$) orbits: $x(\theta) = r_0 \sec(\alpha\theta) \cos(\theta)$ and $y(\theta) = r_0 \sec(\alpha\theta) \sin(\theta)$, for the central-force potential $U(r) = (\ell^2/2\mu)(\alpha^2 - 1) r^{-2}$.

Lastly, the total energy

$$E = \frac{\alpha^2 \ell^2}{2\mu r_0^2} \equiv V(r_0),$$

is determined from the initial conditions $r(0) = r_0$ and $\dot{r}(0) = 0$, i.e., r_0 is the single turning point in the effective potential $V(r) = U(r) + \ell^2/(2\mu r^2) = \alpha^2 \ell^2/(2\mu r^2)$.

4.1.2 *Hamiltonian Formalism*

We now return to the solution of the orbit equation (4.13). First, we write the energy equation

$$\mu E/\ell^2 \equiv \frac{\epsilon}{2} = \frac{s'^2}{2} + \frac{s^2}{2} + \overline{U}(s), \qquad (4.19)$$

in terms of (s, s') and the normalized potential $\overline{U}(s)$, so that we obtain

$$s'(\theta) = \pm \sqrt{\epsilon - 2\overline{U}(s) - s^2}. \qquad (4.20)$$

Hence, for a given central-force potential $U(r)$, we can solve for $r(\theta) = 1/s(\theta)$ by integrating

$$\theta(s) = -\int_{s_0}^{s} \frac{d\sigma}{\sqrt{\epsilon - 2\overline{U}(\sigma) - \sigma^2}}, \qquad (4.21)$$

where s_0 defines $\theta(s_0) = 0$, and performing the inversion $\theta(s) \to s(\theta)$ to obtain the orbit equation $r(\theta) = 1/s(\theta)$.

The evaluation of the integral solution (4.21) requires the determination of the turning points (where $s' = 0$) of the equation

$$\epsilon = s^2 + 2\,\overline{U}(s). \tag{4.22}$$

If two non-vanishing turning points $0 < s_1 < s_2$ exist, the motion is said to be *bounded* in the interval $s_1 < s < s_2$ (or $r_2 < r < r_1$); otherwise, the motion is *unbounded*. If the motion is bounded, the angular period $\Delta\theta$ is defined as

$$\Delta\theta = 2 \int_{s_1}^{s_2} \frac{ds}{\sqrt{\epsilon - 2\,\overline{U}(s) - s^2}}. \tag{4.23}$$

Here, the bounded orbit is *closed* only if $\Delta\theta$ is a rational multiple of 2π.

4.2 Homogeneous Central Potentials*

An important class of central potentials is provided by homogeneous potentials that satisfy the condition $U(\lambda\,\mathbf{r}) = \lambda^n\,U(\mathbf{r})$, where λ denotes a rescaling parameter and n denotes the *order* of the homogeneous potential.

4.2.1 *The Virial Theorem*

The Virial Theorem is an important theorem in Celestial Mechanics and Astrophysics. We begin with the time derivative of $S = \sum_i \mathbf{p}_i \cdot \mathbf{r}_i$:

$$\frac{dS}{dt} = \sum_i \left(\frac{d\mathbf{p}_i}{dt} \cdot \mathbf{r}_i + \mathbf{p}_i \cdot \frac{d\mathbf{r}_i}{dt} \right), \tag{4.24}$$

where $\mathbf{p}_i = m_i\, d\mathbf{r}_i/dt$ denotes the kinetic momentum of the ith particle, and the summation is over all particles in a mechanical system under the influence of a self-interaction potential

$$U = \frac{1}{2} \sum_{i,\,j\neq i} U(\mathbf{r}_i - \mathbf{r}_j) \equiv \frac{1}{2} \sum_{i,\,j\neq i} U_{ij}.$$

We note, however, that S itself can be written as a time derivative

$$S = \sum_i m_i \frac{d\mathbf{r}_i}{dt} \cdot \mathbf{r}_i = \frac{d}{dt} \left(\frac{1}{2} \sum_i m_i\, |\mathbf{r}_i|^2 \right) = \frac{1}{2} \frac{d\mathcal{I}}{dt},$$

where $\mathcal{I}$ denotes the *moment of inertia* of the system. Using Hamilton's equations

$$\frac{d\mathbf{r}_i}{dt} = \frac{\mathbf{p}_i}{m_i} \quad \text{and} \quad \frac{d\mathbf{p}_i}{dt} = -\sum_{j\neq i} \nabla_i U(\mathbf{r}_i - \mathbf{r}_j),$$

Eq. (4.24) can also be written as

$$\frac{1}{2}\frac{d^2\mathcal{I}}{dt^2} = \sum_i \left(\frac{|\mathbf{p}_i|^2}{m_i} - \mathbf{r}_i \cdot \sum_{j\neq i} \nabla_i U_{ij}\right) = 2\,K - \sum_{i,\,j\neq i} \mathbf{r}_i \cdot \nabla_i U_{ij},$$
(4.25)

where K denotes the kinetic energy of the mechanical system. Next, using Newton's Third Law, we write

$$\sum_{i,\,j\neq i} \mathbf{r}_i \cdot \nabla_i U_{ij} = \frac{1}{2}\sum_{i,\,j\neq i}(\mathbf{r}_i - \mathbf{r}_j)\cdot\nabla U(\mathbf{r}_i - \mathbf{r}_j),$$

and, for a homogeneous central potential of order n, we find $\mathbf{r}\cdot\nabla U(\mathbf{r}) = n\,U(\mathbf{r})$, so that

$$\frac{1}{2}\sum_{i,\,j\neq i}(\mathbf{r}_i - \mathbf{r}_j)\cdot\nabla U(\mathbf{r}_i - \mathbf{r}_j) \equiv n\,U.$$

Hence, Eq. (4.25) becomes the *Virial of Clausius* (Rudolph Clausius, 1822-1888)

$$\frac{1}{2}\frac{d^2\mathcal{I}}{dt^2} = 2\,K - n\,U.$$
(4.26)

If we now assume that the mechanical system under consideration is periodic in time (e.g., the system is bounded), then the time average (denoted $\langle\cdots\rangle$) of Eq. (4.26) yields the Virial Theorem

$$\langle K\rangle = \frac{n}{2}\langle U\rangle.$$
(4.27)

Hence, the time-average of the total energy of the mechanical system, $E = K + U$, is expressed as

$$E = (1 + n/2)\langle U\rangle = (1 + 2/n)\langle K\rangle,$$
(4.28)

since $\langle E\rangle = E$, i.e., the time average of a constant of motion is equal to itself. For example, for the Kepler problem ($n = -1$), we find

$$E = \frac{1}{2}\langle U\rangle = -\langle K\rangle < 0,$$
(4.29)

which means that the total energy of a bounded Keplerian orbit is negative (see Sec. 4.3.1).

We note that the Virial Theorem has important applications in astrophysics where the contraction of a self-gravitating cloud (i.e., $\langle U\rangle$ becoming more negative) leads to an increase in its internal energy (i.e., $\langle K\rangle$ becoming more positive).

4.2.2   General Properties of Homogeneous Potentials

We now investigate the dynamical properties of orbits in homogeneous central potentials of the form $U(r) = (k/n)\, r^n$ $(n \neq -2)$, where k denotes a positive constant, which means that the associated central force $\mathbf{F} = -\nabla U = -k\, r^{n-1}\, \hat{r}$ is attractive.

First, the effective potential (4.6):

$$V(r) = \frac{\ell^2}{2\mu\, r^2} + \frac{k}{n}\, r^n$$

has a single extremum point, where

$$V'(r_0) = \frac{1}{r_0^3}\left(k\, r_0^{n+2} - \frac{\ell^2}{\mu}\right) = 0,$$

at a distance $r_0 = 1/s_0$ defined as

$$r_0^{n+2} = \frac{\ell^2}{k\mu} = \frac{1}{s_0^{n+2}}.$$

It is simple to show that this extremum is a maximum if $n < -2$ or a minimum if $n > -2$; we shall, henceforth, focus our attention on the latter case, where the minimum in the effective potential is

$$V_0 = V(r_0) = \left(1 + \frac{n}{2}\right)\frac{k}{n}\, r_0^n = \left(1 + \frac{n}{2}\right)U_0.$$

In the vicinity of this minimum, we can certainly find periodic orbits with turning points ($r_2 = 1/s_2 < r_1 = 1/s_1$) that satisfy the condition $E = V(r)$.

Next, the radial equation (4.13) is written in terms of the potential $\overline{U}(s) = (\mu/\ell^2)\, U(1/s)$ as

$$s'' + s = -\frac{d\overline{U}}{ds} = \frac{s_0^{n+2}}{s^{n+1}},$$

and its solution is given as the orbit integral

$$\theta(s) = \int_s^{s_2} \frac{d\sigma}{\sqrt{\epsilon - (2/n)\, s_0^{n+2}/\sigma^n - \sigma^2}}, \qquad (4.30)$$

where s_2 denotes the upper turning point in the s-coordinate. The solution (4.30) can be expressed in terms of closed analytic expressions obtained by trigonometric substitution only for $n = -1$ or $n = 2$ (when $\epsilon \neq 0$), which we now study in detail below (the cases $n = -3$ and -4, for example, are solved in terms of elliptic functions as discussed in Appendix B).

4.3 Kepler Problem

In this Section, we solve the Kepler problem (Johannes Kepler, 1571-1630) where the attractive central potential $U(r) = -k/r$ is homogeneous with order $n = -1$ and k is a positive constant.[2] The Virial Theorem (4.27) implies that periodic solutions of the Kepler problem have negative total energies $E = -\langle K \rangle = (1/2) \langle U \rangle$.

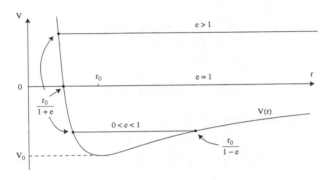

Fig. 4.2 Effective potential (4.32) for the Kepler problem, which has a single minimum $V_0 < 0$ at $r = r_0$. Orbits are labeled by the eccentricity $e \equiv \sqrt{1 - E/V_0}$ and are either bounded: circular ($e = 0$) or elliptical ($e < 1$); or unbounded: parabolic ($e = 1$) or hyperbolic ($e > 1$).

The general solution of the Kepler problem involves solutions for the radial position $r(t)$ and angular position $\theta(t)$

$$\mu \ddot{r} = \frac{\ell^2}{\mu r^3} - \frac{k}{r^2} \equiv -V'(r) \quad \text{and} \quad \dot{\theta} = \frac{\ell}{\mu r^2}, \qquad (4.31)$$

whose orbits $r(\theta)$ are either bounded (periodic) or unbounded (see Fig. 4.2) in the effective potential

$$V(r) = -\frac{k}{r} + \frac{\ell^2}{2\mu r^2}. \qquad (4.32)$$

This potential exhibits a single minimum

$$V_0 = -\frac{k}{2r_0} < 0 \quad \text{at} \quad r = r_0 = \frac{\ell^2}{\mu k}. \qquad (4.33)$$

To obtain an analytic solution $r(\theta)$ for the Kepler problem (4.31), as expressed by the radial force equation (4.13), we use the normalized central potential $\overline{U}(s) = -s_0 s$, where $s_0 = \mu k/\ell^2$, and Eq. (4.13) becomes

$$s'' + s = s_0. \qquad (4.34)$$

---

[2]For problems involving gravitational attraction, we have $k = G m_1 m_2$, while for problems involving electrostatic attraction ($q_1 q_2 < 0$), we have $k = |q_1 q_2|/(4\pi\epsilon_0)$.

The solution of Eq. (4.34) is given by Eq. (4.30), with $n = -1$:

$$\theta(s) = \int_s^{s_2} \frac{d\sigma}{\sqrt{\epsilon + 2\,s_0\,\sigma\, -\, \sigma^2}},$$

and requires finding the turning points $s_1 < s_2$ for the Kepler problem.

The turning points for the Kepler problem are solutions of the quadratic equation

$$s^2 - 2\,s_0\,s - \epsilon = 0,$$

which can be written as $s_{1,2} = s_0 \pm \sqrt{s_0^2 + \epsilon}$:

$$s_1 = s_0\,(1 - \mathrm{e}) \quad \text{and} \quad s_2 = s_0\,(1 + \mathrm{e}),$$

where the eccentricity is defined as

$$\mathrm{e} = \sqrt{1 + \epsilon/s_0^2} = \sqrt{1 - E/V_0}. \tag{4.35}$$

The general solution to Eq. (4.34) is therefore

$$\theta(s) = \int_s^{s_0(1+\mathrm{e})} \frac{d\sigma}{\sqrt{s_0^2\,\mathrm{e}^2 - (\sigma - s_0)^2}} = \arccos\left(\frac{s - s_0}{s_0\,\mathrm{e}}\right), \tag{4.36}$$

which can readily be inverted to give $s(\theta) = s_0\,(1 + \mathrm{e}\cos\theta)$. The general radial orbital solution to the Kepler Problem is

$$r(\theta) = \frac{r_0}{1 + \mathrm{e}\cos\theta}. \tag{4.37}$$

Figure 4.2 shows the types of orbits described by the Kepler solution (4.37), whch is valid for all values of e. We note that motion is bounded (i.e., orbits are periodic between $s_1 \leq s \leq s_2$) when $V_0 \leq E < 0$ ($0 \leq \mathrm{e} < 1$), and the motion is unbounded (i.e., orbits are aperiodic with only s_2 being allowed, since s_1 becomes negative) when $E > 0$ ($\mathrm{e} > 1$). The separatrix solution is defined by $E = 0$ ($\mathrm{e} = 1$) and the motion on the separatrix is also unbounded (with $s_1 = 0$ or $r_1 \to \infty$).

4.3.1 *Bounded Keplerian Orbits*

We first look at the bounded case where $\epsilon < 0$ or $\mathrm{e} = \sqrt{1 - |\epsilon|/s_0} < 1$.

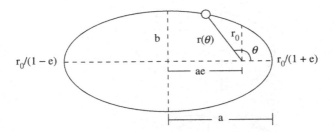

Fig. 4.3 Elliptical orbit for the Kepler problem.

4.3.1.1 *Kepler's First Law*

Equation (4.37) generates an ellipse of semi-major axis (see Fig. 4.3)

$$2\,a \;=\; \frac{r_0}{1+e} + \frac{r_0}{1-e} \;\rightarrow\; a \;=\; \frac{r_0}{1-e^2} \;=\; \frac{k}{2\,|E|} \tag{4.38}$$

and semi-minor axis

$$b \;=\; a\sqrt{1-e^2} \;=\; \sqrt{\ell^2/(2\mu\,|E|)}. \tag{4.39}$$

With these definitions, Eq. (4.37) may be written as

$$\left(\frac{x}{a}+e\right)^2 + \frac{y^2}{b^2} \;=\; 1, \tag{4.40}$$

which yields Kepler's First Law: Planets move around the Sun along elliptical orbits. We note that the eccentricity can also be calculated in terms of the semi-major and semi-minor axes as $e = \sqrt{1-(b/a)^2}$. In addition, because the Moon's eccentricity is $e \simeq 0.055$ (as it moves along an elliptical orbit around Earth), the ratio of the Moon's angular diameter at perigee (when it is closest to Earth) and the Moon's angular diameter at apogee (when it is farthest from Earth) is 1.116 (i.e., the Moon is 11.6 % bigger and nearly 25 % brighter) at perigee than when it is at apogee.

When we plot the positions of the two objects (of mass m_1 and m_2, respectively) by using Kepler's first law (4.37), with the positions $\mathbf{r}_1$ and $\mathbf{r}_2$ determined by Eqs. (2.59), we obtain Fig. 4.4. It is interesting to note that by detecting the small wobble motion of a distant star (with mass m_1), it has been possible to discover extra-solar planets (with masses $m_2 < m_1$).

4.3.1.2 *Kepler's Second Law*

Using the conservation law of angular momentum (4.4), we find

$$dt \;=\; \frac{d\theta}{\dot{\theta}} \;=\; \frac{\mu}{\ell}\, r^2(\theta)\, d\theta \;=\; \frac{2\mu}{\ell}\, dA(\theta),$$

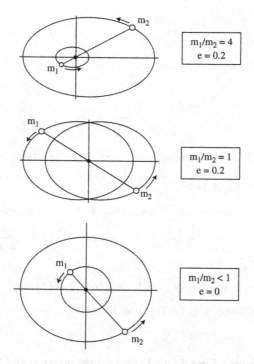

Fig. 4.4   Keplerian two-body orbits for various mass ratios and eccentricities.

where $dA(\theta) = (\int r\, dr)\, d\theta = \frac{1}{2} [r(\theta)]^2\, d\theta$ denotes an infinitesimal area swept by $d\theta$ at radius $r(\theta)$. When integrated, the relation

$$\Delta t = \frac{2\mu}{\ell} \, \Delta A, \tag{4.41}$$

yields Kepler's Second law: Equal areas ΔA are swept in equal times Δt since μ and ℓ are constants.

4.3.1.3   *Kepler's Third Law*

Using Kepler's Second Law (4.41), the orbital period T of a bound system is defined as

$$T = \int_0^{2\pi} \frac{d\theta}{\dot{\theta}} = \frac{\mu}{\ell} \int_0^{2\pi} r^2\, d\theta = \frac{2\mu}{\ell} \, A = \frac{2\pi\,\mu}{\ell} \, a\,b$$

where $A = \pi\,ab$ denotes the area of an ellipse with semi-major axis a and semi-minor axis b. Next, using the expressions (4.38)-(4.39) for a and b, the orbital period becomes

$$T(E) = \frac{2\pi\mu}{\ell} \cdot \frac{k}{2|E|} \cdot \sqrt{\frac{\ell^2}{2\mu|E|}} = 2\pi\sqrt{\frac{\mu k^2}{(2|E|)^3}}. \qquad (4.42)$$

We note that the Keplerian period (4.42) diverges as $E \to 0$. Similar behavior has already been observed with the period (3.46) of bounded motion in the Morse potential.

If we now substitute the expression for $a = k/2|E|$ and square both sides of Eq. (4.42), we obtain Kepler's Third Law:

$$T^2 = \frac{(2\pi)^2\mu}{k} a^3. \qquad (4.43)$$

In Newtonian gravitational theory, where $k/\mu = G(m_1 + m_2)$, Kepler's Third Law states that T^2/a^3 is a constant for all planets in the solar system, which is only an approximation that holds for $m_1 \gg m_2$ (true for all solar planets, e.g., the ratio of Earth's mass to the Sun's is 3×10^{-6}, while it is 9.5×10^{-4} for Jupiter).

Taking $m_1 = M_\odot$ to be the mass of the Sun, we find $k/\mu \simeq G M_\odot = 1.327 \times 10^{20}\,\mathrm{m^3/s^2}$. By introducing the Astronomical Unit[3] $1\,\mathrm{AU} = 1.496 \times 10^{11}\,\mathrm{m}$, defined as the average orbital radius of Earth's orbit (Earth's eccentricity is presently at 0.017), as a unit of distance, and the Year $1\,\mathrm{y} = 3.156 \times 10^7\,\mathrm{s}$, defined as Earth's orbital period around the Sun, as a unit of time, Kepler's Third Law (4.43) for planets and comets orbiting the Sun becomes $T(\mathrm{y})^2 \equiv a(\mathrm{AU})^3$. Hence, at an average distance of $0.387\,\mathrm{AU}$ from the Sun, Mercury has an orbital period of $0.241\,\mathrm{y}$ (or nearly 88 days), while at an average distance of $30.069\,\mathrm{AU}$ from the Sun, Neptune has an orbital period of $165\,\mathrm{y}$. Lastly, using Kepler's Third Law (4.43), the average orbital velocity $v = 2\pi a/T$ is calculated to be $v = \sqrt{k/(\mu a)} = 29.783\,\mathrm{km/s} \times a(\mathrm{AU})^{-1/2}$, so that Mercury, Earth, and Neptune travel at an average orbital velocity of $47.87\,\mathrm{km/s}$, $29.79\,\mathrm{km/s}$, and $5.48\,\mathrm{km/s}$, respectively.

4.3.2 *Unbounded Keplerian Orbits*

We now look at the cases where the total energy is positive or zero, i.e., $\mathsf{e} \geq 1$ in Eq. (4.37). For the case $\mathsf{e} = \sqrt{1 + \epsilon/s_0} > 1$, we redefine $a = r_0/(\mathsf{e}^2 - 1)$ and $b = r_0/\sqrt{\mathsf{e}^2 - 1}$, so that Eq. (4.37) yields the hyperbolic equation

$$\left(\frac{x}{a} - \mathsf{e}\right)^2 - \frac{y^2}{b^2} = 1.$$

---

[3]Data used here are taken from Norton's Star Atlas (Pi Press), Epoch 2000.0 edition.

Furthermore, we may use a parametric form for the coordinates $x = a\,(e - \cosh\psi)$ and $y = b\,\sinh\psi$ and find that the hyperbolic asymptotes (at $\psi \to \pm\infty$) are located at $\pi \pm \Theta$, where $\Theta \equiv \arctan(\sqrt{e^2 - 1}) = \arccos(e^{-1})$. For the separatrix solution ($\epsilon = 0$ or $e = 1$), Eq. (4.37) yields $r + x = r_0$, from which we recover the parabola $x = (r_0^2 - y^2)/2r_0$, with the distance of closest approach reached at $x(0) = r_0/2$.

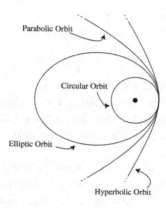

Fig. 4.5   Bounded and unbounded orbits for the Kepler problem.

Figure 4.5 shows the four types of Keplerian orbits: a circular (bounded) orbit for $e = 0$; an elliptical (bounded) orbit for $e < 1$; a hyperbolic (unbounded) orbit for $e > 1$; and the parabolic (unbounded) separatrix solution for $e = 1$.

4.3.3   *Laplace-Runge-Lenz Vector**

The angular period for a bounded Keplerian orbit is calculated from Eq. (4.23) as

$$\Delta\theta = 2 \int_{s_0(1-e)}^{s_0(1+e)} \frac{ds}{\sqrt{s_0^2\,e^2 - (s - s_0)^2}} = 2 \int_0^\pi d\varphi = 2\pi.$$

Since the orientation of the unperturbed Keplerian ellipse is constant (i.e., it does not precess), it turns out there exists a third constant of the motion for the Kepler problem (in addition to energy and angular momentum); we note, however, that only two of these three invariants are independent.

4.3.3.1 *Kepler Problem*

Let us now investigate this additional constant of the motion for the Kepler problem. First, we consider the time derivative of the vector $\mathbf{p} \times \mathbf{L}$, where the linear momentum $\mathbf{p}$ and angular momentum $\mathbf{L}$ are

$$\mathbf{p} = \mu\left(\dot{r}\,\hat{\mathbf{r}} + r\dot{\theta}\,\hat{\theta}\right) \quad \text{and} \quad \mathbf{L} = \ell\,\hat{\mathbf{z}} = \mu r^2 \dot{\theta}\,\hat{\mathbf{z}}.$$

The time derivative of the linear momentum is $\dot{\mathbf{p}} = -\nabla U(r) = -U'(r)\,\hat{\mathbf{r}}$ while the angular momentum $\mathbf{L} = \mathbf{r} \times \mathbf{p}$ is itself a constant of the motion (in a central potential) so that

$$\frac{d}{dt}\,(\mathbf{p} \times \mathbf{L}) = \frac{d\mathbf{p}}{dt} \times \mathbf{L} = -\mu\,\nabla U \times (\mathbf{r} \times \dot{\mathbf{r}})$$
$$= -\mu\dot{\mathbf{r}}\cdot\nabla U\,\mathbf{r} + \mu\mathbf{r}\cdot\nabla U\,\dot{\mathbf{r}}.$$

By re-arranging terms (and using $\dot{\mathbf{r}}\cdot\nabla U = dU/dt$ for time-independent potentials), we find

$$\frac{d}{dt}\,(\mathbf{p} \times \mathbf{L}) = -\frac{d}{dt}\,(\mu U\,\mathbf{r}) + \mu\,(\mathbf{r}\cdot\nabla U + U)\,\dot{\mathbf{r}},$$

or

$$\frac{d\mathbf{A}}{dt} = \mu\,(\mathbf{r}\cdot\nabla U + U)\,\dot{\mathbf{r}}, \tag{4.44}$$

where $\mathbf{A}$ is the Laplace-Runge-Lenz (LRL) vector:

$$\mathbf{A} = \mathbf{p} \times \mathbf{L} + \mu U(r)\,\mathbf{r}. \tag{4.45}$$

We immediately note that the LRL vector (4.45) is a constant of the motion if the potential $U(r)$ satisfies the condition

$$\mathbf{r}\cdot\nabla U(r) + U(r) = \frac{d(r\,U)}{dr} = 0.$$

This condition is satisfied for the Kepler problem, with $U(r) = -k/r$, so that the LRL vector (4.45)

$$\mathbf{A} = \mathbf{p} \times \mathbf{L} - k\mu\hat{\mathbf{r}} = \left(\frac{\ell^2}{r} - k\mu\right)\hat{\mathbf{r}} - \ell\,\mu\dot{r}\,\hat{\theta}, \tag{4.46}$$

is a constant of the motion for the Kepler problem.

Since the vector $\mathbf{A}$ is constant in both magnitude and direction, its constant magnitude

$$|\mathbf{A}|^2 = 2\mu\ell^2\left(\frac{p^2}{2\mu} + U\right) + k^2\mu^2 = k^2\mu^2\left(1 + \frac{2\ell^2 E}{\mu k^2}\right) = k^2\mu^2\mathsf{e}^2$$

is expressed in terms of $\mathsf{e}(E, \ell)$. Next, we choose its direction to be along the x-axis $(\mathbf{A} = k\mu\,\mathsf{e}\,\hat{\mathbf{x}})$ and we can easily show that

$$\left(\frac{\ell^2}{r} - k\mu\right) = \mathbf{A}\cdot\hat{\mathbf{r}} \equiv (k\mu\,\mathsf{e})\cos\theta$$

leads to the Kepler solution (4.37), where $r_0 = \ell^2/k\mu$ and the orbit's eccentricity is $\mathsf{e} = |\mathbf{A}|/k\mu$.

4.3.3.2 Perturbed Kepler Problem

We now use Eq. (4.44) to investigate what happens to a bounded Keplerian orbit when it is perturbed by the introduction of an additional potential term $\delta U(r)$: $U(r) = U_0(r) + \delta U(r)$, where $U_0(r) = -k/r$. In this case, we find $\mathbf{A} = \mathbf{A}_0 + \mu\,\delta U\,\widehat{\mathbf{r}} \equiv \mathbf{A}_0 + \delta\mathbf{A}$ and Eq. (4.44) yields

$$\frac{d\delta\mathbf{A}}{dt} = (\delta U + \mathbf{r}\cdot\nabla\delta U)\,\mathbf{p} \equiv \frac{d}{dr}(r\,\delta U)\,\mathbf{p}.$$

We now show that, under the perturbation potential $\delta U(r)$, the perturbed Keplerian orbit precesses in θ (i.e., $\Delta\theta \neq 2\pi$). First, we obtain the cross product (to lowest order in δU)

$$\mathbf{A}_0 \times \frac{d\delta\mathbf{A}}{dt} = (\delta U + \mathbf{r}\cdot\nabla\delta U)\,\left(p^2 + \mu\,U_0\right)\,\mathbf{L},$$

where we used $\mathbf{A}_0 \times \mathbf{p} = p^2 + \mu\,U_0$. Next, using the expression for the unperturbed total energy

$$E = \frac{p^2}{2\mu} + U_0 = -\frac{k}{2a},$$

we define the precession frequency

$$\omega_{\mathrm{p}}(\theta) \equiv \widehat{\mathbf{z}}\cdot\frac{\mathbf{A}_0}{|\mathbf{A}_0|^2} \times \frac{d\delta\mathbf{A}}{dt} = (\delta U + \mathbf{r}\cdot\nabla\delta U)\,\frac{\ell\mu}{(\mu ke)^2}\,(2E - U_0)$$

$$= (\delta U + \mathbf{r}\cdot\nabla\delta U)\,\frac{\ell\mu k}{(\mu ke)^2}\left(\frac{1}{r} - \frac{1}{a}\right).$$

Hence, using $a = r_0/(1 - e^2)$, the precession frequency becomes

$$\omega_{\mathrm{p}}(\theta) = \ell^{-1}\left(1 + e^{-1}\cos\theta\right)\left[\frac{d}{dr}(r\,\delta U)\right]_{r=r(\theta)}, \qquad (4.47)$$

where the term inside the square brackets is evaluated at the unperturbed Keplerian solution $r(\theta) = r_0/(1 + e\cos\theta)$. We now define the net precession shift $\delta\theta \equiv \int_0^{2\pi} \omega_{\mathrm{p}}(\theta)\,d\theta/\dot{\theta}$ of the perturbed Keplerian orbit over one unperturbed period to be

$$\delta\theta = -\int_0^{2\pi}\left(\frac{1 + e^{-1}\cos\theta}{1 + e\cos\theta}\right)\left[r\,\frac{d}{dr}\left(\frac{\delta U}{U_0}\right)\right]_{r=r(\theta)}d\theta, \qquad (4.48)$$

where we used $\dot{\theta} = (1 + e\cos\theta)\,k/(\ell r)$. For example, if $\delta U = -\beta/r^2$, then $r\,d(\delta U/U_0)/dr = -\beta/kr$ and the net precession shift (4.48) is

$$\delta\theta = \frac{\beta}{kr_0}\int_0^{2\pi}\left(1 + e^{-1}\cos\theta\right)d\theta = 2\pi\,\frac{\beta}{kr_0}.$$

Figure 4.6 shows the numerical solution of the perturbed Kepler problem for the case where $\beta \simeq kr_0/16$.

Fig. 4.6   Perturbed Kepler problem with $\delta U(r) = -\beta/r^2$ and $\beta \simeq kr_0/16$.

4.4   Isotropic Simple Harmonic Oscillator

As a second example of a central potential with closed bounded orbits [see Eq. (4.30) with $n = 2$], we now investigate the case when the central potential is of the form

$$U(r) = \frac{k}{2}\, r^2 \;\rightarrow\; \overline{U}(s) = \frac{\mu k}{2\ell^2\, s^2}. \tag{4.49}$$

This potential corresponds to a two-dimensional simple harmonic oscillator with potential energy $\frac{1}{2}\,(k_x\, x^2 + k_y\, y^2)$ where the spring constants are equal $k_x = k_y = k$. The effective potential for this problem is $V(r) = U(r) + \ell^2/(2\, m\, r^2)$, where m denotes the mass of the particle undergoing isotropic simple harmonic motion, which has a single minimum at $r = r_0 \equiv (\ell^2/k\, m)^{1/4}$. Periodic (bounded) motion is therefore allowed if the total energy $E \geq V_0 = V(r_0) = k\, r_0^2$.

By introducing the dimensionless energy $E \equiv \varepsilon\, kr_0^2$, where $\varepsilon \geq 1$, and the dimensionless radius $\rho = r/r_0$, the turning points for this problem are obtained from solutions of the equation

$$E = \frac{1}{2}\, k\, r^2 + \frac{\ell^2}{2m\, r^2} \;\rightarrow\; \rho^4 - 2\varepsilon\, \rho^2 + 1 = 0,$$

which are easily expressed as $\rho_\pm^2 = \varepsilon \pm \sqrt{\varepsilon^2 - 1} \equiv \varepsilon\,(1 \pm \mathsf{e}) > 0$, where

$$\mathsf{e} \equiv \sqrt{1 - 1/\varepsilon^2} = \sqrt{1 - (V_0/E)^2} \leq 1.$$

Since $r_\pm \equiv r_0\, \rho_\pm$ cannot be negative, the only two turning points are

$$r_1 = r_0 \left(\frac{1 + \mathsf{e}}{1 - \mathsf{e}}\right)^{\frac{1}{4}} = \frac{1}{s_1} \quad\text{and}\quad r_2 = r_0 \left(\frac{1 - \mathsf{e}}{1 + \mathsf{e}}\right)^{\frac{1}{4}} = \frac{1}{s_2},$$

where we used $\varepsilon = 1/\sqrt{1 - \mathsf{e}^2}$ for the dimensionless energy.

Next, using $n = 2$ in Eq. (4.30), we find the integral solution

$$\theta(s) = \int_s^{s_2} \frac{d\sigma}{\sqrt{\epsilon - s_0^4/\sigma^2 - \sigma^2}} = \int_s^{s_2} \frac{\sigma \, d\sigma}{\sqrt{\epsilon \sigma^2 - s_0^4 - \sigma^4}}, \qquad (4.50)$$

where $\epsilon \equiv 2 \, m \, E/\ell^2 = 2 \, \varepsilon \, s_0^2$ [see Eq. (4.19), with $\mu = m$]. Using the change of coordinate $q = \sigma^2/s_0^2$ in Eq. (4.50), we obtain

$$\theta(q) = \frac{1}{2} \int_q^{q_2} \frac{dq}{\sqrt{2 \, \varepsilon \, q - 1 - q^2}}, \qquad (4.51)$$

where $q_2 = \sqrt{(1 + e)/(1 - e)} \equiv \varepsilon \, (1 + e)$. We now complete the square $2 \, \varepsilon \, q - 1 - q^2 = \varepsilon^2 \, e^2 - (q - \varepsilon)^2$ and substitute $q(\varphi) = \varepsilon \, (1 + e \cos \varphi)$ in Eq. (4.51) to obtain

$$\theta(q) = \frac{1}{2} \, \arccos \left(\frac{q - \varepsilon}{\varepsilon \, e} \right),$$

and we easily verify that $\Delta\theta = \pi$ and bounded orbits are closed. The equation $\theta(q)$ can now be inverted (with $r = r_0/\sqrt{q}$) to give

$$r(\theta) = \frac{r_0 \, (1 - e^2)^{1/4}}{\sqrt{1 + e \cos 2\theta}}, \qquad (4.52)$$

which describes the ellipse $x^2/b^2 + y^2/a^2 = 1$, with semi-major axis $a = r_1$ and semi-minor axis $b = r_2$. Note that this solution $x^2/b^2 + y^2/a^2 = 1$ may be obtained from the Cartesian representation for the Lagrangian $L = \frac{1}{2} \, m \, (\dot{x}^2 + \dot{y}^2) - \frac{1}{2} \, k \, (x^2 + y^2)$, which yields the solutions $x(t) = b \, \cos \omega t$ and $y(t) = a \, \sin \omega t$, where $\omega \equiv \sqrt{k/m}$ and the constants a and b are determined from the conservation laws $E = \frac{1}{2} \, m\omega^2 \, (a^2 + b^2)$ and $\ell = m\omega \, a \, b$.

Lastly, the area of the ellipse is $A = \pi \, ab = \pi \, r_0^2$, while the *physical* period is

$$T(E, \ell) = \int_0^{2\pi} \frac{d\theta}{\dot{\theta}} = 2\pi \, \sqrt{\frac{m}{k}}.$$

Note that the *radial* period is $T/2$ since $\Delta\theta = \pi$. We, therefore, find that the period of an isotropic simple harmonic oscillator is independent of the constants of the motion E and ℓ, in analogy with the one-dimensional case.

4.5   Internal Reflection inside a Well

As a last example of bounded motion in a central-force potential, we consider the motion of a particle of mass m in the central potential

$$U(r) = \begin{cases} - U_0 & (r < R) \\ \\ 0 & (r > R) \end{cases} \qquad (4.53)$$

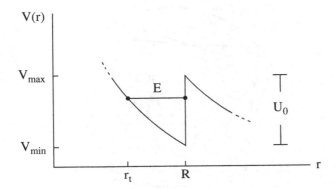

Fig. 4.7   Effective potential for the internal hard sphere.

where U_0 is a constant and R denotes the radius of a sphere. The effective potential $V(r) = \ell^2/(2mr^2) + U(r)$ associated with this potential is shown in Fig. 4.7, and orbits are unbounded when $E > V_{\max} = \ell^2/(2mR^2)$. For energy values

$$V_{\min} = \frac{\ell^2}{2mR^2} - U_0 < E < V_{\max} = \frac{\ell^2}{2mR^2},$$

on the other hand, Fig. 4.7 shows that bounded motion is possible, with turning points

$$r_2 = \sqrt{\frac{\ell^2}{2m\left(E + U_0\right)}} \equiv r_t \quad \text{and} \quad r_1 = R.$$

When $E = V_{\min}$, the left turning point reaches its maximum value $r_t = R$ while it reaches its minimum value $r_t/R = (1 + U_0/E)^{-\frac{1}{2}} < 1$ when $E = V_{\max}$.

We now solve for the bounded motion in the potential (4.53) by assuming that the particle starts at $r = r_t$ at $\theta = 0$. The particle orbit is found by integration by quadrature of $(s')^2 = s_t^2 - s^2$:

$$\theta(s) = \int_s^{s_t} \frac{d\sigma}{\sqrt{s_t^2 - \sigma^2}} = \arccos\left(\frac{s}{s_t}\right),$$

where $s_t = 1/r_t$. We easily invert this relation and find the orbital radial solution

$$r(\theta) = r_t \sec\theta \quad \text{(for } \theta \le \Theta), \tag{4.54}$$

where the maximum angle Θ defines the angle at which the particle hits the turning point R, i.e., $r(\Theta) = R$, where $\Theta = \arccos(r_t/R)$. Subsequent

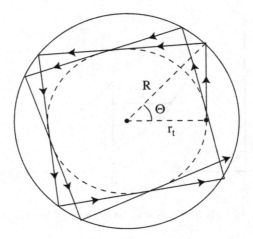

Fig. 4.8　Internal reflections inside a hard sphere.

motion of the particle involves an infinite sequence of *internal reflections* as shown in Fig. 4.8. The case where $E > \ell^2/2mR^2$ involves a single turning point and is discussed in Sec. 5.7.

Lastly, we note that the dynamical radial solution $r(t)$ is obtained by inverting the velocity equation $dt = dr/\dot{r}$:

$$\dot{r} = \frac{\ell}{m} \frac{\sqrt{r^2 - r_t^2}}{r\, r_t} \;\rightarrow\; r(t) = \sqrt{r_t^2 + \left(\frac{\ell\, t}{m\, r_t}\right)^2}, \qquad (4.55)$$

which is valid for $0 \le t \le T$, where

$$T \equiv \frac{m\, r_t}{\ell} \sqrt{R^2 - r_t^2} = \frac{\sin\Theta}{\sqrt{(2/m)\,(E + U_0)}}, \qquad (4.56)$$

and $r(T) = R$. Hence, the radial period for this problem is $2T$. By comparing the orbital radial solution (4.55) with the dynamical radial solution (4.54), we easily obtain the dynamical angular solution $\theta(t) \equiv \arctan(\ell\, t/m\, r_t^2)$, with $\theta(T) = \Theta$. The equations of motion in the (x, y)-plane are, therefore, expressed as $x(t) = r(t)\cos\theta(t) = r_t$ and $y(t) = r(t)\sin\theta(t) = r_t \tan\theta(t) = (\ell/m r_t)\, t$.

4.6　Summary

Chapter 4 presented the general solution of planar motion under the influence of a central-force potential, for which case two constants of motion

Table 4.1   Summary of Chapter 4: Motion in a Central-Force Field.

Topic	Equation
Routhian for Motion in Central-Force Potential	(4.5)
Orbital Radial Motion	(4.13)
Inverse Orbital problem	(4.15)
Orbital Solution	(4.21)
Virial Theorem	(4.28)
Kepler Problem	(4.31)-(4.37)
Kepler's Three Laws of Planetary Motion	(4.40)-(4.43)
Laplace-Runge-Lenz Vector	(4.45)-(4.46)
Perturbed Kepler Orbital Precession	(4.47)-(4.48)
Isotropic Simple Harmonic Oscillator	(4.49)-(4.52)
Internal Reflection inside a Well	(4.53)-(4.56)

(energy E and angular momentum p_θ) exist. With these two invariants, the Energy method enabled the construction of a solution for the radial motion $r(t)$, while the angular motion $\theta(t)$ was obtained by integration of the angular-momentum conservation law. Complete solutions of the Kepler problem, the isotropic simple-harmonic-oscillator problem, and the problem of internal reflection inside a potential well were also given. Table 4.1 presents a summary of the important topics of Chapter 4.

4.7   Problems

1. Consider a comet moving in a parabolic orbit $r(\theta) = r_0/(1 + \cos\theta)$ in the plane of the Earth's orbit (see Fig. 4.9). If the distance of closest approach of the comet to the sun is $r_0/2 = \beta\, r_E$, where $\beta < 1$ and r_E is the radius of the Earth's (assumed) circular orbit, show that the time the comet spends within the orbit of the Earth is given by

$$\Delta t = \sqrt{2\mu} \int_{\beta r_E}^{r_E} \frac{dr}{\sqrt{(k\,r - \ell^2/2\mu)/r^2}} = \sqrt{\frac{2\mu}{k}} \int_{\beta r_E}^{r_E} \frac{r\,dr}{\sqrt{r - \beta\, r_E}}$$

$$= \sqrt{2\,(1-\beta)}\,(1 + 2\,\beta) \times \left(\frac{1\ \text{year}}{3\,\pi}\right),$$

where Eq. (4.9) was used (with $E = 0$ for a parabola) and we used $\mu\, r_E^3/k \simeq (1\ \text{year}/2\pi)^2$.

2. A particle (of mass m) moves in a spiral orbit given by $r(\theta) = k\,\theta^n$, where k is a constant and n is a positive integer. Show that the effective

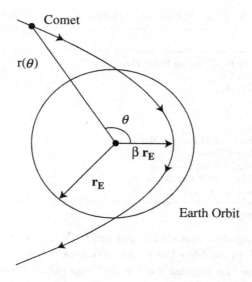

Fig. 4.9　Problem 1.

potential $V(r) = U(r) + \ell^2/(2mr^2)$ that allows this motion is

$$V(r) = -\left(\frac{\ell^2}{2m}\right)\frac{n^2 k^{2/n}}{r^{2+2/n}}.$$

3. A satellite (of mass m) is orbiting Earth along a circular path (at radius r_0) with total energy

$$E = \frac{m}{2}v^2 - \frac{k}{r_0} = -\frac{k}{2r_0},$$

where $v = \sqrt{k/(mr_0)}$ represents the satellite's tangential (angular) speed. During its orbit, the satellite's propulsion system gives an instantaneous boost $v \to v + \Delta v$ to the satellite (with $\Delta v > 0$ tangent to the orbit). The new energy of the satellite is

$$E' \equiv \frac{m}{2}(v + \Delta v)^2 - \frac{k}{r_0} = -\frac{k}{2a},$$

where $a > r_0$ is the semi-major axis of the new elliptical orbit.

(a) Show that the eccentricity of the new elliptical orbit is $e' \equiv 1 - r_0/a$ and that the energy boost is $\Delta E \equiv E' - E = -e'E$.

(b) Show that the velocity boost Δv needed to achieve a desired eccentricity e_1 is

$$\Delta v(e_1) \equiv v\left(\sqrt{1 + e_1} - 1\right).$$

(c) Find the minimum boost value $\Delta v_{\min}$ necessary for the satellite to escape Earth's gravity (i.e., the satellite's after-boost orbit is a parabola).

4. Show that the radial "fall-in" time of a particle of mass m in a potential $U = -k/r$ starting from radius R without angular momentum is

$$\Delta t = \int_0^R \frac{dr}{|\dot{r}|} = \pi \sqrt{\frac{m R^3}{8k}}.$$

5. Consider the perturbed Kepler problem in which a particle of mass m, energy $E < 0$, and angular momentum ℓ is moving in the central-force potential

$$U(r) = -\frac{k}{r} + \frac{\alpha}{r^2},$$

where the perturbation potential α/r^2 is considered small in the sense that the dimensionless parameter $\epsilon = 2m\alpha/\ell^2 \ll 1$ is small.

(a) Show that the energy equation for this problem can be written using $s = 1/r$ as

$$E = \frac{\ell^2}{2m} \left[(s')^2 + \gamma^2 s^2 - 2 s_0 s \right],$$

where $s_0 = mk/\ell^2$ and $\gamma^2 = 1 + \epsilon$.

(b) Show that the turning points are

$$s_1 = \frac{s_0}{\gamma^2} (1 - \mathbf{e}) \quad \text{and} \quad s_2 = \frac{s_0}{\gamma^2} (1 + \mathbf{e}),$$

where $\mathbf{e} = \sqrt{1 + 2\gamma^2 \ell^2 E/mk^2}$.

(c) By solving the integral

$$\theta(s) = -\int_{s_2}^s \frac{d\sigma}{\sqrt{(2mE/\ell^2) + 2 s_0 \sigma - \gamma^2 \sigma^2}},$$

where $\theta(s_2) = 0$, show that

$$r(\theta) = \frac{\gamma^2 r_0}{1 + \mathbf{e} \cos(\gamma\theta)},$$

where $r_0 = 1/s_0$. Hence, the ellipse precesses with an angular step $\Delta\theta = 2\pi/\gamma$.

6. Consider a particle of mass m moving in the potential $U(\mathbf{r}) = -k/r - \mathbf{r} \cdot \mathbf{F}$, where $\mathbf{F}$ is a constant force vector. Show that, while the angular momentum $\mathbf{L}$ is no longer conserved, the quantity

$$\mathbf{p} \times \mathbf{L} \cdot \mathbf{F} - \frac{mk}{r} \mathbf{r} \cdot \mathbf{F} + \frac{m}{2} |\mathbf{r} \times \mathbf{F}|^2$$

is a constant of the motion.

7. A Keplerian elliptical orbit, described by the relation $r(\theta) = r_0/(1 + e\cos\theta)$, undergoes a precession motion when perturbed by the perturbation potential $\delta U(r)$, with precession frequency (4.47). Show that, if $\delta U(r) = -\alpha/r^3$ (where α is a constant), the net precession shift $\delta\theta$ of the Keplerian orbit over one unperturbed period is

$$\delta\theta = \int_0^{2\pi} \omega_p(\theta) \, \frac{d\theta}{\dot{\theta}} = 6\pi \, \frac{\alpha}{kr_0^2},$$

where we used $\dot{\theta} = \ell/\mu r^2 = (\ell/\mu r_0^2)(1 + e\cos\theta)^2$.

Fig. 4.10   Problem 8.

8. In Kepler's work, angles are referred to as *anomalies*. In Fig. 4.10, an ellipse (with eccentricity $e < 1$) of semi-major axis a and semi-minor axis b is inscribed by a circle of radius a.

(a) Show that the orbit of the planet (at point P in Fig. 4.10) is described in terms of the *eccentric anomaly* ψ as

$$r(\psi) = a(1 - e\cos\psi),$$

and the *true* anomaly θ is defined in terms of ψ as

$$\cos\theta(\psi) = \left(\frac{\cos\psi - e}{1 - e\cos\psi}\right).$$

Note that by using the eccentric anomaly angle ψ, we find $a\cos\psi = ae + r\cos\theta$ from which we obtain $\cos\psi = (e + \cos\theta)/(1 + e\cos\theta)$ or $\cos\theta = (\cos\psi - e)/(1 - e\cos\psi)$. By substituting this last expression into Kepler's First Law (4.37), we obtain $r(\psi) = a(1 - e\cos\psi)$.

(b) Show that the time from perihelion ($\psi = 0$) is given by Kepler's Equation:

$$t(\psi) \equiv \int_0^{\theta(\psi)} \frac{d\theta}{|\dot\theta|} = \frac{\mu}{\ell}\int_0^\psi a^2(1 - e\cos\psi)^2\,\frac{d\theta(\psi)}{d\psi}\,d\psi$$

$$= \frac{\tau}{2\pi}\,(\psi - e\sin\psi), \tag{4.57}$$

where $\tau \equiv 2\pi\sqrt{\mu a^3/k}$ denotes the orbital period (i.e., 1 year for Earth).

(c) If the Earth's orbit is divided in two by the *latus rectum* (i.e., the vertical line drawn through the Sun), show that the times spent in the inner and outer halves (in fractions of a year) are

$$t_{\text{inner}} = \frac{1}{\pi}\left[\cos^{-1}e - e\sqrt{1 - e^2}\right],$$

$$t_{\text{outer}} = \frac{1}{\pi}\left[(\pi - \cos^{-1}e) + e\sqrt{1 - e^2}\right],$$

and that the difference between the times is

$$\Delta t \equiv t_{\text{outer}} - t_{\text{inner}} = \frac{2}{\pi}\left[\sin^{-1}e + e\sqrt{1 - e^2}\right].$$

(d) Using $t(\psi)$ and $r(\psi)$, show that the average orbital radius is

$$\bar{r} \equiv \frac{2}{\tau}\int_0^\pi r(\psi)\,\frac{dt(\psi)}{d\psi}\,d\psi = a\left(1 + \frac{e^2}{2}\right).$$

9. A particle of unit mass moves from infinity along a straight line that, if continued would allow it to pass a distance $b\sqrt{2}$ from a point P. If the particle is attracted toward point P with a force varying as k/r^5, and if the angular momentum about the point P is $\sqrt{k}/b$, show that the trajectory is given by [*Hint: Show that* $s'' + s = b^2 s^3$.]

$$r(\theta) = b\,\coth(\theta/\sqrt{2}).$$

10. An Earth satellite moves in an elliptical orbit with a period τ, eccentricity e, and semi-major axis $a = ((\tau/2\pi)^2 k/\mu)^{1/3}$. Show that the maximum radial velocity of the satellite is

$$\dot{r}_{\text{max}} = \sqrt{\frac{2}{\mu}[E - V(r_0)]} = \frac{2\pi\,ae}{\tau\sqrt{1 - e^2}}.$$

11. (a) Show that if a particle describes a circular orbit under the influence of an attractive central force directed toward a point on the circle,

$$r(\theta) = 2R\,|\cos\theta| \equiv \sqrt{2}\,R\,\sqrt{1 + \cos 2\theta},$$

then the force varies as the inverse-fifth power of the distance.

(b) Show that for the orbit described the total energy of the particle is zero.

(c) Find the period of the motion.

(d) Find $\dot{x}$, $\dot{y}$, and v as a function of angle around the circle and show that all three quantities are inifinite as the particle goes the center of force.

12*. At perigee of an elliptic gravitational orbit, a particle experiences an impulse in the radial direction, sending the particle into another elliptic orbit. Determine the new semi-major axis, eccentricity, and orientation in terms of the old.

13. Show that for elliptical motion in a gravitational field the radial speed can be written as

$$\dot{r} = \omega\,\frac{a}{r}\,\sqrt{a^2 e^2 - (r-a)^2}.$$

By replacing the radial coordinate r with the eccentric anomaly angle ψ, show that the resulting differential equation can be integrated immediately to give Kepler's Equation (4.57).

14. Discuss the possible types of orbit for a particle moving under the central potential $k/2r^2$. (a) For the repulsive case ($k > 0$), show that the orbit equation is $r(\theta) = b \sec n(\theta - \theta_0)$, where n, b, and θ_0 are constants.

(b) For the attractive case ($k < 0$), the nature of the orbit depends on the sign of $\ell^2 + mk$ and E. Find the orbit equation for each possible type.

15. The *Hohmann* transfer orbit (H) represents the passage from a circular orbit at radius r_A to another circular orbit at radius $r_B > r_A$ (see inset in Fig. 4.11).

The transfer orbit requires two boosts at points 1 and 2, with energy changes

$$\Delta E_1 = \frac{k}{2}\left(\frac{1}{r_A} - \frac{1}{a}\right) \quad \text{and} \quad \Delta E_2 = \frac{k}{2}\left(\frac{1}{a} - \frac{1}{r_B}\right),$$

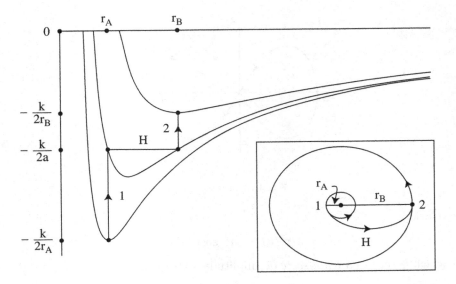

Fig. 4.11   Problem 15.

where $a = (r_A + r_B)/2$ is the semi-major axis for the Hohmann (H) transfer orbit. For each boost ($j = 1, 2$), compute the velocity change Δv_j.

16. Consider a binary-star system composed of two stars of masses m_1 and $m_2 < m_1$ undergoing circular motion about their common center of mass (see Fig. 4.4).

(a) Use Kepler's Third Law to show that the masses m_1 and m_2 are expressed in terms of the orbital radii r_1 and r_2 according to the relations

$$m_1 = \frac{\omega^2 a^2}{G} r_2 \quad \text{and} \quad m_2 = \frac{\omega^2 a^2}{G} r_1,$$

where $\omega \equiv 2\pi/T$ is the orbital frequency of the two stars and $a = r_1 + r_2$ is the constant separation between the two stars.

(b) Compute the masses of Sirius A and Sirius B (in units of solar mass) if their separation is 20 AU ($1\,\text{AU} = 1.5 \times 10^{11}\,\text{m}$), with $r_B/r_A = 2.3$, and the orbital period is $T = 50$ years.

17. A particle of mass m and angular momentum ℓ is observed to undergo periodic motion with its distance $r(\theta)$ from the center of force given by the

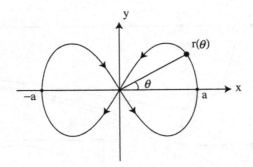

Fig. 4.12   Problem 17.

relation

$$r^2(\theta) = a^2 \cos(2\theta),$$

which describes a *lemniscate* of amplitude a (see Fig. 4.12).

(a) Show that the potential energy $U(r)$ leading to this periodic motion is given by the attractive potential

$$U(r) = -\frac{\ell^2}{2m}\frac{a^4}{r^6}.$$

[*Hint: Show that* $s'' + s = 3 a^4 s^5 \equiv -\overline{U}'(s)$.]

(b) Solve the differential equation for $\theta(t)$ given by

$$\dot\theta = \ell/mr^2 = (\ell/ma^2)\sec(2\theta)$$

for $\theta < \pi/4$ subject to the initial condition $\theta = 0$ at time $t = 0$.

18. Show that for a particle moving along an elliptical orbit in an inverse-square-law potential, the eccentricity of the orbit can be written as

$$e = \frac{\sqrt{n}-1}{\sqrt{n}+1},$$

where $n = \dot\theta_{\max}/\dot\theta_{\min}$ is the ratio of the maximum angular velocity to the minimum angular velocity.

Chapter 5

Collisions and Scattering Theory

In Chapter 4, we investigated two types of orbits (bounded and unbounded) for two-particle systems evolving under the influence of a central-force potential. In the present Chapter, we focus our attention on unbounded orbits within the context of *elastic* collision theory (i.e., a collision for which energy and momentum are conserved). In this context, a collision between two interacting particles involves a three-step process (see Fig. 5.1): Step I – two particles (with masses m_1, m_2 and momenta $\mathbf{p}_1$, $\mathbf{p}_2$) are initially infinitely far apart (in which case, the total energy of the two-particle system is assumed to be strictly kinetic: $|\mathbf{p}_1|^2/2m_1 + |\mathbf{p}_2|^2/2m_2$); Step II – as the two particles approach each other, their interacting potential (repulsive or attractive) causes them to reach a distance of closest approach (where the interaction force is strongest); and Step III – the two particles then move progressively farther apart (with momenta $\mathbf{q}_1$ and $\mathbf{q}_2$), eventually reaching a point where the total energy is once again strictly kinetic: $|\mathbf{q}_1|^2/2m_1 + |\mathbf{q}_2|^2/2m_2$).

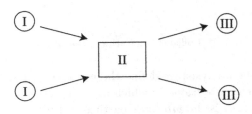

Fig. 5.1   Collision kinematics ($I \to III$) and dynamics (II).

These three steps form the foundations of Collision *Kinematics* and Collision *Dynamics*. The topic of Collision Kinematics, which describes the collision in terms of the conservation laws of momentum and energy:

$$\mathbf{p}_1 + \mathbf{p}_2 = \mathbf{q}_1 + \mathbf{q}_2, \tag{5.1}$$

$$\frac{|\mathbf{p}_1|^2}{2m_1} + \frac{|\mathbf{p}_2|^2}{2m_2} = \frac{|\mathbf{q}_1|^2}{2m_1} + \frac{|\mathbf{q}_2|^2}{2m_2}, \tag{5.2}$$

deals with Steps I and III; here, the incoming particles define the initial state of the two-particle system while the outgoing particles define the final state. The topic of Collision Dynamics, on the other hand, deals with Step II, in which the particular nature of the interaction is taken into account.

5.1   Two-Particle Collisions in the LAB Frame

Consider the collision of two particles (labeled 1 and 2) of masses m_1 and m_2, respectively. Let us denote the velocities of particles 1 and 2 *before* the collision as $\mathbf{u}_1$ and $\mathbf{u}_2$, respectively, while the velocities *after* the collision are denoted $\mathbf{v}_1$ and $\mathbf{v}_2$. Hence, the particle momenta before and after the collision are denoted $\mathbf{p}_i = m_i\,\mathbf{u}_i$ and $\mathbf{q}_i = m_i\,\mathbf{v}_i$, respectively.

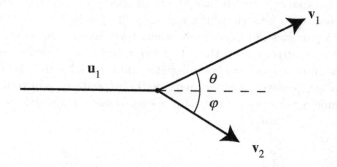

Fig. 5.2   Collision kinematics in the LAB frame.

To simplify the analysis, we define the laboratory (LAB) frame to correspond to the reference frame in which m_2 is at rest (i.e., $\mathbf{u}_2 = 0$); in this collision scenario, m_1 is the *projectile* particle and m_2 is the *target* particle. We now write the velocities $\mathbf{u}_1$, $\mathbf{v}_1$, and $\mathbf{v}_2$ as

$$\left.\begin{aligned}
\mathbf{u}_1 &= u\,\widehat{\mathbf{x}} \\
\mathbf{v}_1 &= v_1\,(\cos\theta\,\widehat{\mathbf{x}} + \sin\theta\,\widehat{\mathbf{y}}) \\
\mathbf{v}_2 &= v_2\,(\cos\varphi\,\widehat{\mathbf{x}} - \sin\varphi\,\widehat{\mathbf{y}})
\end{aligned}\right\}, \tag{5.3}$$

where the *deflection* angle θ of the projectile particle and the *recoil* angle φ of the target particle are defined in Fig. 5.2. The conservation laws of momentum and energy (5.1)-(5.2) yield

$$m_1\mathbf{u}_1 = m_1\mathbf{v}_1 + m_2\mathbf{v}_2 \quad \text{and} \quad \frac{m_1}{2}\,u^2 = \frac{m_1}{2}\,|\mathbf{v}_1|^2 + \frac{m_2}{2}\,|\mathbf{v}_2|^2,$$

which can be written in terms of the mass ratio $\alpha = m_1/m_2$ of the projectile mass to the target mass as

$$\alpha\,v_1\cos\theta = \alpha\,u - v_2\cos\varphi, \tag{5.4}$$

$$\alpha\,v_1\sin\theta = v_2\sin\varphi, \tag{5.5}$$

$$\alpha\,v_1^2 = \alpha\,u^2 - v_2^2. \tag{5.6}$$

Since the **three** equations (5.4)-(5.6) are expressed in terms of **four** unknown quantities $(v_1, \theta;\ v_2, \varphi)$, for given incident velocity u and mass ratio α, we must choose **one** post-collision coordinate as an independent variable to get a closed solution for the three remaining post-collision variables.

Here, we choose the recoil angle φ of the target particle, and proceed with finding expressions for $v_1(u, \varphi; \alpha)$, $v_2(u, \varphi; \alpha)$ and $\theta(u, \varphi; \alpha)$; other choices lead to similar formulas (see problems at the end of the Chapter). First, adding the square of the momentum components (5.4) and (5.5), we obtain

$$\alpha^2\,v_1^2 = \alpha^2\,u^2 - 2\,\alpha\,uv_2\cos\varphi + v_2^2. \tag{5.7}$$

Next, using the energy equation (5.6), we find

$$\alpha^2\,v_1^2 = \alpha\left(\alpha\,u^2 - v_2^2\right) = \alpha^2\,u^2 - \alpha\,v_2^2, \tag{5.8}$$

so that these two equations combine to give

$$v_2(u, \varphi; \alpha) = 2\left(\frac{\alpha}{1+\alpha}\right)u\cos\varphi. \tag{5.9}$$

After substituting Eq. (5.9) into Eq. (5.8), we find

$$v_1(u, \varphi; \alpha) = u\,\sqrt{1 - 4\,\frac{\mu}{M}\,\cos^2\varphi} \equiv u\,\nu(\varphi), \tag{5.10}$$

where $\mu/M = \alpha/(1+\alpha)^2$ is the ratio of the reduced mass $\mu = m_1 m_2/M$ and the total mass $M = m_1 + m_2$.

Lastly, we take the ratio of the momentum components (5.5) over (5.4) in order to eliminate the unknown v_1 and we find

$$\tan\theta = \frac{v_2\sin\varphi}{\alpha\,u - v_2\cos\varphi}.$$

If we substitute Eq. (5.9), we easily obtain

$$\tan \theta = \frac{2 \sin \varphi \cos \varphi}{1 + \alpha - 2 \cos^2 \varphi} \equiv \frac{\sin 2\varphi}{\alpha - \cos 2\varphi},$$

or

$$\theta(\varphi; \alpha) = \arctan \left(\frac{\sin 2\varphi}{\alpha - \cos 2\varphi} \right). \tag{5.11}$$

In the limit $\alpha = 1$ (i.e., a collision involving identical particles), we find

$$v_2 = u \cos \varphi \quad \text{and} \quad v_1 = u \sin \varphi \tag{5.12}$$

from Eqs. (5.9) and (5.10), respectively, and

$$\tan \theta = \cot \varphi \quad \rightarrow \quad \varphi = \frac{\pi}{2} - \theta, \tag{5.13}$$

from Eq. (5.11). Hence, the angular sum $\theta + \varphi$ for like-particle collisions is always 90° (for $\varphi \neq 0$).

We summarize by stating that, after the collision, the momenta $\mathbf{q}_1$ and $\mathbf{q}_2$ in the LAB frame (where m_2 is initially at rest) are functions of the initial momentum $p = m_1 u$ and the angles θ and φ:

$$\mathbf{q}_1 = p \nu(\varphi) \left(\cos \theta \, \widehat{\mathbf{x}} + \sin \theta \, \widehat{\mathbf{y}} \right), \tag{5.14}$$

$$\mathbf{q}_2 = \frac{2 p \cos \varphi}{1 + \alpha} \left(\cos \varphi \, \widehat{\mathbf{x}} - \sin \varphi \, \widehat{\mathbf{y}} \right), \tag{5.15}$$

where $\nu(\varphi)$ is defined in Eq. (5.10). We note that, by using the relations obtained from Eq. (5.11): $\nu(\varphi) \cos \theta = (\alpha - \cos 2\varphi)/(1 + \alpha)$ and $\nu(\varphi) \sin \theta = \sin 2\varphi/(1 + \alpha)$, Eqs. (5.14)-(5.15) satisfy the momentum conservation law (5.1):

$$\mathbf{q}_1 + \mathbf{q}_2 = \frac{p}{1 + \alpha} \left[\left((\alpha - \cos 2\varphi) + 2 \cos^2 \varphi \right) \widehat{\mathbf{x}} \right.$$
$$\left. + \left(\sin 2\varphi - 2 \cos \varphi \sin \varphi \right) \widehat{\mathbf{y}} \right]$$
$$= p \, \widehat{\mathbf{x}} \equiv \mathbf{p}_1 + \mathbf{p}_2,$$

and the energy conservation law (5.2):

$$\frac{|\mathbf{q}_1|^2}{2 \, m_1} + \frac{|\mathbf{q}_2|^2}{2 \, m_2} = \frac{p^2}{2 m_1} \left(1 - 4 \, \frac{\mu}{M} \, \cos^2 \varphi \right) + \frac{p^2}{2 m_1} \left(4 \, \frac{\mu}{M} \, \cos^2 \varphi \right)$$
$$= \frac{p^2}{2 \, m_1} \equiv \frac{|\mathbf{p}_1|^2}{2 \, m_1} + \frac{|\mathbf{p}_2|^2}{2 \, m_2}.$$

5.2 Two-Particle Collisions in the CM Frame

In the center-of-mass (CM) frame (see Sec. 2.6), the elastic collision between particles 1 and 2 is described quite simply; the CM velocities and momenta are, henceforth, denoted with a prime. The simplicity of the CM collisional kinematics has important theoretical advantages when we investigate the CM collisional dynamics (Step II in Fig. 5.1) in Sec. 5.4.2.

We begin with the momentum of the center-of-mass $\mathbf{P} \equiv \mathbf{p}_1 + \mathbf{p}_2 = m_1 u \widehat{\mathbf{x}}$ which, by momentum conservation, is a constant of motion. Note that we are still using the assumption that $\mathbf{u}_2 = 0$ (i.e., the target is at rest in the LAB frame). Before the collision takes place, the CM momenta of particles 1 and 2 are equal in magnitude but with opposite directions (see Fig. 5.3): $\mathbf{p}'_1 + \mathbf{p}'_2 \equiv 0$. We now introduce the relations $\mathbf{p}_1 = \mathbf{p}'_1 + \gamma \mathbf{P}$ and $\mathbf{p}_2 = \mathbf{p}'_2 + (1 - \gamma) \mathbf{P}$, where γ is a dimensionless mass parameter that is determined as follows. First, in the limit $m_2 \to \infty$, it is clear that $\mathbf{p}'_1 = \mathbf{p}_1$ and thus $\gamma = 0$ in that limit. Hence, the ratio is $\gamma \equiv m_1/M$ and $1 - \gamma = m_2/M$. Next, since $\mathbf{p}_2 = 0$ in the LAB frame, we then find

$$\mathbf{p}'_2 = -(1 - \gamma)\mathbf{P} = -\mu u \widehat{\mathbf{x}} \equiv -\mathbf{p}'_1, \tag{5.16}$$

where μ is the reduced mass of the two-particle system and $\mu u = p/(1+\alpha)$.

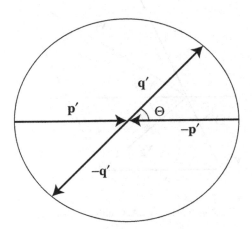

Fig. 5.3 Collision kinematics in the CM frame.

After the collision, conservation of energy-momentum dictates that

$$\mathbf{q}'_1 = \mu u \, (\cos\Theta \widehat{\mathbf{x}} + \sin\Theta \widehat{\mathbf{y}}) = -\mathbf{q}'_2, \tag{5.17}$$

where Θ is the scattering angle in the CM frame. Thus the particle velocities after the collision in the CM frame are

$$\mathbf{v}_1' = \frac{\mathbf{q}_1'}{m_1} = \frac{u}{1+\alpha}\left(\cos\Theta\,\widehat{\mathbf{x}} + \sin\Theta\,\widehat{\mathbf{y}}\right) \quad \text{and} \quad \mathbf{v}_2' = \frac{\mathbf{q}_2'}{m_2} = -\,\alpha\,\mathbf{v}_1'.$$

It is quite clear, thus, that the initial and final kinematic states lie on the same circle in CM momentum space and the single variable defining the outgoing two-particle state is represented by the CM scattering angle Θ.

5.3   Connection between the CM and LAB Frames

We now establish the connection between the momenta (5.14)-(5.15) in the LAB frame and the momenta (5.17) in the CM frame (see Fig. 5.4). First, we denote the velocity of the CM as

$$\mathbf{w} = \frac{m_1\,\mathbf{u}_1 + m_2\,\mathbf{u}_2}{m_1 + m_2} = \frac{\alpha\,u}{1+\alpha}\,\widehat{\mathbf{x}},$$

so that $w = |\mathbf{w}| = \alpha\,u/(1+\alpha)$ and $|\mathbf{v}_2'| = w = \alpha\,|\mathbf{v}_1'|$.

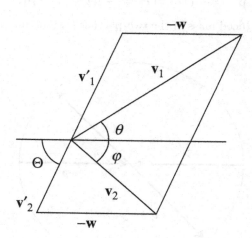

Fig. 5.4   CM and LAB collision geometries.

The connection between $\mathbf{v}_1'$ and $\mathbf{v}_1$ is expressed as

$$\mathbf{v}_1 = \mathbf{v}_1' + \mathbf{w} \quad \rightarrow \quad \begin{cases} v_1\cos\theta = w\left(1 + \alpha^{-1}\cos\Theta\right) \\[2mm] v_1\sin\theta = w\,\alpha^{-1}\sin\Theta \end{cases}$$

so that

$$\tan\theta = \frac{\sin\Theta}{\alpha + \cos\Theta}, \tag{5.18}$$

and

$$v_1 = \frac{u}{(1+\alpha)}\sqrt{1 + \alpha^2 + 2\alpha\cos\Theta}. \tag{5.19}$$

Likewise, the connection between $\mathbf{v}_2'$ and $\mathbf{v}_2$ is expressed as

$$\mathbf{v}_2 = \mathbf{v}_2' + \mathbf{w} \quad \rightarrow \quad \begin{cases} v_2 \cos\varphi = w(1 - \cos\Theta) \\[2mm] v_2 \sin\varphi = w \sin\Theta \end{cases}$$

so that $\tan\varphi = \sin\Theta/(1 - \cos\Theta) = \cot\Theta/2$, which yields

$$\varphi = \frac{1}{2}(\pi - \Theta), \tag{5.20}$$

and

$$v_2 = \frac{2\alpha u}{(1+\alpha)}\sin\frac{\Theta}{2}. \tag{5.21}$$

It is interesting to note that Eq. (5.20) is true for all values of $\alpha = m_1/m_2$. Hence, once the recoil angle φ is known (i.e., measured), then the CM deflection angle is $\Theta = \pi - 2\varphi$.

5.4 Scattering Cross Sections

In the previous Section, we investigated the connection between the initial and final kinematic states of an elastic collision described by Steps I and III, respectively, introduced earlier (see Fig. 5.1). Here, the initial kinematic state is described in terms of the speed u of the projectile particle in the LAB frame (assuming that the target particle is at rest), while the final kinematic state is described in terms of the velocity coordinates for the deflected projectile particle (v_1, θ) and the recoiled target particle (v_2, φ).

In the present Section, we shall investigate Step II, namely, how the distance of closest approach influences the deflection angles (θ, φ) in the LAB frame and the deflection angle Θ in the CM frame.

5.4.1 *Definitions*

First, we consider for simplicity the case of a projectile particle of mass m being deflected by a repulsive central-force potential $U(r) > 0$ whose center is at rest at the origin (i.e., $\alpha = 0$). As the projectile particle approaches from the right (at $r = \infty$ and $\theta = 0$) moving with speed u, it is progressively deflected until it reaches a minimum radius ρ at $\theta = \chi$ after which the projectile particle moves away from the repulsion center until it reaches $r = \infty$ at a deflection angle $\theta = \Theta$ and again moving with speed u.

Fig. 5.5 Scattering geometry: the incoming particle is entering the interaction region from the right and is leaving on an asymptote after having been deflected by Θ.

Figure 5.5 shows that the scattering process is symmetric about the line of closest approach (i.e., $2\chi = \pi - \Theta$). The angle of closest approach

$$\chi = \frac{1}{2}(\pi - \Theta) \tag{5.22}$$

is a function of the distance of closest approach ρ, the total energy E, and the angular momentum ℓ.[1] The distance ρ is, of course, a turning point ($\dot{r} = 0$) and is the only positive root of the energy equation

$$E = U(\rho) + \frac{\ell^2}{2m\,\rho^2}, \tag{5.23}$$

where $E = m u^2/2$ is the total initial energy of the projectile particle.

The path of the projectile particle in Fig. 5.5 is labeled by the *impact parameter* b (the distance of closest approach in the non-interacting case: $U = 0$). A simple calculation (using $\mathbf{r} = -ut\,\widehat{\mathbf{x}} + b\,\widehat{\mathbf{y}}$ and $\mathbf{u} = -u\,\widehat{\mathbf{x}}$) shows that the angular momentum $\mathbf{L} = \ell\,\widehat{\mathbf{z}}$ is

$$\ell = \widehat{\mathbf{z}}\cdot(m\mathbf{r}\times\mathbf{u}) = m u b = \sqrt{2mE}\,b. \tag{5.24}$$

---

[1] The sign convention for scattering by an attractive force is $\chi = \frac{1}{2}(\pi+\Theta)$; see Eq. (5.59) for an example.

It is, thus, quite clear that ρ is a function of $E \equiv \ell^2/(2m\,b^2)$, m, and b. The angle χ in Fig. 5.5 is now defined as

$$\chi = \int_\rho^\infty \frac{(\ell/r^2)\,dr}{\sqrt{2m\,[E - U(r)] - (\ell^2/r^2)}}$$

$$\equiv \int_0^{b/\rho} \frac{dx}{\sqrt{1 - x^2 - 2\,b^2\,\overline{U}(x/b)}}, \tag{5.25}$$

where we used the substitution $x = b/r$ to obtain the second integral, with the definition $b^2\,\overline{U}(x/b) \equiv U(b/x)/(2E)$.

Once an expression $\Theta(b) \equiv \pi - 2\,\chi(b)$ is obtained from Eq. (5.25), we may invert it to obtain $b(\Theta)$. We note that as the impact parameter b increases, we generally see that the angle of closest approach increases (see Fig. 5.5) and, thus, the scattering angle Θ decreases according to Eq. (5.22).

5.4.2   *Cross Sections in CM and LAB Frames*

We are now ready to discuss the *likelihood* of the outcome of a collision (for a given impact parameter b) by introducing the concept of differential cross section $\sigma'(\Theta)$ in the CM frame. The infinitesimal cross section $d\sigma'$ in the CM frame is defined in terms of $b(\Theta)$ as $d\sigma'(\Theta) = 2\pi\,b(\Theta)\,db(\Theta)$. Physically, $d\sigma'/d\Omega$ measures the ratio of the number of incident particles per unit time scattered into a solid angle $d\Omega$.

Using Eqs. (5.22) and (5.25), the differential cross section in the CM frame is defined as

$$\sigma'(\Theta) = \frac{d\sigma'}{2\pi\sin\Theta\,d\Theta} = \frac{b(\Theta)}{\sin\Theta}\left|\frac{db(\Theta)}{d\Theta}\right|, \tag{5.26}$$

where, since the quantity $db/d\Theta$ is negative, we must take its absolute value to ensure that $\sigma'(\Theta)$ is positive. The total cross section in the CM frame is defined as

$$\sigma_T' = 2\pi \int_0^\pi \sigma'(\Theta)\,\sin\Theta\,d\Theta. \tag{5.27}$$

The differential cross section can also be written in the LAB frame in terms of the deflection angle θ as

$$\sigma(\theta) = \frac{d\sigma}{2\pi\sin\theta\,d\theta} = \frac{b(\theta)}{\sin\theta}\left|\frac{db(\theta)}{d\theta}\right|. \tag{5.28}$$

Since the infinitesimal cross section $d\sigma = d\sigma'$ is the same in both frames (i.e., the likelihood of a collision should not depend on the choice of a frame of reference), we find

$$\sigma(\theta)\,\sin\theta\,d\theta = \sigma'(\Theta)\,\sin\Theta\,d\Theta,$$

from which we obtain

$$\sigma(\theta) = \sigma'(\Theta) \frac{\sin \Theta}{\sin \theta} \frac{d\Theta}{d\theta}. \tag{5.29}$$

This relation ensures that the total cross section in the LAB frame

$$\sigma_T = 2\pi \int_0^\pi \sigma(\theta) \sin \theta \, d\theta = \sigma_T'$$

is the same as the cross section in the CM frame (5.27). The inversion of Eq. (5.29) yields

$$\sigma'(\Theta) = \sigma(\theta) \frac{\sin \theta}{\sin \Theta} \frac{d\theta}{d\Theta}, \tag{5.30}$$

which gives the differential cross section in the LAB frame $\sigma(\theta)$ once the differential cross section in the CM frame $\sigma'(\Theta)$ and an explicit formula for $\Theta(\theta)$ are known.

We point out that, whereas the CM differential cross section $\sigma'(\Theta)$ is naturally associated with theoretical calculations, the LAB differential cross section $\sigma(\theta)$ is naturally associated with experimental measurements. Hence, the transformation (5.29) is used to translate a theoretical prediction into an observable experimental cross section, while the transformation (5.30) is used to translate experimental measurements into a format suitable for theoretical analysis.

We note that these transformations (5.29)-(5.30) rely on finding relations between the LAB deflection angle θ and the CM deflection angle Θ given by Eq. (5.18), which can be converted into

$$\sin(\Theta - \theta) = \alpha \sin \theta. \tag{5.31}$$

Using these relations, we now show how to obtain an expression for Eq. (5.29) by using Eqs. (5.18) and (5.31). First, we use Eq. (5.31) to obtain $(d\Theta - d\theta) \cos(\Theta - \theta) = \alpha \cos \theta \, d\theta$, which yields

$$\frac{d\Theta}{d\theta} = \frac{\alpha \cos \theta + \cos(\Theta - \theta)}{\cos(\Theta - \theta)}, \tag{5.32}$$

where $\cos(\Theta - \theta) = \sqrt{1 - \alpha^2 \sin^2 \theta}$ requires that $\alpha \sin \theta < 1$. In the case $\alpha < 1$, the maximum deflection angle is therefore $\theta_{max} = \pi$; in the case $\alpha > 1$, on the other hand, $\theta_{max} = \arcsin(\alpha^{-1})$.

Next, using Eqs. (5.18) and (5.31), we obtain

$$\frac{\sin \Theta}{\sin \theta} = \frac{\alpha + \cos \Theta}{\cos \theta} = \frac{\alpha + [\cos(\Theta - \theta) \cos \theta - \sin(\Theta - \theta) \sin \theta]}{\cos \theta}$$

$$= \frac{\alpha (1 - \sin^2 \theta) + \cos(\Theta - \theta) \cos \theta}{\cos \theta}$$

$$= \alpha \cos \theta + \sqrt{1 - \alpha^2 \sin^2 \theta}. \tag{5.33}$$

Thus, by combining Eqs. (5.32) and (5.33), we find

$$\frac{\sin\Theta}{\sin\theta}\frac{d\Theta}{d\theta} = \frac{[\alpha\cos\theta + \sqrt{1 - \alpha^2\sin^2\theta}]^2}{\sqrt{1 - \alpha^2\sin^2\theta}}$$

$$= 2\alpha\cos\theta + \frac{1 + \alpha^2\cos 2\theta}{\sqrt{1 - \alpha^2\sin^2\theta}}. \tag{5.34}$$

Lastly, noting from Eq. (5.31) that the CM deflection angle is defined as

$$\Theta(\theta) = \theta + \arcsin(\alpha\sin\theta),$$

the transformation $\sigma'(\Theta) \to \sigma(\theta)$ is now complete. Similar manipulations yield the transformation $\sigma(\theta) \to \sigma'(\Theta)$. We now show that the LAB-frame cross section $\sigma(\theta)$ are generally difficult to obtain for arbitrary α.

5.5 Hard-Sphere Scattering

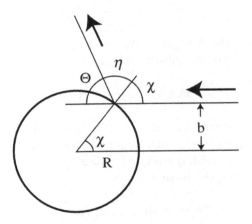

Fig. 5.6 Hard-sphere scattering geometry.

Explicit calculations of differential cross sections tend to be very complex for general central potentials. In this Section, we look at the simplest scattering problem involving the collision of a point-like particle of mass m_1 with a hard sphere of mass m_2 and radius R. The hard-sphere central-force potential is

$$U(r) = \begin{cases} \infty & (\text{for } r < R) \\ 0 & (\text{for } r > R) \end{cases} \tag{5.35}$$

and the collision is shown in Fig. 5.6, where we see that the impact parameter

$$b = R \sin\chi, \tag{5.36}$$

depends simply on the angle of incidence χ.

We note that, in Fig. 5.6, the angle of reflection η is shown to be different from the angle of incidence χ for the case of arbitrary mass ratio $\alpha = m_1/m_2$. To show this, we decompose the velocities in terms of components perpendicular and tangential to the surface of the sphere at the point of impact, i.e., we respectively find

$$\alpha u \cos\chi = v_2 - \alpha v_1 \cos\eta$$
$$\alpha u \sin\chi = \alpha v_1 \sin\eta.$$

From these expressions we obtain

$$\tan\eta = \frac{\alpha u \sin\chi}{v_2 - \alpha u \cos\chi}.$$

From Fig. 5.6, we also find the deflection angle $\theta = \pi - (\chi + \eta)$ and the recoil angle $\varphi = \chi$. Hence, substituting Eq. (5.9), we find

$$\tan\eta = \left(\frac{1+\alpha}{1-\alpha}\right) \tan\chi. \tag{5.37}$$

We, therefore, easily see that $\eta = \chi$ (the standard form of the Law of Reflection) only if $\alpha = 0$ (i.e., the target particle is infinitely massive).

In the CM frame (with $\alpha = 0$), the collision is symmetric with a deflection angle $\chi = \frac{1}{2}(\pi - \Theta)$, so that

$$b = R \sin\chi = R \cos\frac{\Theta}{2}.$$

The scattering cross section (5.26) in the CM frame is

$$\sigma'(\Theta) = \frac{b(\Theta)}{\sin\Theta} \left|\frac{db(\Theta)}{d\Theta}\right| = \frac{R\cos(\Theta/2)}{\sin\Theta} \cdot \frac{R}{2}\sin(\Theta/2) = \frac{R^2}{4}, \tag{5.38}$$

and the total cross section is

$$\sigma_T = 2\pi \int_0^\pi \sigma'(\Theta) \sin\Theta \, d\Theta = \pi R^2, \tag{5.39}$$

i.e., the total cross section for the hard-sphere problem is equal to the effective area of the sphere.

The scattering cross section in the LAB frame (we consider the case $\alpha < 1$) can also be obtained from (5.38) using Eqs. (5.29) and (5.34):

$$\sigma(\theta) = \frac{R^2}{4} \left(2\alpha \cos\theta + \frac{1 + \alpha^2 \cos 2\theta}{\sqrt{1 - \alpha^2 \sin^2\theta}} \right). \qquad (5.40)$$

The integration of this formula yields the total cross section

$$\sigma_T = \frac{\pi}{2} R^2 \int_0^\pi \left(2\alpha \cos\theta + \frac{1 + \alpha^2 \cos 2\theta}{\sqrt{1 - \alpha^2 \sin^2\theta}} \right) \sin\theta \, d\theta$$

$$= \frac{\pi}{2} R^2 \int_0^\pi \left[\alpha \sin(2\theta) - \frac{(1 - \alpha^2) \sin\theta}{\sqrt{1 - \alpha^2 \sin^2\theta}} + 2\sqrt{1 - \alpha^2 \sin^2\theta} \, \sin\theta \right] d\theta$$

$$= \pi R^2 \int_0^1 \left[2\sqrt{(1 - \alpha^2) + \alpha^2 x^2} - \frac{(1 - \alpha^2)}{\sqrt{(1 - \alpha^2) + \alpha^2 x^2}} \right] dx,$$

where the first integral in the second line vanishes while the substitution $x = \cos\theta$ was introduced in the second and third integrals. Next, we use the hyperbolic-trigonometric substitution $\alpha x = \sqrt{1 - \alpha^2} \sinh y$ in both remaining integrals, so that we find

$$\sigma_T = \pi R^2 \left[\left(\frac{1}{\alpha} - \alpha \right) \int_0^Y (2 \cosh^2 y - 1) \, dy \right]$$

$$= \pi R^2 \left[\left(\frac{1}{\alpha} - \alpha \right) \int_0^Y \cosh(2y) \, dy \right]$$

$$= \pi R^2 \left[\left(\frac{1}{\alpha} - \alpha \right) \cosh Y \cdot \sinh Y \right] = \pi R^2, \qquad (5.41)$$

where $Y \equiv \operatorname{arcsinh}(\alpha/\sqrt{1 - \alpha^2})$, so that $\cosh Y \cdot \sinh Y = \alpha/(1 - \alpha^2)$. Thus the total cross section (5.41) in the LAB frame is that same as the total cross section (5.39) in the CM frame, as expected.

5.6 Rutherford Scattering

5.6.1 *Classical Rutherford Scattering*

We now turn our attention to the scattering of a charged particle of mass m_1 and charge q_1 by another charged particle of mass $m_2 \gg m_1$ and charge q_2 such that $q_1 q_2 > 0$ and $\mu \simeq m_1$. This situation is described by a repulsive force produced by the Coulomb potential

$$U(r) = \frac{k}{r}, \qquad (5.42)$$

where $k = q_1 q_2 / (4\pi\,\varepsilon_0) > 0$. The problem of the electrostatic repulsive interaction between a positively-charged alpha particle (i.e., the nucleus of a helium atom) and positively-charged nucleus of a gold atom was first studied by Rutherford (Ernest Rutherford, 1871-1937) and the scattering cross section for this problem is known as the Rutherford cross section.

Using the definition $\ell^2 / 2\,m_1 \equiv E\,b^2$, the turning-point equation in this case is

$$E = E\,\frac{b^2}{\rho^2} + \frac{k}{\rho} \;\to\; \rho^2 - 2\,r_0\,\rho - b^2 = 0,$$

where $2\,r_0 = k/E$ is the distance of closest approach for a *head-on* collision (i.e., $b = 0$). The physical solution for the distance of closest approach is

$$\rho = r_0 + \sqrt{r_0^2 + b^2} = b\left(\varepsilon + \sqrt{1 + \varepsilon^2}\right), \qquad (5.43)$$

where $\varepsilon = r_0/b$. Note that the second radial solution $r_0 - \sqrt{r_0^2 + b^2}$ to the turning-point equation is negative and, therefore, is not allowed.

The angle χ at which the distance of closest approach is reached is calculated from Eq. (5.25) as

$$\chi = \int_0^{b/\rho} \frac{dx}{\sqrt{1 - x^2 - 2\,\varepsilon\,x}} = \int_0^{b/\rho} \frac{dx}{\sqrt{(1 + \varepsilon^2) - (x + \epsilon)^2}},$$

where $b^2\,\overline{U}(x/b) = (k/2E)\,x/b = \varepsilon\,x$ was used in Eq. (5.25) and the upper integration boundary is

$$\frac{b}{\rho} = \frac{1}{\varepsilon + \sqrt{1 + \varepsilon^2}} = -\varepsilon + \sqrt{1 + \varepsilon^2}.$$

Making use of the trigonometric substitution $x = -\varepsilon + \sqrt{1 + \varepsilon^2}\,\cos\chi$, we find that

$$\chi = \arccos\left(\frac{\varepsilon}{\sqrt{1 + \varepsilon^2}}\right) \;\to\; \varepsilon = \cot\chi = r_0/b, \qquad (5.44)$$

which becomes $b = r_0\,\tan\chi$. Using the relation (5.22), we now find

$$b(\Theta) = r_0\,\cot(\Theta/2), \qquad (5.45)$$

with $db(\Theta)/d\Theta = -(r_0/2)\,\csc^2(\Theta/2)$. Hence, using the definition (5.26), the CM Rutherford cross section becomes

$$\sigma'(\Theta) = \frac{r_0^2}{4\,\sin^4(\Theta/2)} = \left(\frac{k}{4E\,\sin^2(\Theta/2)}\right)^2. \qquad (5.46)$$

Note that the Rutherford scattering cross section (5.46) does not depend on the sign of k and is thus valid for both repulsive and attractive interactions.

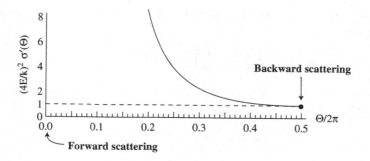

Fig. 5.7   Rutherford scattering cross section.

Figure 5.7 shows that the Rutherford scattering cross section (5.46) becomes very large in the forward direction $\Theta \to 0$ (where $\sigma' \to \Theta^{-4}$) while the differential cross section exhibits the hard-sphere limit

$$\lim_{\Theta \to \pi} \sigma'(\Theta) = \frac{r_0^2}{4}, \qquad (5.47)$$

corresponding to the hard-sphere scattering problem (5.38) for a sphere of radius r_0. Thus, the probability of backward scattering is not zero, which led to the development of the "planetary" model of the atom, in which the small positively-charged nucleus (where most of the atomic mass resides) is surrounded by a "cloud" of electrons. Note that the forward-scattering divergence of the Rutherford formula (5.46) can be eliminated by a slight modification of the Coulomb potential $U(r)$.

5.6.2   *Modified Rutherford Scattering*

In an attempt to remove the divergence of the classical Rutherford scattering cross section (5.46), we now consider the scattering of a particle of mass m by the modified Coulomb potential

$$U(r) = \begin{cases} k\,(1/r - 1/R) & (r \leq R) \\ 0 & (r > R) \end{cases} \qquad (5.48)$$

where R denotes the radial distance beyond which the Coulomb repulsive force is set at zero. The classical Coulomb potential (5.42) is recovered in the limit $R \to \infty$.

The distance of closest approach ρ is the single positive root of the equation

$$E = E\,\frac{b^2}{\rho^2} + k\left(\frac{1}{\rho} - \frac{1}{R}\right) \;\to\; \alpha^2\rho^2 - 2\,r_0\,\rho - b^2 = 0,$$

where $r_0 = k/2E$ and $\alpha = \sqrt{1 + 2\,r_0/R}$. The positive root of this equation yields the distance of closest approach

$$\rho = \frac{b}{\alpha^2}\left(\varepsilon + \sqrt{\varepsilon^2 + \alpha^2}\right), \tag{5.49}$$

where $\varepsilon = r_0/b$ (the negative root for ρ is, of course, not physical). We note that Eq. (5.49) is a modified version of Eq. (5.43) in which α^2 is replaced by 1.

The angle χ at which the distance of closest approach is reached is calculated from Eq. (5.25) as

$$\chi = \int_0^{b/R} \frac{dx}{\sqrt{1 - x^2}} + \int_{b/R}^{b/\rho} \frac{dx}{\sqrt{1 - x^2 - 2\,\varepsilon\,(x - b/R)}}$$

$$= \beta + \int_{\sin\beta}^{b/\rho} \frac{dx}{\sqrt{(\varepsilon^2 + \alpha^2) - (x + \varepsilon)^2}},$$

where $b^2\,\overline{U}(x/b) = (k/2E)\,(x/b - 1/R) = (\varepsilon\,x - r_0/R)$ was used in Eq. (5.25), the angle β is defined as $\beta \equiv \arcsin(b/R)$, and

$$\frac{b}{\rho} = \frac{\alpha^2}{\varepsilon + \sqrt{\varepsilon^2 + \alpha^2}} = \sqrt{\varepsilon^2 + \alpha^2} - \varepsilon.$$

By using the trigonometric substitution $x = -\,\varepsilon + \sqrt{\varepsilon^2 + \alpha^2}\,\cos\psi$, we easily find the angle of closest approach

$$\chi = \beta + \arccos\left(\frac{\varepsilon + \sin\beta}{\sqrt{\varepsilon^2 + \alpha^2}}\right), \tag{5.50}$$

where the right side is a complicated function of the impact parameter b. We obtain the function $b(\Theta)$ by following a few simple steps. First, we write Eq. (5.50) as $\chi \equiv \beta + \arccos(\Delta)$, where $\Delta = (\varepsilon + \sin\beta)/\sqrt{\varepsilon^2 + \alpha^2}$. Next, we derive the pair of equations

$$\cos\chi = \sin(\Theta/2) = \cos\beta\,\Delta - \sin\beta\,\sqrt{1 - \Delta^2},$$

$$\sin\chi = \cos(\Theta/2) = \sin\beta\,\Delta + \cos\beta\,\sqrt{1 - \Delta^2},$$

from which we obtain the relation

$$\cot(\Theta/2) = \frac{\sin\beta\,\Delta + \cos\beta\,\sqrt{1 - \Delta^2}}{\cos\beta\,\Delta - \sin\beta\,\sqrt{1 - \Delta^2}} = \frac{\varepsilon\sin\beta + 1}{\varepsilon\cos\beta}, \tag{5.51}$$

where we used

$$1 - \Delta^2 = \frac{\alpha^2 - 2\varepsilon \sin\beta - \sin^2\beta}{\varepsilon^2 + \alpha^2} = \frac{\cos^2\beta}{\varepsilon^2 + \alpha^2},$$

with $\alpha^2 = 1 + r_0/R = 1 + 2\varepsilon \sin\beta$. Lastly, we solve Eq. (5.51) for the impact parameter $b = r_0/\varepsilon$ and we find

$$b(\Theta) = r_0 \left[\cot(\Theta/2) \cos\beta - \sin\beta\right] = \frac{R \cos(\Theta/2)}{\sqrt{1 + \lambda \sin^2(\Theta/2)}}, \qquad (5.52)$$

where $\beta \equiv \arcsin(b/R)$ and $\lambda \equiv (1 + R/r_0)^2 - 1$. We can easily verify that, in the limit $R \to \infty$, we recover the classical Rutherford formula (5.45) from Eq. (5.52).

We are now ready to calculate the differential cross section (5.26) for the modified Coulomb potential (5.48). Using the impact-parameter function (5.52), we readily find

$$\sigma'(\Theta) = \frac{R^2}{4} \frac{(1+\lambda)}{[1 + \lambda \sin^2(\Theta/2)]^2} = \frac{r_0^2}{4} \frac{[(\sqrt{\lambda+1} - 1)^2 (1+\lambda)]}{[1 + \lambda \sin^2(\Theta/2)]^2}, \qquad (5.53)$$

where we used $R/r_0 = -1 + \sqrt{\lambda + 1}$. We again recover the classical Rutherford formula (5.46) from Eq. (5.53) in the limit $R \to \infty$. The modified Rutherford cross section (5.53) no longer diverges as $\Theta \to 0$, since

$$\sigma'(0) = \frac{R^2}{4}(1+\lambda) = \frac{R^2}{4}\left(1 + \frac{R}{r_0}\right)^2,$$

which scales as R^4 when $R \gg r_0$. We note that $\sigma'(\pi) = (R^2/4)/(1+\lambda)$, which yields the hard-sphere limit (5.47): $\sigma'(\pi) \to r_0^2/4$ when $R \to \infty$.

Lastly, the total cross section (5.27) for the modified Coulomb potential (5.48) is also finite:

$$\sigma_T = \frac{\pi}{2}R^2 \int_0^\pi \frac{(1+\lambda)\sin\Theta \, d\Theta}{[1 + \lambda \sin^2(\Theta/2)]^2} = \frac{\pi}{2}R^2 \int_{-1}^1 \frac{(1+\lambda) \, dx}{(1 + \lambda/2 - \lambda x/2)^2}$$

$$= \pi R^2 \left(\frac{1}{\lambda} + 1\right) \int_1^{1+\lambda} \frac{du}{u^2} = \pi R^2, \qquad (5.54)$$

where the result follows from simple integration (after using the substitution $x = \cos\Theta$). It is interesting to note that the total cross section for a scattering problem that involves a central-force potential that is confined within a radius R is $\sigma_T = \pi R^2$, i.e., the same as for the hard-sphere potential (5.35).

5.7   Soft-Sphere Scattering

We now proceed with another example of scattering by a confined central-force potential, by considering a modified version of the hard-sphere scattering problem. We introduce the attractive potential considered in Sec. 4.5:

$$U(r) = \begin{cases} -U_0 & (\text{for } r < R) \\ \\ 0 & \text{for } r > R \end{cases} \tag{5.55}$$

where the constant U_0 denotes the depth of the attractive potential well and the condition $E > \ell^2/2\mu R^2$ involves a single turning point. We denote β the angle at which the incoming particle enters the *soft-sphere* potential (see Fig. 5.8), and thus the impact parameter b of the incoming particle is $b = R\sin\beta$. For the case of a repulsive soft-sphere (see problem 10), we replace $-U_0$ with U_0 in Eq. (5.55).

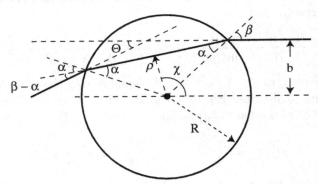

Fig. 5.8   Soft-sphere scattering geometry.

The particle enters the soft-sphere potential region ($r < R$) and reaches a distance of closest approach ρ, defined from the turning-point equation

$$E = -U_0 + E\,\frac{b^2}{\rho^2} \;\rightarrow\; \rho = \frac{b}{\sqrt{1 + U_0/E}} = \frac{R}{n}\sin\beta, \tag{5.56}$$

where

$$n = \sqrt{1 + U_0/E} \equiv b/\rho > 1 \tag{5.57}$$

denotes the *index of refraction* of the attractive soft-sphere potential. From Fig. 5.8, we note that an optical analogy helps us determine that, through Snell's law, we find

$$\sin\beta = n\,\sin\left(\beta - \frac{\Theta}{2}\right), \tag{5.58}$$

where the *transmission* angle α is given in terms of the *incident* angle β and the CM scattering angle $-\Theta$ is defined as $\Theta = 2\,(\beta - \alpha)$.

The distance of closest approach is reached at the angle

$$
\chi = \beta + \int_{\rho}^{R} \frac{b\,dr}{r\,\sqrt{n^2 r^2 - b^2}}
$$

$$
= \beta + \arccos\left(\frac{b}{nR}\right) - \underbrace{\arccos\left(\frac{b}{n\rho}\right)}_{=\,0}
$$

$$
= \beta + \arccos\left(\frac{b}{nR}\right) \equiv \frac{1}{2}\,(\pi + \Theta), \tag{5.59}
$$

where we used Eq. (5.57) in the second line, and the sign convention for $\chi = \frac{1}{2}\,(\pi + \Theta)$ corresponds to the case of scattering by an attractive potential. Using Snell's Law (5.58), the impact parameter $b(\Theta)$ can be expressed as

$$
b(\Theta) = nR\,\sin\left(\beta(b) - \frac{\Theta}{2}\right),
$$

which can be solved for $b(\Theta)$ as

$$
b(\Theta) = \frac{nR\,\sin(\Theta/2)}{\sqrt{1 + n^2 - 2\,n\,\cos(\Theta/2)}}. \tag{5.60}
$$

Its derivative with respect to Θ yields

$$
\frac{db}{d\Theta} = \frac{nR}{2}\,\frac{[n\,\cos(\Theta/2) - 1]\,[n - \cos(\Theta/2)]}{[1 + n^2 - 2\,n\,\cos(\Theta/2)]^{3/2}},
$$

and the scattering cross section (5.26) in the CM frame is

$$
\sigma'(\Theta) = \frac{n^2 R^2}{4}\,\frac{|[n\,\cos(\Theta/2) - 1]\,[n - \cos(\Theta/2)]|}{\cos(\Theta/2)\,[1 + n^2 - 2\,n\,\cos(\Theta/2)]^2}. \tag{5.61}
$$

Note that, on the one hand, when $\beta = 0$, we find $\chi = \pi/2$ and $\Theta_{\min} = 0$ while, on the other hand, when $\beta = \pi/2$, we find $b = R$ and

$$
1 = n\,\sin\left(\frac{\pi}{2} - \frac{\Theta_{\max}}{2}\right) = n\,\cos(\Theta_{\max}/2),
$$

which yields the maximum angle

$$
\Theta_{\max} = 2\,\arccos\left(n^{-1}\right).
$$

Moreover, when $\Theta = \Theta_{\max}$, we find that $db/d\Theta$ vanishes and, therefore, the differential cross section vanishes $\sigma'(\Theta_{\max}) = 0$, while at $\Theta = 0$, we find

$$
\sigma'(0) = \frac{R^2}{4}\left(\frac{n}{n-1}\right)^2.
$$

Fig. 5.9   Soft-sphere scattering cross section in the soft-sphere limit $(n \to 1)$ and the hard-sphere limit $(n \gg 1)$; here, note that $\bar{\sigma}(0) = n^2/(n-1)^2$.

Figure 5.9 shows the soft-sphere scattering cross section $\bar{\sigma}(\Theta) \equiv (4/R^2)\,\sigma'(\Theta)$ as a function of Θ for four cases: $n = (1.1, 1.15)$ in the *soft-sphere limit* $(n - 1 \ll 1)$ and $n = (10, 20, 50, 1000)$ in the *hard-sphere limit* $(n \gg 1)$. We clearly see the strong forward-scattering behavior as $n \to 1$ (or $U_0 \to 0$) in the soft-sphere limit and the hard-sphere limit $\bar{\sigma} \to 1$ as $n \to \infty$. We note that, using the substitution $x = n \cos \Theta/2$, the total scattering cross section associated with Eq. (5.61)

$$\sigma_T = 2\pi \int_0^{\Theta_{max}} \sigma'(\Theta)\,\sin\Theta\,d\Theta = 2\pi R^2 \int_1^n \frac{(x-1)(n^2-x)\,dx}{(1+n^2-2x)^2}$$

$$= \frac{\pi}{4} R^2 \int_{(n-1)^2}^{n^2-1} \left[\frac{(n^2-1)^2}{y^2} - 1 \right] dy = \pi R^2, \tag{5.62}$$

where we used the substitution $x = \frac{1}{2}(n^2 + 1 - y)$. Note that the total cross section for the attractive soft-sphere potential (5.55) is independent of the index of refraction n and is equal to the hard-sphere total cross section (5.39), as expected for scattering by a confined central-force potential [see Eq. (5.54)].

5.8   Elastic Scattering by a Hard Surface

We now generalize the hard-sphere scattering problem by considering scattering by a smooth hard surface of revolution[2] $\rho(z)$ with maximal radial extent R (see Fig. 5.10). Here, a particle of mass m, initially traveling along

---

[2] Adapted from J. L. Brun and A. F. Pacheco, Euro. J. Phys. **26**, 747 (2005).

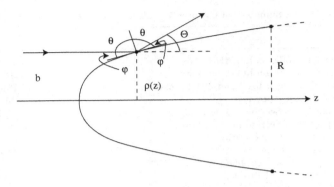

Fig. 5.10   Scattering by a hard surface $\rho(z)$.

the z-axis with velocity u with an impact parameter $b < R$, collides with the hard surface and is scattered with deflection angle Θ. The particle hits the surface (assumed to be infinitely massive) at a distance $b = \rho(z)$ from its axis of symmetry and the angle of incidence $\theta = \pi/2 - \varphi$ (measured from the normal to the surface) is defined in terms of the complementary angle φ, where $\cos\varphi = [1 + (\rho')^2]^{-1/2}$. Since the deflection angle Θ is defined in terms of φ as $\Theta = \pi - 2\theta = 2\varphi$, we find

$$\tan\varphi = \rho'(z) = \tan\frac{\Theta}{2}. \tag{5.63}$$

By using the identity $b(\Theta) = \rho(z)$, we can solve for $z(\Theta)$ [or $\Theta(z)$], and we can thus calculate the differential cross section (5.26).

First, we use the identity

$$\frac{db}{d\Theta} = \rho'\frac{dz}{d\Theta} = \rho'\left(\frac{d\Theta}{dz}\right)^{-1},$$

where, by inverting Eq. (5.63), we obtain $\Theta(z) = 2\arctan(\rho')$, which yields

$$\frac{d\Theta}{dz} = \frac{2\rho''}{[1 + (\rho')^2]},$$

so that we obtain

$$\left|\frac{db}{d\Theta}\right| = \frac{\rho'}{2|\rho''|}\left[1 + (\rho')^2\right].$$

Lastly, using the relation

$$\sin\Theta = 2\cos\varphi\sin\varphi = 2\frac{\tan\varphi}{\sec^2\varphi} \equiv \frac{2\rho'}{[1 + (\rho')^2]},$$

Table 5.1   Summary of Chapter 5: Collisions and Scattering Theory.

Topic	Equation
Two-particle Collision in LAB Frame	(5.9)-(5.11)
Two-particle Collision in CM Frame	(5.16)-(5.17)
Connection between LAB and CM Frames	(5.18)-(5.21)
Angle of Closest Approach	(5.25)
Scattering Cross Section in CM Frame	(5.26)
Scattering Cross Section in LAB Frame	(5.28)-(5.30)
Total Scattering Cross Section	(5.27)
Hard-sphere Problem	(5.38)-(5.41)
Classical Rutherford Problem	(5.46)
Modified Rutherford Problem	(5.53)-(5.54)
Soft-sphere Problem	(5.61)-(5.62)
Scattering by Hard Surface	(5.64)

we find the differential scattering cross section

$$\sigma(\Theta(z)) = \frac{b(z)}{\sin \Theta(z)} \left| \frac{db(z)}{d\Theta(z)} \right| = \frac{\rho}{4 |\rho''|} \left[1 + (\rho')^2 \right]^2$$

$$\equiv \left(\frac{\rho}{4 \kappa} \right) \sec \frac{\Theta}{2}, \tag{5.64}$$

where $\kappa \equiv |\rho''|/[1 + (\rho')^2]^{3/2}$ denotes the Frenet-Serret curvature of the curve $\rho(z)$ in the (ρ, z)-plane.

For example, we revisit the hard-sphere scattering problem studied in Sec. 5.5, with $\rho(z) = \sqrt{R^2 - z^2}$ for $-R \leq z \leq 0$. Here, the Frenet-Serret curvature is simply $\kappa = 1/R$ and

$$z = -\rho \tan \frac{\Theta}{2} \quad \rightarrow \quad \rho = R \cos \frac{\Theta}{2},$$

so that the differential cross section (5.64) yields the standard hard-sphere result (5.38):

$$\sigma(\Theta(z)) = \left(\frac{R^2}{4} \cos \frac{\Theta}{2} \right) \sec \frac{\Theta}{2} = \frac{R^2}{4}.$$

Note that it is possible to invert the relation $\rho(z) \to \sigma(\Theta(z))$, given by Eq. (5.64), to obtain the shape of a surface from its scattering data $\sigma(\Theta) \to \rho(z(\Theta))$.

5.9   Summary

Chapter 5 investigated the kinematics and dynamics of the collision involving two particles that interact through a central-force potential field.

In planar collision kinematics, the conservation laws of energy and momentum are not sufficient to determine a unique outgoing set of momenta. With the help of a single post-collision measurement (in the LAB frame or the CM frame), however, a unique set of post-collision momenta are obtained. The collision dynamics was shown to be easily expressed in the CM frame in terms of a distance and an angle of closest approach, which yielded a differential cross section that described the likelihood of an outgoing momentum state. Table 5.1 presents a summary of the important topics of Chapter 5.

5.10   Problems

1. A particle of mass m_1 traveling in a straight line at velocity v_1 has a head-on elastic collision with a particle of mass m_2 traveling at velocity $-v_2$ (along the same line but in the opposite direction). Show that after the collision, the masses m_1 and m_2 are assumed to travel at velocities $-v_1'$ and v_2' defined as

$$v_1' = \frac{(1-\alpha)\,v_1 + 2\,v_2}{(1+\alpha)} \quad \text{and} \quad v_2' = \frac{2\alpha\,v_1 - (1-\alpha)\,v_2}{(1+\alpha)},$$

where $\alpha \equiv m_1/m_2$.

2. Consider the transfer of momentum from a particle of mass $M_L = M$, initially traveling to the right at velocity u, toward another of the same mass $M_R = M$, mediated by a third particle of mass $m < M$. The particles are arranged in a straight line, with the lighter particle placed in the middle (initially at rest), so that the collision process begins when the heavier particle M_L collides head-on with the lighter particle m, which then collides with the third particle M_R (also initially at rest). Show that the fraction of the initial momentum that is finally transferred to particle M_R from M_L is $u_R = (15/16)\,u$ for $\alpha \equiv m/M = 1/3 > \sqrt{5} - 2$.

[*Hint: For $\alpha \geq \sqrt{5}-2$, only two collisions between m and M_R are involved; for $\alpha = \sqrt{5} - 2$, M_L is at rest after the second collision.*]

3. (a) Using Eq. (5.3) and the conservation laws of energy and momentum, solve for $v_1(u, \theta; \beta)$, where $\beta = m_2/m_1$.

(b) Discuss the number of physical solutions for $v_1(u, \theta; \beta)$ for $\beta < 1$ and

$\beta > 1$.

(c) For $\beta < 1$, show that physical solutions for $v_1(u, \theta; \beta)$ exist for $\theta <$ arcsin$(\beta) = \theta_{max}$.

4. Show that the momentum transfer $\Delta \mathbf{p}'_1 = \mathbf{q}'_1 - \mathbf{p}'_1$ of the projectile particle in the CM frame has a magnitude

$$|\Delta \mathbf{p}'_1| = 2\,\mu u\,\sin\frac{\Theta}{2},$$

where μ, u, and Θ are the reduced mass, initial projectile LAB speed, and CM scattering angle, respectively.

5. Show that the differential cross section $\sigma'(\Theta)$ for the elastic scattering of a particle of mass m from the repulsive central-force potential $U(r) = k/r^2$ with a fixed force-center at $r = 0$ (or an infinitely massive target particle) is

$$\sigma'(\Theta) = \frac{2\pi^2 k}{m\,u^2}\,\frac{(\pi - \Theta)}{[\Theta\,(2\pi - \Theta)]^2\,\sin\Theta},$$

where u is the speed of the incoming projectile particle at $r = \infty$.

Hint :   *Show that* $b(\Theta) = \dfrac{r_0\,(\pi - \Theta)}{\sqrt{2\pi\,\Theta - \Theta^2}},$ *where* $r_0^2 = \dfrac{2\,k}{m\,u^2}.$

6. By using the relations $\tan\theta = \sin\Theta/(\alpha + \cos\Theta)$ and/or $\sin(\Theta - \theta) = \alpha\sin\theta$, where $\alpha = m_1/m_2$, show that the relation between the differential cross section in the CM frame, $\sigma'(\Theta)$, and the differential cross section in the LAB frame, $\sigma(\theta)$, is

$$\sigma'(\Theta) = \sigma(\theta) \cdot \frac{1 + \alpha\cos\Theta}{(1 + 2\alpha\cos\Theta + \alpha^2)^{3/2}}.$$

7. Consider the scattering of a particle of mass m by the localized repulsive central potential

$$U(r) = \begin{cases} -kr^2/2 & (r \leq R) \\ 0 & (r > R) \end{cases}$$

where the radius R denotes the range of the interaction.

(a) Show that for a particle of energy $E > 0$ moving towards the center of attraction with impact parameter $b = R\sin\beta$, the distance of closest approach ρ for this problem is

$$\rho = \sqrt{\frac{E}{k}} \, (e - 1) = b \sqrt{\frac{2}{1 + e}}, \quad \text{where} \quad e = \sqrt{1 + \frac{2 \, k b^2}{E}}$$

(b) Show that the angle χ at closest approach is

$$\chi = \beta + \int_{\sin \beta}^{(1+e)/2} \frac{dx}{\sqrt{1 - x^2 + (e^2 - 1)/x^2}}$$

$$= \beta + \frac{1}{2} \arccos \left(\frac{2 \sin^2 \beta - 1}{e} \right)$$

(c) Using the relation $\chi = \frac{1}{2} (\pi - \Theta)$ between χ and the CM scattering angle Θ (since the scattering involves a repulsive potential), show that

$$e = \frac{\cos 2\beta}{\cos(2\beta + \Theta)} > 1.$$

8. Consider the scattering of a particle of mass m by the potential

$$U(r) = \frac{k}{r} - \frac{\ell^2 \, \beta}{2m \, r^2},$$

where $0 < \beta < 1$ is a constant.

(a) Show that the distance of closest approach is

$$\rho = r_0 + \sqrt{r_0^2 + b^2 \gamma^2},$$

where $b = \ell / \sqrt{2m \, E}$ is the impact parameter, $2 \, r_0 = k/E$ is the distance of closest approach for a head-on collision, and $\gamma^2 \equiv 1 - \beta$.

(b) Show that the angle of closest approach

$$\chi = \int_0^{b/\rho} \frac{dx}{\sqrt{1 - \gamma^2 x^2 - 2 \epsilon \gamma x}} = \frac{1}{\gamma} \arccos \left(\frac{\epsilon}{\sqrt{1 + \epsilon^2}} \right),$$

where $\epsilon \equiv r_0/(b\gamma)$, yields

$$b = \frac{r_0}{\gamma} \tan(\gamma \chi) = \frac{r_0}{\gamma} \tan \left(\gamma \frac{\pi}{2} - \gamma \frac{\Theta}{2} \right).$$

We can easily recover the Rutherford expression (5.45) when $\gamma = 1$ (i.e., $\beta = 0$).

(c) Show that the differential cross section (5.26) for this problem is

$$\sigma'(\Theta) = \frac{r_0^2 \, \tan(\gamma \chi) \, \sec^2(\gamma \chi)}{2\gamma \, \sin(\Theta)},$$

from which we recover the classical Rutherford formula (5.46) when $\gamma = 1$.

(d) While the function $\sigma'(\Theta)$ diverges as $\Theta \to 0$, verify that the total cross section is

$$\sigma_T = 2\pi \int_0^\pi \sigma'(\Theta)\, \sin(\Theta)\, d\Theta = \frac{\pi\, r_0^2}{\gamma^2} \tan^2(\gamma\,\pi/2),$$

which is finite for $\gamma < 1$, but diverges as $\gamma \to 1$.

9*. Consider the scattering of a particle by the repulsive potential[3]

$$U(r) = \frac{k}{r^4}.$$

(a) Show that the distance of closest approach is

$$\rho = b\,\sqrt{\frac{1+e}{2}} = r_0 \left(\frac{e+1}{e-1}\right)^{1/4}, \quad \text{where} \quad e = \sqrt{1 + \frac{4k}{E\,b^4}} \geq 1,$$

and the distance of closest approach for a head-on collision ($b = 0$, $e \to \infty$) is $r_0 = (k/E)^{1/4}$.

(b) Show that the angle of closest approach is

$$\chi = \int_0^{b/\rho} \frac{dx}{\sqrt{1 - x^2 - (k/E b^4)\, x^4}}$$

$$= \int_0^{b/\rho} \frac{dx}{\sqrt{(b^2/\rho^2 - x^2)\,[(k/E b^4)\, x^2 + \rho^2/b^2]}}.$$

Using the substitution $x = (b/\rho) \sin\phi$, show that χ is expressed as

$$\chi = \frac{b^2}{\rho^2} \int_0^{\pi/2} \frac{d\phi}{\sqrt{1 + (k/E\rho^4)\, \sin^2\phi}} = \frac{2}{(1+e)}\, \mathsf{K}\left(\frac{1-e}{1+e}\right),$$

where $k/(E\rho^4) = (e-1)/(e+1)$ and $\mathsf{K}(m) \equiv \int_0^{\pi/2} d\phi/\sqrt{1 - m\,\sin^2\phi}$ denotes the complete elliptic integral of the first kind (defined for $m < 1$).

(c) The CM scattering angle Θ is now parameterized by e (see Fig. 5.11):

$$\Theta(e) = \pi - \frac{4}{(1+e)}\, \mathsf{K}\left(\frac{1-e}{1+e}\right),$$

where $\Theta(1) = 0$ and $\Theta(\infty) = \pi$. Using the relation [14]

$$\frac{d\mathsf{K}(m)}{dm} = \frac{\mathsf{E}(m) - (1-m)\,\mathsf{K}(m)}{2\,m\,(1-m)},$$

---

[3] See also Section 7.7 of *Classical Mechanics with Applications* by P. W. Johnson (World Scientific, 2010).

Fig. 5.11 Scattering angle $\Theta(e)$ as a function of the parameter e.

show that

$$\frac{d\Theta(e)}{de} = \frac{4\,e}{(e^2 - 1)\,(e + 1)} \left[\mathsf{K}\left(\frac{1-e}{1+e}\right) - \left(\frac{1+e}{2\,e^2}\right) \mathsf{E}\left(\frac{1-e}{1+e}\right) \right],$$

and

$$\frac{db}{d\Theta} = \frac{db/de}{d\Theta/de} = -\frac{b}{8}\,(e+1) \left[\mathsf{K}\left(\frac{1-e}{1+e}\right) - \left(\frac{1+e}{2\,e^2}\right) \mathsf{E}\left(\frac{1-e}{1+e}\right) \right]^{-1}.$$

(d) Show that the differential cross section is

$$2\pi\,\sigma'(\Theta)\,\sin\Theta\,d\Theta = \pi\,r_0^2\,\frac{2\,e\,de}{(e^2 - 1)^{3/2}},$$

which diverges at $e = 1$ (i.e., $\Theta = 0$).

10. Consider elastic scattering by a hard ellipsoid $\rho(z) = \rho_0\,\sqrt{1 - (z/z_0)^2}$ $(-z_0 \le z \le 0)$, where $\rho_0 = z_0\,\sqrt{1 - e^2} \le z_0$ and $0 \le e < 1$ denotes the eccentricity of the ellipse in the (ρ, z)-plane.

(a) Show that the differential scattering cross section is expressed as

$$\sigma'(\Theta) = \frac{\rho_0^2\,(1 - e^2)}{4\,(1 - e^2\,\cos^2\frac{\Theta}{2})}.$$

(b) Show that the total cross section σ_T is

$$\sigma_T = 2\pi \int_0^\pi \sigma'(\Theta)\,\sin\Theta\,d\Theta = \pi\,\rho_0^2 \left[\left(\frac{1}{e^2} - 1\right) \ln\left(\frac{1}{1 - e^2}\right) \right].$$

(c) Show that, when $e \to 0$, we recover the hard-sphere result $\sigma_T = \pi\,\rho_0^2$.

11. An electron is moving to the right (from $-\infty$), with speed u and impact parameter b, and collides with another electron initially at rest. According to collision kinematics in the LAB frame, the deflection angle θ of the projectile electron and the recoil angle φ of the target electron satisfy the identity (5.13): $\tan\theta = \cot\varphi$. According collision kinematics in the CM frame, on the other hand, we can use the relation (5.20): $\varphi = \pi/2 - \Theta/2$, to obtain $\cot\varphi = \tan(\Theta/2)$.

By making use of the classical Rutherford formula $b = r_0 \cot(\Theta/2)$, where $r_0 = k/E$ for the case of equal-mass Coulomb scattering (we ignore all quantum effects, of course), show that the speeds v_1 (for the deflected electron) and v_2 (for the recoil electron) after the collision are

$$v_1 = u \sin\varphi = \frac{b\,u}{\sqrt{r_0^2 + b^2}} \quad \text{and} \quad v_2 = u\cos\varphi = \frac{r_0\,u}{\sqrt{r_0^2 + b^2}},$$

and $\tan\theta = \cot\varphi = r_0/b$.

12. Consider the scattering problem associated with a repulsive soft-sphere potential, where $-U_0$ is replaced with U_0 in Eq. (5.55). By replacing $n = (1 + U_0/E)^{\frac{1}{2}}$ with $n = (1 - U_0/E)^{-\frac{1}{2}}$, show that Eq. (5.60) is replaced with $b(\Theta) = n^{-1} R \sin(\beta(b) + \Theta/2)$, or

$$b(\Theta) = \frac{R \sin(\Theta/2)}{\sqrt{1 + n^2 - 2n \cos(\Theta/2)}},$$

while Snell's law (5.58) is replaced with

$$\sin\left(\beta + \frac{\Theta}{2}\right) = n \sin\beta.$$

13. Using Eqs. (5.63)-(5.64), show that the elastic scattering of a particle by the hard surface $\rho(z) = 2\sqrt{R\,z}$, where R is a constant, yields the Rutherford formula

$$\sigma = R^2/\sin^4\frac{\Theta}{2},$$

where $z = R \cot^2\frac{\Theta}{2}$.

Chapter 6

Motion in a Non-Inertial Frame

A reference frame is said to be an *inertial* frame if the motion of particles in that frame is subject only to physical forces (e.g., forces that are derivable from a physical potential U such that $m\ddot{\mathbf{x}} = -\nabla U$). The Principle of Galilean Relativity (Sec. 2.5.3) states that the laws of physics are the same in all inertial frames and that all reference frames moving at constant velocity with respect to an inertial frame are also inertial frames. Hence, physical accelerations are identical in all inertial frames.

In contrast, a reference frame is said to be *non-inertial* if the motion of particles in that frame of reference violates the Principle of Galilean Relativity. Such non-inertial frames include all rotating frames and accelerated reference frames.

6.1 Time Derivatives in Rotating Frames

To investigate the relationship between inertial and non-inertial frames, we consider the time derivative of an arbitrary vector $\mathbf{A}(t)$ in two reference frames. The first reference frame is called the *fixed* (inertial) frame and is expressed in terms of the Cartesian coordinates $\mathbf{r}' = (x', y', z')$. The second reference frame is called the *rotating* (non-inertial) frame and is expressed in terms of the Cartesian coordinates $\mathbf{r} = (x, y, z)$. In Fig. 6.1, the rotating frame shares the same origin as the fixed frame (we remove this condition later) and the rotation angular velocity $\boldsymbol{\omega}$ of the rotating frame (with respect to the rotating frame) has components $(\omega_x, \omega_y, \omega_z)$.

Since observations can also be made in a rotating frame of reference, we decompose the vector $\mathbf{A}$ in terms of components A_i in the rotating frame (with unit vectors $\widehat{\mathbf{x}}^i$). Thus, $\mathbf{A} = A_i\,\widehat{\mathbf{x}}^i$ (using the summation rule) and

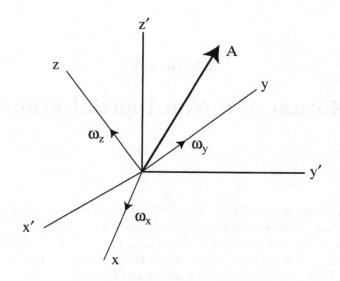

Fig. 6.1 Rotating and fixed frames.

the time derivative of **A** as observed in the fixed frame is

$$\frac{d\mathbf{A}}{dt} = \frac{dA_i}{dt}\,\widehat{\mathsf{x}}^i + A_i\,\frac{d\widehat{\mathsf{x}}^i}{dt}. \tag{6.1}$$

The interpretation of the first term is that of the time derivative of **A** as observed in the rotating frame (where the unit vectors $\widehat{\mathsf{x}}^i$ are constant) while the second term involves the time-dependence of the relation between the fixed and rotating frames. By construction, the vector $d\widehat{\mathsf{x}}^i/dt$ is simply expressed in terms of the angular velocity $\boldsymbol{\omega}$ of the rotating frame as

$$\frac{d\widehat{\mathsf{x}}^i}{dt} = \boldsymbol{\omega} \times \widehat{\mathsf{x}}^i, \tag{6.2}$$

which automatically guarantees that $\widehat{\mathsf{x}}^i \cdot d\widehat{\mathsf{x}}^i/dt = 0$. Hence, the second term in Eq. (6.1) becomes

$$A_i\,\frac{d\widehat{\mathsf{x}}^i}{dt} = \boldsymbol{\omega} \times \mathbf{A}, \tag{6.3}$$

and the time derivative of an arbitrary rotating-frame vector **A** in a fixed frame is, therefore, expressed as

$$\left(\frac{d\mathbf{A}}{dt}\right)_f = \left(\frac{d\mathbf{A}}{dt}\right)_r + \boldsymbol{\omega} \times \mathbf{A}. \tag{6.4}$$

Here, $(d/dt)_f$ denotes the time derivative as observed in the fixed (f) frame while $(d/dt)_r$ denotes the time derivative as observed in the rotating (r)

frame. An important application of this formula relates to the time derivative of the rotation angular velocity ω itself, where one can easily see that

$$\left(\frac{d\omega}{dt}\right)_f = \dot{\omega} = \left(\frac{d\omega}{dt}\right)_r,$$

since the second term in Eq. (6.4) vanishes for $\mathbf{A} = \omega$. The time derivative of ω is, therefore, the same in both frames of reference and is denoted $\dot{\omega}$ in what follows.

6.2 Accelerations in Rotating Frames

We now consider the general case of a rotating frame and fixed frame being related by translation and rotation. The position of a point P according to the fixed frame of reference is labeled $\mathbf{r}'$, while the position of the same point according to the rotating frame of reference is labeled $\mathbf{r}$, and

$$\mathbf{r}' = \mathbf{R} + \mathbf{r}, \tag{6.5}$$

where $\mathbf{R}$ denotes the position of the origin of the rotating frame (e.g., the center of mass) according to the fixed frame. Since the velocity of the point P involves the rate of change of position, we must now be careful in defining which time-derivative operator, $(d/dt)_f$ or $(d/dt)_r$, is used.

The velocities of point P as observed in the fixed and rotating frames are defined as

$$\mathbf{v}_f = \left(\frac{d\mathbf{r}'}{dt}\right)_f \quad \text{and} \quad \mathbf{v}_r = \left(\frac{d\mathbf{r}}{dt}\right)_r, \tag{6.6}$$

respectively. Using Eq. (6.4), the relation between the fixed-frame and rotating-frame velocities is expressed as

$$\mathbf{v}_f = \dot{\mathbf{R}} + \left(\frac{d\mathbf{r}}{dt}\right)_f = \dot{\mathbf{R}} + \mathbf{v}_r + \omega \times \mathbf{r}, \tag{6.7}$$

where $\dot{\mathbf{R}}$ denotes the translation velocity of the rotating-frame origin (as observed in the fixed frame).

Using Eq. (6.7), we are now in a position to evaluate expressions for the acceleration of point P as observed in the fixed and rotating frames of reference, which are defined as

$$\mathbf{a}_f = \left(\frac{d\mathbf{v}_f}{dt}\right)_f \quad \text{and} \quad \mathbf{a}_r = \left(\frac{d\mathbf{v}_r}{dt}\right)_r, \tag{6.8}$$

respectively. Hence, using Eq. (6.7), we find

$$\mathbf{a}_f = \ddot{\mathbf{R}} + \left(\frac{d\mathbf{v}_r}{dt}\right)_f + \left(\frac{d\boldsymbol{\omega}}{dt}\right)_f \times \mathbf{r} + \boldsymbol{\omega} \times \left(\frac{d\mathbf{r}}{dt}\right)_f$$

$$= \ddot{\mathbf{R}} + (\mathbf{a}_r + \boldsymbol{\omega} \times \mathbf{v}_r) + \dot{\boldsymbol{\omega}} \times \mathbf{r} + \boldsymbol{\omega} \times (\mathbf{v}_r + \boldsymbol{\omega} \times \mathbf{r}),$$

or

$$\mathbf{a}_f = \ddot{\mathbf{R}} + \mathbf{a}_r + 2\boldsymbol{\omega} \times \mathbf{v}_r + \dot{\boldsymbol{\omega}} \times \mathbf{r} + \boldsymbol{\omega} \times (\boldsymbol{\omega} \times \mathbf{r}), \qquad (6.9)$$

where $\ddot{\mathbf{R}}$ denotes the translational acceleration of the rotating-frame origin (as observed in the fixed frame of reference). We can write an expression for the acceleration of point P as observed in the rotating frame as

$$\mathbf{a}_r = \left(\mathbf{a}_f - \ddot{\mathbf{R}}\right) - \boldsymbol{\omega} \times (\boldsymbol{\omega} \times \mathbf{r}) - 2\boldsymbol{\omega} \times \mathbf{v}_r - \dot{\boldsymbol{\omega}} \times \mathbf{r}, \qquad (6.10)$$

which represents the sum of the net *inertial* acceleration $(\mathbf{a}_f - \ddot{\mathbf{R}})$, the centrifugal acceleration $-\boldsymbol{\omega} \times (\boldsymbol{\omega} \times \mathbf{r})$ and the *Coriolis* acceleration $-2\boldsymbol{\omega} \times \mathbf{v}_r$ (see Fig. 6.2) and an angular acceleration term $-\dot{\boldsymbol{\omega}} \times \mathbf{r}$ that depends explicitly on the time dependence of the rotation angular velocity $\boldsymbol{\omega}$. The centrifugal acceleration $\mathbf{a}_{\mathrm{Cf}} = -\boldsymbol{\omega} \times (\boldsymbol{\omega} \times \mathbf{r}) = \omega^2 \mathbf{r} - (\boldsymbol{\omega} \cdot \mathbf{r})\boldsymbol{\omega}$ (which is directed outwardly from the rotation axis) represents a familiar *non-inertial* effect in physics.

A less familiar *non-inertial* effect is the Coriolis acceleration $\mathbf{a}_{\mathrm{Co}} \equiv -2\boldsymbol{\omega} \times \dot{\mathbf{r}}$ discovered in 1831 by Gaspard Gustave de Coriolis (1792-1843). Figure 6.2 shows that an object *falling* inwardly (toward Earth), for example, also experiences an *eastward* acceleration. It is also quite clear that, since the Coriolis acceleration does not change the kinetic energy of a particle (i.e., $\dot{\mathbf{r}} \cdot \mathbf{a}_{\mathrm{Co}} \equiv 0$), it only changes the direction of the particle's motion (if $\dot{\mathbf{r}}$ is not directed along $\boldsymbol{\omega}$).

6.3   Lagrangian Formulation of Non-Inertial Motion

We can recover the expression (6.10) for the acceleration in a rotating (non-inertial) frame from a Lagrangian formulation as follows. The Lagrangian for a particle of mass m moving in a non-inertial rotating frame (with its origin coinciding with the fixed-frame origin) in the presence of the potential $U(\mathbf{r})$ is expressed as

$$L(\mathbf{r}, \dot{\mathbf{r}}) = \frac{m}{2} \left|\dot{\mathbf{r}} + \boldsymbol{\omega} \times \mathbf{r}\right|^2 - U(\mathbf{r}) \qquad (6.11)$$

$$= \frac{m}{2} \left|\dot{\mathbf{r}}\right|^2 + m\,\dot{\mathbf{r}} \cdot (\boldsymbol{\omega} \times \mathbf{r}) + \frac{m}{2} \left|\boldsymbol{\omega} \times \mathbf{r}\right|^2 - U(\mathbf{r}),$$

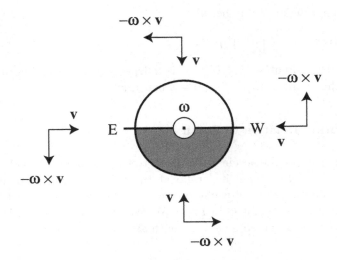

Fig. 6.2 Coriolis acceleration for a falling object on Earth (the shaded area shows the night-side of Earth and the rotation angular velocity is pointing out of the page).

where $\boldsymbol{\omega}$ is the angular velocity vector. Using the Lagrangian (6.11), we now derive the general Euler-Lagrange equations for $\mathbf{r}$. First, we derive an expression for the canonical momentum

$$\mathbf{p} = \frac{\partial L}{\partial \dot{\mathbf{r}}} = m\,(\dot{\mathbf{r}} + \boldsymbol{\omega} \times \mathbf{r})\,, \tag{6.12}$$

so that the time derivative of the canonical momentum is

$$\frac{d}{dt}\left(\frac{\partial L}{\partial \dot{\mathbf{r}}}\right) = m\,(\ddot{\mathbf{r}} + \dot{\boldsymbol{\omega}} \times \mathbf{r} + \boldsymbol{\omega} \times \dot{\mathbf{r}})\,.$$

Next, we derive the generalized force

$$\frac{\partial L}{\partial \mathbf{r}} = -\,\nabla U(\mathbf{r}) - m\,[\,\boldsymbol{\omega} \times \dot{\mathbf{r}} + \boldsymbol{\omega} \times (\boldsymbol{\omega} \times \mathbf{r})\,]\,,$$

so that the Euler-Lagrange equations are

$$m\,\ddot{\mathbf{r}} = -\,\nabla U(\mathbf{r}) - m\,[\,\dot{\boldsymbol{\omega}} \times \mathbf{r} + 2\,\boldsymbol{\omega} \times \dot{\mathbf{r}} + \boldsymbol{\omega} \times (\boldsymbol{\omega} \times \mathbf{r})\,]\,. \tag{6.13}$$

Here, the potential energy term generates the fixed-frame acceleration, $-\nabla U = m\,\mathbf{a}_f$, and, thus, the Euler-Lagrange equation (6.13) yields Eq. (6.10) for $\mathbf{a}_r = \ddot{\mathbf{r}}$.

It is interesting to note that if we adopt cylindrical coordinates (ρ, φ, z), with the z-axis aligned along the angular velocity $\boldsymbol{\omega}$ vector (i.e., $\boldsymbol{\omega} = \omega\,\hat{\mathbf{z}}$), the vector $\dot{\mathbf{r}} + \boldsymbol{\omega} \times \mathbf{r}$ becomes

$$\dot{\mathbf{r}} + \boldsymbol{\omega} \times \mathbf{r} = \dot{\rho}\,\hat{\rho} + \rho\,(\dot{\varphi} + \omega)\,\hat{\varphi} + \dot{z}\,\hat{\mathbf{z}}\,,$$

and the Lagrangian (6.11) becomes

$$L(\mathbf{r}, \dot{\mathbf{r}}) = \frac{m}{2}\left[\dot{\rho}^2 + \rho^2\,(\dot{\varphi} + \omega)^2 + \dot{z}^2\right] - U(\rho, \varphi, z). \tag{6.14}$$

Hence, if the potential $U = U(\rho, z)$ is independent of the azimuthal angle φ, then the quantity $m\,\rho^2(\dot{\varphi} + \omega)$ is a constant of the motion.

6.4 Motion Relative to Earth

We can now apply the acceleration (6.10) to the important case of the fixed frame of reference having its origin at the center of Earth (point O' in Fig. 6.3) and the rotating frame of reference having its origin at latitude λ and longitude ψ (point O in Fig. 6.3). We note that the rotation of the Earth is now represented as $\dot{\psi} = \omega$ (with $\dot{\boldsymbol{\omega}} = 0$).

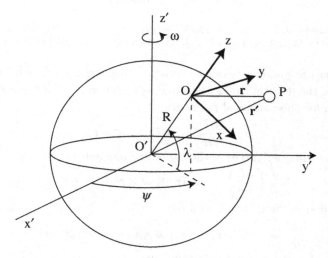

Fig. 6.3 Earth frame: $\hat{\mathbf{x}}$ = southward (northern hemisphere), $\hat{\mathbf{y}}$ = eastward, and $\hat{\mathbf{z}}$ = (radially) upward.

We arrange the (x, y, z)-axes of the rotating frame so that the z-axis is a continuation of the position vector $\mathbf{R}$ of the rotating-frame origin, i.e., $\mathbf{R} = R\hat{\mathbf{z}}$ in the rotating frame (where $R = 6378$ km is the radius of a *spherical* Earth). When expressed in terms of the fixed-frame latitude angle λ and the azimuthal angle ψ, the unit vector $\hat{\mathbf{z}}$ is

$$\hat{\mathbf{z}} = \cos\lambda\,(\cos\psi\,\hat{\mathbf{x}}' + \sin\psi\,\hat{\mathbf{y}}') + \sin\lambda\,\hat{\mathbf{z}}'.$$

Likewise, we choose the x-axis to be tangent to a *great circle* passing through the North and South poles, so that

$$\widehat{x} \equiv -\frac{\partial \widehat{z}}{\partial \lambda} = \sin \lambda \ (\cos \psi \, \widehat{x}' + \sin \psi \, \widehat{y}') - \cos \lambda \, \widehat{z}',$$

i.e., $\widehat{x}$ points southward. Lastly, the y-axis is chosen such that

$$\widehat{y} = \widehat{z} \times \widehat{x} = -\sin \psi \, \widehat{x}' + \cos \psi \, \widehat{y}' \equiv \sec \lambda \, \frac{\partial \widehat{z}}{\partial \psi},$$

i.e., $\widehat{y}$ points eastward.

We now consider the acceleration of a point P as observed in the rotating frame O by writing Eq. (6.10) as

$$\frac{d^2 \mathbf{r}}{dt^2} = \mathbf{g}_0 - \ddot{\mathbf{R}} - \boldsymbol{\omega} \times (\boldsymbol{\omega} \times \mathbf{r}) - 2\boldsymbol{\omega} \times \frac{d\mathbf{r}}{dt}. \tag{6.15}$$

The first term represents the *pure* gravitational acceleration due to the gravitational pull of the Earth on point P (as observed in the fixed frame located at Earth's center)

$$\mathbf{g}_0 = -\frac{GM}{|\mathbf{r}'|^3} \, \mathbf{r}', \tag{6.16}$$

where $\mathbf{r}' = \mathbf{R} + \mathbf{r}$ is the position of point P in the fixed frame and $\mathbf{r}$ is the location of P in the rotating frame. When expressed in terms of rotating-frame spherical coordinates (r, θ, φ):

$$\mathbf{r} = r\,[\,\sin \theta \ (\cos \varphi \, \widehat{x} + \sin \varphi \, \widehat{y}) + \cos \theta \, \widehat{z}\,],$$

the fixed-frame position $\mathbf{r}'$ is written as

$$\mathbf{r}' = (R + r \cos \theta)\, \widehat{z} + r \sin \theta \ (\cos \varphi \, \widehat{x} + \sin \varphi \, \widehat{y}),$$

and thus

$$|\mathbf{r}'|^3 = \left(R^2 + 2\,R r \cos \theta + r^2\right)^{3/2}.$$

The pure gravitational acceleration (6.16) is, therefore, expressed in the rotating frame of the Earth as

$$\mathbf{g}_0 = -g_0 \left[\frac{(1 + \epsilon \cos \theta)\, \widehat{z} + \epsilon \sin \theta \ (\cos \varphi \, \widehat{x} + \sin \varphi \, \widehat{y})}{(1 + 2\epsilon \cos \theta + \epsilon^2)^{3/2}} \right] \tag{6.17}$$

$$= -g_0 \left[(1 - 2\epsilon \cos \theta)\, \widehat{z} + \epsilon \sin \theta \ (\cos \varphi \, \widehat{x} + \sin \varphi \, \widehat{y}) + \cdots\right],$$

where $g_0 = GM/R^2 = 9.789$ m/s^2 and $\epsilon = r/R \ll 1$ (e.g., $\epsilon \sim 10^{-6}$ at $r \sim 10$ m).

The angular velocity in the fixed frame is $\boldsymbol{\omega} = \omega \widehat{z}'$, where

$$\omega = \frac{2\pi \text{ rad}}{24 \times 3600 \text{ sec}} = 7.27 \times 10^{-5} \text{ rad/s}$$

is the angular rotation speed of Earth about its axis. In the rotating frame, we find

$$\boldsymbol{\omega} = \omega \left(\sin\lambda\,\widehat{\mathbf{z}} - \cos\lambda\,\widehat{\mathbf{x}}\right). \tag{6.18}$$

Because the position vector $\mathbf{R}$ rotates with the origin of the rotating frame, its time derivatives yield

$$\dot{\mathbf{R}} = \boldsymbol{\omega} \times \mathbf{R} = (\omega\,R\,\cos\lambda)\,\widehat{\mathbf{y}},$$
$$\ddot{\mathbf{R}} = \boldsymbol{\omega} \times \dot{\mathbf{R}}_f = \boldsymbol{\omega} \times (\boldsymbol{\omega} \times \mathbf{R}) = -\,\omega^2\,R\,\cos\lambda\,(\cos\lambda\,\widehat{\mathbf{z}} + \sin\lambda\,\widehat{\mathbf{x}}),$$

and thus the centrifugal acceleration due to $\mathbf{R}$ is

$$-\,\ddot{\mathbf{R}} = -\,\boldsymbol{\omega} \times (\boldsymbol{\omega} \times \mathbf{R}) = \alpha\,g_0\,\cos\lambda\,(\cos\lambda\,\widehat{\mathbf{z}} + \sin\lambda\,\widehat{\mathbf{x}}), \tag{6.19}$$

where $\omega^2 R = 0.0337$ m/s^2 can be expressed in terms of the *pure* gravitational acceleration g_0 as $\omega^2 R = \alpha\,g_0$, where $\alpha = 3.4 \times 10^{-3}$ is the normalized centrifugal acceleration. We now define the physical gravitational acceleration as

$$\mathbf{g} = \mathbf{g}_0 - \boldsymbol{\omega} \times (\boldsymbol{\omega} \times \mathbf{R})$$
$$= -\,g_0\left[\left(1 - \alpha\,\cos^2\lambda\right)\widehat{\mathbf{z}} - (\alpha\,\cos\lambda\,\sin\lambda)\,\widehat{\mathbf{x}}\right], \tag{6.20}$$

where terms of order $\epsilon = r/R$ have been neglected (since $\epsilon \ll \alpha$). For example, a plumb line experiences a small angular deviation $\delta(\lambda)$ (southward) from the true vertical given as

$$\tan\delta(\lambda) = \frac{g_x}{|g_z|} = \frac{\alpha\,\sin 2\lambda}{(2 - \alpha) + \alpha\,\cos 2\lambda}.$$

This function exhibits a maximum at a latitude $\overline{\lambda}$ defined as $\cos 2\overline{\lambda} = -\alpha/(2 - \alpha)$, so that

$$\tan\overline{\delta} = \frac{\alpha\,\sin 2\overline{\lambda}}{(2 - \alpha) + \alpha\,\cos 2\overline{\lambda}} = \frac{\alpha}{2\sqrt{1 - \alpha}} \simeq 1.7 \times 10^{-3},$$

or

$$\overline{\delta} \simeq 5.86\,\text{arcmin} \quad \text{at} \quad \overline{\lambda} \simeq \left(\frac{\pi}{4} + \frac{\alpha}{4}\right)\,\text{rad} = 45.05^{\circ}.$$

6.4.1   *Coriolis-corrected Projectile Motion*

We now return to Eq. (6.15), which is written to lowest order in ϵ and α as

$$\frac{d^2\mathbf{r}}{dt^2} = -\,g\,\widehat{\mathbf{z}} - 2\,\boldsymbol{\omega} \times \frac{d\mathbf{r}}{dt}, \tag{6.21}$$

where g denotes the effective (constant) gravitational acceleration and the Coriolis acceleration is

$$-2\,\boldsymbol{\omega}\times\frac{d\mathbf{r}}{dt} \;=\; -2\omega\,[\,(\dot{x}\sin\lambda\,+\,\dot{z}\cos\lambda)\widehat{\mathbf{y}}\,-\,\dot{y}\,(\sin\lambda\,\widehat{\mathbf{x}}\,+\,\cos\lambda\,\widehat{\mathbf{z}})\,].$$

Thus, we find the three components of Eq. (6.21) written explicitly as

$$\left.\begin{aligned}
\ddot{x} &= 2\,\omega\,\sin\lambda\,\dot{y}\\
\ddot{y} &= -\,2\,\omega\,(\sin\lambda\,\dot{x}\,+\,\cos\lambda\,\dot{z})\\
\ddot{z} &= -\,g\,+\,2\,\omega\,\cos\lambda\,\dot{y}
\end{aligned}\right\}. \tag{6.22}$$

An interesting comment can be made concerning horizontal motion ($z = \dot{z} = 0$) of a floating object at sea in the presence of the Coriolis acceleration $(\ddot{x}, \ddot{y}) = 2\omega\sin\lambda\,(\dot{y}, -\dot{x})$. By calculating the Frenet-Serret curvature [see Eq. (A.25)] for this planar motion, we find

$$\kappa \equiv \frac{\dot{y}\,\ddot{x} - \dot{x}\,\ddot{y}}{(\dot{x}^2 + \dot{y}^2)^{3/2}} \;=\; \frac{2\omega}{v}\,\sin\lambda,$$

where v is a constant. Hence, the Coriolis acceleration generates an *inertia circle* with a radius equal to $v/(2\,\omega\sin\lambda)$. For example, a particle drifting horizontally at sea, with speed 10 cm/s at latitude $\lambda = 45^o$, performs an inertia circle with a radius of approximately 1 km.

A first integration of Eq. (6.22) yields

$$\left.\begin{aligned}
\dot{x} &= 2\,\omega\,\sin\lambda\,y\,+\,V_x\\
\dot{y} &= -\,2\,\omega\,(\sin\lambda\,x\,+\,\cos\lambda\,z)\,+\,V_y\\
\dot{z} &= -\,g\,t\,+\,2\,\omega\,\cos\lambda\,y\,+\,V_z
\end{aligned}\right\}, \tag{6.23}$$

where (V_x, V_y, V_z) are constants defined from initial conditions (x_0, y_0, z_0) and $(\dot{x}_0, \dot{y}_0, \dot{z}_0)$:

$$\left.\begin{aligned}
V_x &= \dot{x}_0\,-\,2\,\omega\,\sin\lambda\,y_0\\
V_y &= \dot{y}_0\,+\,2\,\omega\,(\sin\lambda\,x_0\,+\,\cos\lambda\,z_0)\\
V_z &= \dot{z}_0\,-\,2\,\omega\,\cos\lambda\,y_0
\end{aligned}\right\}. \tag{6.24}$$

A second integration of Eq. (6.23) yields

$$x(t) = x_0\,+\,V_x\,t\,+\,2\,\omega\,\sin\lambda\int_0^t y(\tau)\,d\tau,$$

$$y(t) = y_0\,+\,V_y\,t\,-\,2\,\omega\,\sin\lambda\int_0^t x(\tau)\,d\tau\,-\,2\,\omega\,\cos\lambda\int_0^t z(\tau)\,d\tau,$$

$$z(t) = z_0\,+\,V_z\,t\,-\,\frac{1}{2}\,g\,t^2\,+\,2\,\omega\,\cos\lambda\int_0^t y(\tau)\,d\tau,$$

which can also be rewritten as

$$\left.\begin{array}{l} x(t) = x_0 + V_x\,t + \delta x(t) \\ y(t) = y_0 + V_y\,t + \delta y(t) \\ z(t) = z_0 + V_z\,t - \frac{1}{2}\,g\,t^2 + \delta z(t) \end{array}\right\}, \tag{6.25}$$

where the Coriolis *drifts* $(\delta x, \delta y, \delta z)$ are

$$\delta x(t) = 2\,\omega\,\sin\lambda\left(y_0\,t + \frac{1}{2}\,V_y\,t^2 + \int_0^t \delta y(\tau)\,d\tau\right) \tag{6.26}$$

$$\delta y(t) = -\,2\,\omega\,\sin\lambda\left(x_0\,t + \frac{1}{2}\,V_x\,t^2 + \int_0^t \delta x(\tau)\,d\tau\right) \tag{6.27}$$

$$-\,2\,\omega\,\cos\lambda\left(z_0\,t + \frac{1}{2}\,V_z\,t^2 - \frac{1}{6}\,g\,t^3 + \int_0^t \delta z(\tau)\,d\tau\right)$$

$$\delta z(t) = 2\,\omega\,\cos\lambda\left(y_0\,t + \frac{1}{2}\,V_y\,t^2 + \int_0^t \delta y(\tau)\,d\tau\right). \tag{6.28}$$

Note that each Coriolis drift can be expressed as an infinite series in powers of ω and that all Coriolis effects vanish when $\omega = 0$ (i.e., a fixed Earth).

We can investigate Coriolis effects in the problem of projectile motion by writing the equations (6.25) to first order in Coriolis effects (i.e., first order in ω):

$$x(t) = x_0 + \dot{x}_0\,t + \dot{y}_0\,(\omega\,\sin\lambda)\,t^2, \tag{6.29}$$

$$y(t) = y_0 + \dot{y}_0\,t - \dot{x}_0\,(\omega\,\sin\lambda)\,t^2 - \omega\,\cos\lambda\left(\dot{z}_0\,t^2 - \frac{g}{3}\,t^3\right), \tag{6.30}$$

$$z(t) = z_0 + \dot{z}_0\,t - \frac{1}{2}\,g\,t^2 + \dot{y}_0\,(\omega\,\cos\lambda)\,t^2. \tag{6.31}$$

These equations can be expressed in vector form as

$$\mathbf{r} = \mathbf{r}_0 + \mathbf{v}_0\,t + \frac{1}{2}\,\mathbf{g}\,t^2 - \boldsymbol{\omega}\times\left(\mathbf{v}_0\,t^2 + \frac{1}{3}\,\mathbf{g}\,t^3\right), \tag{6.32}$$

where $\mathbf{g} \equiv -\,g\,\widehat{\mathbf{z}}$, $\mathbf{v}_0 = \dot{x}_0\,\widehat{\mathbf{x}} + \dot{y}_0\,\widehat{\mathbf{y}} + \dot{z}_0\,\widehat{\mathbf{z}}$ denotes the initial velocity, and $\boldsymbol{\omega}$ is defined in Eq. (6.18) as a function of λ.

6.4.2   *Frenet-Serret-Coriolis Formulas*

Using the Frenet-Serret formulas presented in Sec. A.3 of App. A, we now derive the Frenet-Serret formulation of the equation (6.32) for Coriolis-corrected projectile motion, where the velocity $\dot{\mathbf{r}}$, acceleration $\ddot{\mathbf{r}}$, and jerk $\dddot{\mathbf{r}}$

are

$$\dot{\mathbf{r}} = (\mathbf{v}_0 + \mathbf{g}\,t) \ - \ 2\,\boldsymbol{\omega} \times \left(\mathbf{v}_0\,t + \mathbf{g}\,t^2/2\right), \tag{6.33}$$

$$\ddot{\mathbf{r}} = \mathbf{g} \ - \ 2\,\boldsymbol{\omega} \times (\mathbf{v}_0 + \mathbf{g}\,t), \tag{6.34}$$

$$\dddot{\mathbf{r}} = -\,2\,\boldsymbol{\omega} \times \mathbf{g} \ = \ 2\,g\,\omega\,\cos\lambda\,\widehat{\mathbf{y}}, \tag{6.35}$$

where the Coriolis jerk (6.35) is seen to be eastward, when Eq. (6.18) is used. We now wish to use Eqs. (6.33)-(6.35) to calculate the Frenet-Serret curvature

$$\kappa \equiv \frac{|\dot{\mathbf{r}} \times \ddot{\mathbf{r}}|}{|\dot{\mathbf{r}}|^3} \tag{6.36}$$

and the Frenet-Serret torsion

$$\tau \equiv \frac{\dot{\mathbf{r}} \cdot (\ddot{\mathbf{r}} \times \dddot{\mathbf{r}})}{\kappa^2\,|\dot{\mathbf{r}}|^6} \ = \ \frac{(\dot{\mathbf{r}} \times \ddot{\mathbf{r}}) \cdot \dddot{\mathbf{r}}}{|\dot{\mathbf{r}} \times \ddot{\mathbf{r}}|^2} \tag{6.37}$$

for the Coriolis-corrected equations for projectile motion (6.32).

First, we note that, in the absence of Coriolis effects (i.e., $\omega = 0$), the initial velocity $\mathbf{v}_0$ and the gravitational acceleration $\mathbf{g}$ define a constant plane of motion (i.e., the perpendicular vector $\mathbf{v} \times \mathbf{g} = \mathbf{v}_0 \times \mathbf{g}$ is a constant of motion) and, thus, projectile motion in the absence of Coriolis effects corresponds to planar (two-dimensional) motion. In this planar case, with $\mathbf{r} = (v_0\,\cos\theta\,t)\,\widehat{\mathbf{x}} + (v_0\,\sin\theta\,t - g\,t^2/2)\,\widehat{\mathbf{z}}$, the Frenet-Serret curvature (6.36) is

$$\kappa(t) \ = \ \frac{|\mathbf{v}_0 \times \mathbf{g}|}{|\mathbf{v}_0 + \mathbf{g}\,t|^3} \ = \ \frac{(g/v_0^2)\,\cos\theta}{[\cos^2\theta + (\sin\theta - g\,t/v_0)^2]^{3/2}}. \tag{6.38}$$

The curvature (6.38) has a minimum $\kappa_{\min} = (g/v_0^2)\,\cos\theta$ at $t = 0$ and $t = 2\,(v_0/g)\,\sin\theta$ (when the projectile is on the ground at $z = 0$) and a maximum $\kappa_{\max} = (g/v_0^2)\,\sec^2\theta$ at $t = (v_0/g)\,\sin\theta$ (when the projectile reaches its maximum height). For a planar curve, however, the Frenet-Serret torsion (6.37) is zero since the Coriolis jerk (6.35) vanishes in the absence of Coriolis effects.

Next, in the presence of Coriolis effects ($\omega \neq 0$), Eqs. (6.33)-(6.34) yield

$$\dot{\mathbf{r}} \times \ddot{\mathbf{r}} = \mathbf{v}_0 \times (\mathbf{g} - 2\,\boldsymbol{\omega} \times \mathbf{v}_0) \ + \ \left(\mathbf{v}_0\,t + \mathbf{g}\,t^2/2\right) \times \dddot{\mathbf{r}},$$

$$|\dot{\mathbf{r}}|^2 \simeq |\mathbf{v}_0 + \mathbf{g}\,t|^2 \ - \ \mathbf{v}_0\,t^2 \times \dddot{\mathbf{r}},$$

where we used the Coriolis jerk (6.35). Hence, the Frenet-Serret curvature (6.36) is weakly modified by Coriolis effects. The Frenet-Serret torsion (6.37), on the other hand, is

$$\tau \ \simeq \ -\,2 \left(\frac{\mathbf{v}_0 \times \mathbf{g}}{|\mathbf{v}_0 \times \mathbf{g}|}\right) \cdot \frac{\boldsymbol{\omega} \times \mathbf{g}}{|\mathbf{v}_0 \times \mathbf{g}|}, \tag{6.39}$$

to lowest order in Coriolis effects (i.e., we neglected terms of order ω^2 and higher). If we now introduce the initial velocity conditions

$$\mathbf{v}_0 \equiv v_0 \left[\sin\theta_0 \,(\cos\varphi_0\,\widehat{\mathbf{x}} \,+\, \sin\varphi_0\,\widehat{\mathbf{y}}) \,+\, \cos\theta_0\,\widehat{\mathbf{z}}\right],$$

which are expressed in terms of spherical coordinates $(v_0, \theta_0, \varphi_0)$, we find $\mathbf{v}_0 \times \mathbf{g} = v_0 g \,(\cos\varphi_0\,\widehat{\mathbf{y}} - \sin\varphi_0\,\widehat{\mathbf{x}})$, with $|\mathbf{v}_0 \times \mathbf{g}| = v_0 g\,\sin\theta_0$, and thus the Frenet-Serret Coriolis torsion is $\tau \simeq -2\,(\omega/v_0)\,\cos\varphi_0 / \sin\theta_0$.

6.4.3   *Free-Fall Problem Revisited*

To demonstrate the importance of the Coriolis effects in describing motion relative to Earth, we consider the simple *free-fall* problem, where

$$(x_0, y_0, z_0) \;=\; (0, 0, h) \quad \text{and} \quad (\dot{x}_0, \dot{y}_0, \dot{z}_0) \;=\; (0, 0, 0).$$

Substituting these initial conditions into Eqs. (6.29)-(6.31), we obtain

$$x(t) = 0, \tag{6.40}$$

$$y(t) = \frac{g}{3}\,(\omega\,\cos\lambda)\,t^3, \tag{6.41}$$

$$z(t) = h \,-\, \frac{1}{2}\,gt^2. \tag{6.42}$$

Hence, a free-falling object starting from rest at height h touches the ground $z(T) = 0$ after a time $T = \sqrt{2h/g}$, after which time the object has drifted eastward by a distance of

$$y(T) \;=\; \frac{g}{3}\,(\omega\,\cos\lambda)\,T^3 \;=\; \frac{1}{3}\,\omega\,\cos\lambda\,\sqrt{\frac{8h^3}{g}}.$$

This eastward Coriolis drift is maximum at the equator $(\lambda = 0)$. At a height of 100 m and latitude 45°, for example, we find an eastward drift of 15.5 mm, which is easily measurable.

6.4.4   *Foucault Pendulum*

In 1851, Jean Bernard Léon Foucault (1819-1868) was able to demonstrate, in a classic experiment demonstrating Earth's rotation, the role played by the Coriolis acceleration in his investigations of the motion of a pendulum (of length ℓ and mass m) in the rotating frame of the Earth. His analysis showed that, because of the Coriolis acceleration associated with the rotation of the Earth, the motion of the pendulum exhibits a precession motion whose period depends on the latitude at which the pendulum is located.

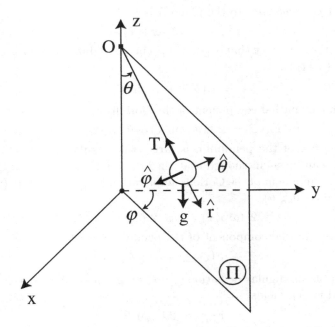

Fig. 6.4   Plane of oscillation of the Foucault pendulum (in the absence of Coriolis effects) spanned by the unit vectors $\widehat{r}$ and $\widehat{\theta}$.

The equation of motion for the pendulum is given as

$$\ddot{\mathbf{r}} = \mathbf{a}_f - 2\,\boldsymbol{\omega} \times \dot{\mathbf{r}}, \qquad (6.43)$$

where $\mathbf{a}_f = \mathbf{g} + \mathbf{T}/m$ is the net fixed-frame acceleration of the pendulum expressed in terms of the gravitational acceleration $\mathbf{g}$ and the string tension $\mathbf{T}$ (see Fig. 6.4). Note that the vectors $\mathbf{g}$ and $\mathbf{T}$ span a plane Π in which the pendulum moves in the absence of the Coriolis acceleration $- 2\boldsymbol{\omega} \times \dot{\mathbf{r}}$. Using spherical coordinates (r, θ, φ) in the rotating frame and placing the origin O of the pendulum system at its pivot point (see Fig. 6.4), the position of the pendulum bob is

$$\mathbf{r} = \ell\,[\,\sin\theta\,(\sin\varphi\,\widehat{\mathbf{x}} + \cos\varphi\,\widehat{\mathbf{y}}) - \cos\theta\,\widehat{\mathbf{z}}\,] = \ell\,\widehat{r}(\theta, \varphi). \qquad (6.44)$$

From this definition, we construct the unit vectors $\widehat{\theta}$ and $\widehat{\varphi}$ as

$$\frac{\partial\widehat{r}}{\partial\theta} = \widehat{\theta}, \quad \frac{\partial\widehat{r}}{\partial\varphi} = \sin\theta\,\widehat{\varphi}, \quad \text{and} \quad \frac{\partial\widehat{\theta}}{\partial\varphi} = \cos\theta\,\widehat{\varphi}. \qquad (6.45)$$

Note that, whereas the unit vectors $\widehat{r}$ and $\widehat{\theta}$ lie on the plane Π, the unit vector $\widehat{\varphi}$ is perpendicular to it and, thus, the equation of motion of the pendulum *perpendicular* to the plane Π is

$$\ddot{\mathbf{r}} \cdot \widehat{\varphi} = -2 \, (\boldsymbol{\omega} \times \dot{\mathbf{r}}) \cdot \widehat{\varphi}, \tag{6.46}$$

where we used the fact that $\widehat{\varphi} \cdot \mathbf{a}_f = 0$. The pendulum velocity is obtained from Eq. (6.44) as

$$\dot{\mathbf{r}} = \ell \left(\dot{\theta} \, \widehat{\theta} + \dot{\varphi} \sin\theta \, \widehat{\varphi} \right), \tag{6.47}$$

so that the azimuthal component of the Coriolis acceleration is

$$-2 \, (\boldsymbol{\omega} \times \dot{\mathbf{r}}) \cdot \widehat{\varphi} = 2 \, \ell \omega \dot{\theta} \, (\sin\lambda \, \cos\theta + \cos\lambda \, \sin\theta \, \sin\varphi).$$

If the length ℓ of the pendulum is large, the angular deviation θ of the pendulum can be small enough that $\sin\theta \ll 1$ and $\cos\theta \simeq 1$ and, thus, the azimuthal component of the Coriolis acceleration is approximately (ignoring $\dot{\theta} \sin\theta \ll \dot{\theta} \cos\theta \simeq \dot{\theta}$)

$$-2 \, (\boldsymbol{\omega} \times \dot{\mathbf{r}}) \cdot \widehat{\varphi} \simeq 2 \, \ell \, (\omega \sin\lambda) \, \dot{\theta}. \tag{6.48}$$

Next, the azimuthal component of the pendulum acceleration is

$$\ddot{\mathbf{r}} \cdot \widehat{\varphi} = \ell \left(\ddot{\varphi} \sin\theta + 2 \, \dot{\theta}\dot{\varphi} \cos\theta \right),$$

which, for small angular deviations ($\theta \ll 1$) and assuming that $\ddot{\varphi} = 0$ (to be verified later), yields

$$\ddot{\mathbf{r}} \cdot \widehat{\varphi} \simeq 2 \, \ell \, (\dot{\varphi}) \, \dot{\theta}. \tag{6.49}$$

By combining these expressions into Eq. (6.46), we obtain an expression for the precession angular frequency of the Foucault pendulum

$$\dot{\varphi} = \omega \sin\lambda \tag{6.50}$$

as a function of latitude λ. As expected, the constant precession motion is clockwise in the Northern Hemisphere and reaches a maximum at the North Pole ($\lambda = 90°$). Note that the precession period of the Foucault pendulum is $(1 \text{ day}/\sin\lambda)$ so that the period is 1.41 days at a latitude of $45°$ or 2 days at a latitude of $30°$.

The more traditional approach to describing the precession motion of the Foucault pendulum makes use of Cartesian coordinates (x, y, z). The motion of the Foucault pendulum in the (x, y)-plane is described in terms of Eqs. (6.43) as

$$\left. \begin{array}{l} \ddot{x} + \omega_0^2 \, x = 2\omega\sin\lambda \, \dot{y} \\ \ddot{y} + \omega_0^2 \, y = -2\omega\sin\lambda \, \dot{x} \end{array} \right\}, \tag{6.51}$$

where $\omega_0^2 = T/m\ell \simeq g/\ell$ and $\dot{z} \simeq 0$ if ℓ is very large. Figure 6.5 shows the numerical solution of Eqs. (6.51) for the Foucault pendulum starting from rest at $(x_0, y_0) = (0, 1)$ with $2 \, (\omega/\omega_0) \sin\lambda = 0.05$ at $\lambda = 45°$. Figure 6.6 shows that, over a finite period of time, the pendulum motion progressively moves from the East-West axis to the North-South axis.

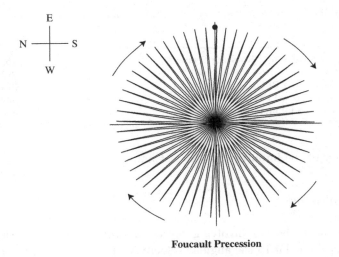

Foucault Precession

Fig. 6.5   Numerical solution of the Foucault-pendulum equations (6.51). The precession motion is clockwise and the initial plane of oscillation is vertical (East-West axis).

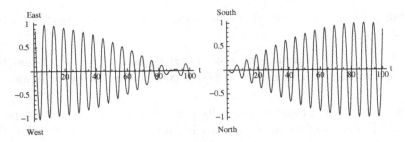

Fig. 6.6   Projection of the Foucault pendulum along East-West and North-South directions. The initial plane of oscillation is along the East-West axis.

6.5   Summary

Chapter 6 studied accelerated motion in a rotating frame, where non-inertial effects due to centrifugal acceleration and the Coriolis acceleration are included. The inclusion of first-order Coriolis corrections in the equations of projectile motion yielded a non-vanishing Frenet-Serret torsion that caused the projectile motion to become non-planar. The Coriolis acceleration associated with Earth's rotation was also shown to play a crucial role in explaining the precession of the Foucault pendulum. Table 6.1 presents

Table 6.1   Summary of Chapter 6: Motion in a Non-Inertial Frame.

Topic	Equation
Time Derivative in a Rotating Frame	(6.4)
Acceleration in a Rotating Frame	(6.10)
Lagrangian in a Rotating Frame	(6.11)
Coriolis-corrected Projectile Motion	(6.32)
Foucault-pendulum Precession	(6.50)

a summary of the important topics of Chapter 6.

6.6   Problems

1. (a) Consider the case involving motion in a rotating frame on the (x, y)-plane perpendicular to the angular velocity vector $\boldsymbol{\omega} = \omega\,\widehat{\mathbf{z}}$ with the potential energy

$$U(\mathbf{r}) \;=\; \frac{1}{2}\,k\,\left(x^2 \,+\, y^2\right).$$

Using the Euler-Lagrange equations (6.13), derive the equations of motion for x and y.

(b) By using the equations of motion derived in Part (a), show that the canonical angular momentum $\ell = \widehat{\mathbf{z}} \cdot (\mathbf{r} \times \mathbf{p})$ is a constant of the motion.

2. If a particle is projected vertically upward to a height h above a point on the Earth's surface at a northern latitude λ, show that it strikes the ground at a point

$$\frac{4\omega}{3}\,\cos\lambda\,\sqrt{\frac{8\,h^3}{g}}$$

to the west.   (Neglect air resistance, and consider only small vertical heights.)

3. For the potential

$$U(\mathbf{r}, \dot{\mathbf{r}}) \;=\; V(r) \,+\, \boldsymbol{\sigma} \cdot \mathbf{r} \times \dot{\mathbf{r}},$$

where $V(r)$ denotes an arbitrary central potential and $\boldsymbol{\sigma}$ denotes an arbitrary constant vector, derive the Euler-Lagrange equations of motion in terms of spherical coordinates.

4. The Lagrangian for the Foucault-pendulum equations (6.51) is

$$L(x, y; \dot{x}, \dot{y}) = \frac{1}{2}\left(\dot{x}^2 + \dot{y}^2\right) - \frac{\omega_0^2}{2}\left(x^2 + y^2\right) + \omega \sin\lambda \, (x\dot{y} - \dot{x}y).$$

(a) By using the polar transformation

$$x(t) = \rho(t)\sin\varphi(t) \quad \text{and} \quad y(t) = \rho(t)\cos\varphi(t),$$

derive the new Lagrangian $L(\rho; \dot{\rho}, \dot{\varphi})$.

(b) Since the new Lagrangian $L(\rho; \dot{\rho}, \dot{\varphi})$ is independent of φ, derive an expression for the conserved momentum $p_\varphi \equiv \partial L/\partial\dot{\varphi}$ and find the Routhian $R(\rho, \dot{\rho}; p_\varphi)$ and the Routh-Euler-Lagrange equation for ρ.

5. We define the complex-valued function $q = y + ix = \ell \sin\theta \, e^{i\varphi}$, so that Eq. (6.51) becomes

$$\ddot{q} + \omega_0^2 q - 2i\,\omega\sin\lambda\,\dot{q} = 0.$$

(a) Insert the eigenfunction $q(t) = \rho\exp(i\Omega t)$ into this equation and find that the solution for the eigenfrequency Ω is

$$\Omega = \omega\sin\lambda \pm \sqrt{\omega^2\sin^2\lambda + \omega_0^2},$$

so that the eigenfunction is

$$q = \rho\exp(i\omega\sin\lambda t)\sin\left(\sqrt{\omega^2\sin^2\lambda + \omega_0^2}\,t\right).$$

(b) Verify that

$$\rho\sin\left(\sqrt{\omega^2\sin^2\lambda + \omega_0^2}\,t\right) = \ell\sin\theta \simeq \ell\,\theta(t),$$

and

$$\varphi(t) = (\omega\sin\lambda)\,t,$$

from which we recover the Foucault pendulum precession frequency (6.50).

6. The equations of motion for a sphere of mass m traveling in air with velocity $\mathbf{v}$ in a constant gravitational field $\mathbf{g}$ are

$$m\,\frac{d\mathbf{v}}{dt} = m\mathbf{g} + \mu\,\boldsymbol{\omega}\times\mathbf{v} - (\beta\,\mathbf{v} + \gamma\,v\mathbf{v}), \tag{6.52}$$

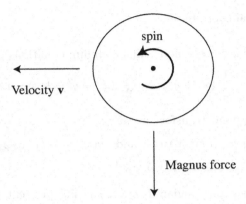

Fig. 6.7 Magnus force acting on a spinning sphere (with spin axis directed out of the page) moving with velocity **v**.

where the second term is the Magnus force (see Fig. 6.7), and the last two terms represent the effects of linear and quadratic air resistance.

The air-resistance coefficients for a sphere of diameter D are $\beta = 1.6 \times 10^{-4} \times D$ and $\gamma = 0.25 \times D^2$ in SI units. The Magnus coefficient for a sphere of diameter D traveling in air (with density 1.168 kg/m^3 at 25 $^{\circ}$C and 1 atm) is $\mu = (\pi^2/8)\, 1.168 \times D^3$. Previous studies have shown that the torque experienced by the sphere during its trajectory is negligible and, thus, the angular velocity $\boldsymbol{\omega} = \omega\,\widehat{\boldsymbol{\omega}}$ is treated as constant in Eq. (6.52).

(a) Show that the Magnus force is energy-conserving (i.e., it does no work on the sphere) and, thus, its sole purpose is to change the direction of motion of the sphere.

(b) For a sphere of diameter $D = 0.07$ m and mass $m = 0.145$ kg (e.g., a baseball) spinning at $\omega = 30$ rev/sec and traveling horizontally at $v = 44$ m/s, compare the magnitudes of the Magnus force (assume that $\boldsymbol{\omega} \perp \mathbf{v}$) and the linear and quadratic air-resistance drag forces with the sphere's weight.

7. To analyze the three-dimensional motion of a baseball described by Eq. (6.52), we use Cartesian coordinates (x, y, z) with the origin located at the pitcher's mound, the x-axis is directed toward home-plate (located approximately 18 m away), the y-axis is directed toward first base, and the z-axis is directed upward (i.e., $\mathbf{g} = -g\,\widehat{z}$). The standard pitches in the arsenal of a baseball pitcher are the fast-ball ($\widehat{\boldsymbol{\omega}} = -\widehat{y}$), the curve-ball

($\widehat{\omega} = \widehat{y}$), and the slider ($\widehat{\omega} = \widehat{z}$).

(a) In general, we may write the rotation unit vector $\widehat{\omega}$ for fast-balls, curve-balls, and sliders in terms of an angle ϕ in the (y, z)-plane as $\widehat{\omega} = \cos\phi\, \widehat{y} + \sin\phi\, \widehat{z}$, where $\phi = \pi$ for a fast-ball, $\phi = \pi/2$ for a slider, and $\phi = 0$ for a curve-ball. Find the general expression for the Magnus-force vector component $\widehat{\omega} \times \mathbf{v}$ for these standard pitches.

(b) For a fast-ball and a curve-ball, i.e., the front end of the baseball is rotating upward (downward) for a fast-ball (curve-ball), determine the direction of the Magnus-force vector component $\widehat{\omega} \times \mathbf{v}$ and discuss qualitatively which ball appears to rise or sink. In addition, if the fast-ball (curve-ball) is initially released horizontally ($\mathbf{v}_0 = v_0\,\widehat{x}$), discuss how its downward motion under the action of gravity ($\dot{z} < 0$) causes it to accelerate (decelerate) along the x-axis under the action of the Magnus force.

(c) Discuss qualitatively the dominant Magnus-force effect on a slider.

Chapter 7

Rigid Body Motion

So far in this textbook, objects have been considered as point-like particles. In the present Chapter, we consider objects known as *rigid bodies* defined as non-deformable discrete collections of massive particles or continuous mass distributions. The inertial properties of such objects are described not only in terms of their masses (i.e., their translational inertia) but also in terms of how their masses are distributed about their instantaneous axis of rotation (i.e., their rotational inertia).

7.1 Inertia Tensor of a Rigid Body

The motion of a rigid body is described in terms of six degrees of freedom. Three degrees of freedom are associated with the translational motion of the center of mass of the rigid body, and three degrees of freedom are associated with the rotational motion about the center of mass. Two of these rotational degrees of freedom are associated with the rotation of an arbitrary point P about the center of mass O, while the third rotational degree of freedom is associated with the rotation of a third point Q about the axis defined by the line OP. Hence, the rotational inertia is described in terms of a 3×3 matrix known as the inertia tensor.

7.1.1 *Discrete Particle Distribution*

We begin our description of rigid body motion by considering the case of a rigid *discrete* particle distribution in which the inter-particle distances are constant. The position of each particle α $(= 1, \cdots, N)$ as measured from a *fixed* laboratory (LAB) frame (using primed coordinates) is

$$\mathbf{r}'_\alpha = \mathbf{R} + \mathbf{r}_\alpha,$$

where $\mathbf{R} \equiv \sum_\alpha (m_\alpha/M)\,\mathbf{r}'_\alpha$ is the position of the center of mass (CM) in the LAB and $\mathbf{r}_\alpha$ is the position of the αth particle in the CM frame, with

$$\left. \begin{array}{c} \sum_\alpha m_\alpha \equiv M \\[2mm] \sum_\alpha m_\alpha \mathbf{r}_\alpha \equiv 0 \end{array} \right\}. \tag{7.1}$$

Using Eq. (6.7), the velocity of particle α in the LAB frame is

$$\mathbf{v}'_\alpha = \dot{\mathbf{R}} + \boldsymbol{\omega} \times \mathbf{r}_\alpha, \tag{7.2}$$

where $\boldsymbol{\omega}$ is the angular velocity vector associated with the rotation of the particle distribution about an axis of rotation which passes through the CM, the velocity $\mathbf{v}_\alpha = \dot{\mathbf{r}}_\alpha = 0$ for each particle of a discrete rigid (non-deformable) body, and $\dot{\mathbf{R}}$ is the CM velocity in the LAB frame. The total linear momentum in the LAB frame is equal to the momentum of the center of mass since

$$\mathbf{P}' = \sum_\alpha m_\alpha \mathbf{v}'_\alpha = M\dot{\mathbf{R}} + \boldsymbol{\omega} \times \left(\sum_\alpha m_\alpha \mathbf{r}_\alpha \right) = M\dot{\mathbf{R}},$$

i.e., the total momentum of a rigid body in its CM frame is zero.

Next, the total angular momentum in the LAB frame is expressed as

$$\mathbf{L}' = \sum_\alpha m_\alpha\,\mathbf{r}'_\alpha \times \mathbf{v}'_\alpha = M\,\mathbf{R} \times \dot{\mathbf{R}} + \sum_\alpha m_\alpha\,\mathbf{r}_\alpha \times (\boldsymbol{\omega} \times \mathbf{r}_\alpha), \tag{7.3}$$

and the kinetic energy of particle α (with mass m_α) in the LAB frame is

$$K'_\alpha = \frac{m_\alpha}{2}\,|\mathbf{v}'_\alpha|^2 = \frac{m_\alpha}{2}\left(|\dot{\mathbf{R}}|^2 + 2\,\dot{\mathbf{R}} \cdot \boldsymbol{\omega} \times \mathbf{r}_\alpha + |\boldsymbol{\omega} \times \mathbf{r}_\alpha|^2 \right).$$

The total kinetic energy $K' = \sum_\alpha K'_\alpha$ of the particle distribution is thus

$$K' = \frac{M}{2}\,|\dot{\mathbf{R}}|^2 + \frac{1}{2}\left\{ \omega^2 \left(\sum_\alpha m_\alpha r_\alpha^2 \right) - \left(\sum_\alpha m_\alpha\,(\boldsymbol{\omega} \cdot \mathbf{r}_\alpha)^2 \right) \right\}, \tag{7.4}$$

where we used Eq. (7.1).

Looking at Eqs. (7.3) and (7.4), we introduce the *inertia tensor* of the particle distribution calculated in the CM frame:[1]

$$\mathbf{I} = \sum_\alpha m_\alpha \left(r_\alpha^2\,\mathbf{1} - \mathbf{r}_\alpha \mathbf{r}_\alpha \right), \tag{7.5}$$

---

[1] We use the dyadic notation $\mathbf{a}\,\mathbf{b} = a_i\,b_j\,\widehat{\mathsf{x}}^i\,\widehat{\mathsf{x}}^j$ to denote a 3×3 matrix constructed out of two 3-dimensional vectors $\mathbf{a}$ and $\mathbf{b}$. Thus the i,j-component of $\mathbf{a}\,\mathbf{b}$ is simply $(\mathbf{a}\,\mathbf{b})_{ij} \equiv a_i\,b_j$, and the transpose of $\mathbf{a}\,\mathbf{b}$ is $(\mathbf{a}\,\mathbf{b})^\mathsf{T} \equiv \mathbf{b}\,\mathbf{a}$.

where $\underline{1}$ denotes the unit tensor (i.e., $\underline{1} = \hat{x}\hat{x} + \hat{y}\hat{y} + \hat{z}\hat{z}$ in Cartesian coordinates). In terms of the inertia tensor (7.5), the angular momentum of a rigid body in the CM frame and its rotational kinetic energy are

$$\mathbf{L} = \mathbf{I}\cdot\boldsymbol{\omega} \quad \text{and} \quad K_{rot} = \frac{1}{2}\,\boldsymbol{\omega}\cdot\mathbf{I}\cdot\boldsymbol{\omega}. \tag{7.6}$$

The inertia tensor (7.5) can also be represented as a matrix

$$\mathbf{I} = \sum_\alpha \begin{pmatrix} m_\alpha\left(y_\alpha^2 + z_\alpha^2\right) & -m_\alpha\left(x_\alpha y_\alpha\right) & -m_\alpha\left(x_\alpha z_\alpha\right) \\ -m_\alpha\left(y_\alpha x_\alpha\right) & m_\alpha\left(x_\alpha^2 + z_\alpha^2\right) & -m_\alpha\left(y_\alpha z_\alpha\right) \\ -m_\alpha\left(z_\alpha x_\alpha\right) & -m_\alpha\left(z_\alpha y_\alpha\right) & m_\alpha\left(x_\alpha^2 + y_\alpha^2\right) \end{pmatrix},$$

where the symmetry property of the inertia tensor ($I^{ji} = I^{ij}$) is readily apparent.

7.1.2 *Parallel-Axes Theorem*

A translation of the origin from which the inertia tensor (7.5) is calculated leads to a different inertia tensor. Let $\mathbf{Q}_\alpha$ denote the position of particle α in a new frame of reference and let $\boldsymbol{\rho} = \mathbf{r}_\alpha - \mathbf{Q}_\alpha$ be the displacement from the old origin to the new origin. The new inertia tensor

$$\mathbf{J} = \sum_\alpha m_\alpha\left(Q_\alpha^2\,\underline{1} - \mathbf{Q}_\alpha\mathbf{Q}_\alpha\right)$$

can be expressed as

$$\mathbf{J} = \sum_\alpha m_\alpha\left(\rho^2\,\underline{1} - \boldsymbol{\rho}\boldsymbol{\rho}\right) + \sum_\alpha m_\alpha\left(r_\alpha^2\,\underline{1} - \mathbf{r}_\alpha\mathbf{r}_\alpha\right)$$
$$+ \left\{ \boldsymbol{\rho}\left(\sum_\alpha m_\alpha\mathbf{r}_\alpha\right) + \left(\sum_\alpha m_\alpha\mathbf{r}_\alpha\right)\boldsymbol{\rho} \right\}$$
$$- \left\{ \boldsymbol{\rho}\cdot\left(\sum_\alpha m_\alpha\mathbf{r}_\alpha\right) \right\}\underline{1}.$$

Using Eq. (7.1), we obtain the Parallel-Axes Theorem:

$$\mathbf{J} = M\left(\rho^2\,\underline{1} - \boldsymbol{\rho}\boldsymbol{\rho}\right) + \mathbf{I}. \tag{7.7}$$

Hence, once the inertia tensor (7.5) is calculated in the CM frame, it can be calculated anywhere else using Eq. (7.7).

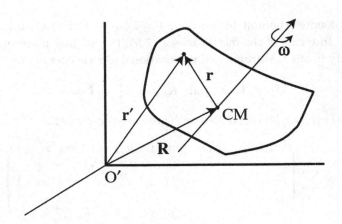

Fig. 7.1   Continuous distribution of mass.

7.1.3   *Continuous Particle Distribution*

For a continuous particle distribution the CM inertia tensor (7.5) becomes

$$\mathbf{I} = \int dm \left(r^2 \, \underline{\mathbf{1}} - \mathbf{r}\,\mathbf{r} \right), \tag{7.8}$$

where $dm(\mathbf{r}) = \rho(\mathbf{r}) \, d^3 r$ is the infinitesimal mass element at point $\mathbf{r}$, with mass density $\rho(\mathbf{r})$.

Consider, for example, the case of a uniform cube of mass M and volume b^3, with $dm = (M/b^3) \, dx \, dy \, dz$. The inertia tensor (7.8) in the LAB frame (with the origin placed at one of its corners) has the components

$$J_{xx} = \frac{M}{b^3} \int_0^b dx \int_0^b dy \int_0^b dz \cdot (y^2 + z^2) = \frac{2}{3} \, M b^2 = J_{yy} = J_{zz}$$

$$J_{xy} = -\frac{M}{b^3} \int_0^b dx \int_0^b dy \int_0^b dz \cdot x\,y = -\frac{1}{4} \, M b^2 = J_{yz} = J_{zx}$$

and thus the inertia matrix for the uniform cube is

$$\mathbf{J} = \frac{M\,b^2}{12} \begin{pmatrix} 8 & -3 & -3 \\ -3 & 8 & -3 \\ -3 & -3 & 8 \end{pmatrix}, \tag{7.9}$$

when the origin is chosen as one of the cube's corners. On the other hand, the inertia tensor calculated in the CM frame (computed with the axes parallel to the axes of the cube) has the components

$$I_{xx} = \frac{M}{b^3} \int_{-b/2}^{b/2} dx \int_{-b/2}^{b/2} dy \int_{-b/2}^{b/2} dz \cdot (y^2 + z^2)$$

$$= \frac{1}{6} M b^2 = I_{yy} = I_{zz},$$

$$I_{xy} = -\frac{M}{b^3} \int_{-b/2}^{b/2} dx \int_{-b/2}^{b/2} dy \int_{-b/2}^{b/2} dz \cdot x\, y$$

$$= 0 = I_{yz} = I_{zx},$$

and thus the CM inertia matrix for the cube is

$$\mathbf{I} = \frac{M b^2}{6} \begin{pmatrix} 1 & 0 & 0 \\ 0 & 1 & 0 \\ 0 & 0 & 1 \end{pmatrix}. \tag{7.10}$$

The displacement vector $\boldsymbol{\rho}$ from the CM point $(b/2, b/2, b/2)$ to the corner $(0, 0, 0)$ of the cube is given as

$$\boldsymbol{\rho} = -\frac{1}{2} \left(b\widehat{\mathbf{x}} + b\widehat{\mathbf{y}} + b\widehat{\mathbf{z}} \right),$$

so that $\rho^2 = 3\, b^2/4$. By using the Parallel-Axis Theorem (7.7), the inertia tensor

$$M \left(\rho^2\, \underline{\mathbf{1}} - \boldsymbol{\rho}\boldsymbol{\rho} \right) = \frac{M b^2}{4} \begin{pmatrix} 2 & -1 & -1 \\ -1 & 2 & -1 \\ -1 & -1 & 2 \end{pmatrix}$$

is added to the CM inertia tensor (7.10) and yields the inertia tensor (7.9).

7.1.4 *Principal Axes of Inertia*

In general, the CM inertia tensor $\mathbf{I}$ can be transformed into a *diagonal* tensor with components given by the eigenvalues I_1, I_2, and I_3 of the inertia tensor. These components (known as principal moments of inertia) are the three roots of the cubic polynomial

$$I^3 - \text{Tr}(\mathbf{I})\, I^2 + \text{Ad}(\mathbf{I})\, I - \text{Det}(\mathbf{I}) = 0, \tag{7.11}$$

obtained from $\text{Det}(\mathbf{I} - I\underline{1}) = 0$, with coefficients

$$\text{Tr}(\mathbf{I}) = I^{11} + I^{22} + I^{33},$$

$$\text{Ad}(\mathbf{I}) = \text{ad}_{11} + \text{ad}_{22} + \text{ad}_{33},$$

$$\text{Det}(\mathbf{I}) = I^{11}\,\text{ad}_{11} - I^{12}\,\text{ad}_{12} + I^{13}\,\text{ad}_{13},$$

where ad_{ij} is the determinant of the two-by-two matrix obtained from $\mathbf{I}$ by removing the ith-row and jth-column from the inertia matrix $\mathbf{I}$. (See Appendix A for additional details on Linear Algebra.)

Each principal moment of inertia I_i represents the moment of inertia calculated about the principal axis of inertia with unit vector $\widehat{e}_i$. Since the inertia tensor $\mathbf{I}$ is real and symmetric, the eigenvalues (I_1, I_2, I_3) are necessarily real and the eigenvectors $(\widehat{e}_1, \widehat{e}_2, \widehat{e}_3)$ form a new orthonormal frame of reference known as the *Body* frame. The unit vectors $(\widehat{e}_1, \widehat{e}_2, \widehat{e}_3)$ are related by a sequence of rotations to the Cartesian CM unit vectors $(\widehat{x}^1, \widehat{x}^2, \widehat{x}^3)$ by the relation

$$\widehat{e}_i = R_{ij}\,\widehat{x}^j, \tag{7.12}$$

where R_{ij} are components of the rotation matrix R. Note that a general rotation matrix has the form

$$\mathsf{R}_n(\alpha) = \widehat{n}\,\widehat{n} + \cos\alpha\,(\underline{1} - \widehat{n}\,\widehat{n}) - \sin\alpha\,\widehat{n} \times \underline{1}, \tag{7.13}$$

where the unit vector $\widehat{n}$ defines the axis of rotation about which an angular rotation of angle α is performed according to the right-hand-rule. The general rotation matrix (7.13) has the following properties. First, the matrix $\mathsf{R}_n(-\alpha)$ is the inverse matrix of $\mathsf{R}_n(\alpha)$, i.e., $\mathsf{R}_n(-\alpha) \cdot \mathsf{R}_n(\alpha) = \underline{1}$. Next, the determinant of $\mathsf{R}_n(\alpha)$ is $+1$ and the eigenvalues of $\mathsf{R}_n(\alpha)$ are $+1$ and $\exp(\pm\,i\alpha)$ (see Appendix A.4 for further details).

A rigid body can be classified into one of three different categories (see Table 7.1) depending on its principal moments of inertia (I_1, I_2, I_3). By denoting as $\mathbf{I}'$ the diagonal inertia tensor calculated in the *body* frame of reference (along the principal axes), we find

$$\mathbf{I}' = \mathsf{R} \cdot \mathbf{I} \cdot \mathsf{R}^\mathsf{T} = \begin{pmatrix} I_1 & 0 & 0 \\ 0 & I_2 & 0 \\ 0 & 0 & I_3 \end{pmatrix}, \tag{7.14}$$

where R^T denotes the transpose of R, i.e., $(\mathsf{R}^\mathsf{T})_{ij} = R_{ji}$. In the body frame, the inertia tensor is, therefore, expressed in dyadic form as

$$\mathbf{I}' = I_1\,\widehat{e}_1\,\widehat{e}_1 + I_2\,\widehat{e}_2\,\widehat{e}_2 + I_3\,\widehat{e}_3\,\widehat{e}_3, \tag{7.15}$$

Table 7.1   Three Categories of Rigid Bodies.

Rigid Body	Principal Moments of Inertia	Example
Asymmetric Top	$I_1 > I_2 > I_3$	textbook
Symmetric Top	$I_1 = I_2 > I_3$	oblate spheroid (pancake)
	$I_1 = I_2 < I_3$	prolate spheroid (football)
Spherical Top	$I_1 = I_2 = I_3$	cube

the angular velocity is defined as

$$\boldsymbol{\omega} = \omega_1 \,\widehat{\mathbf{e}}_1 + \omega_2 \,\widehat{\mathbf{e}}_2 + \omega_3 \,\widehat{\mathbf{e}}_3, \tag{7.16}$$

and the rotational kinetic energy (7.6) is

$$K'_{rot} = \frac{1}{2}\,\boldsymbol{\omega} \cdot \mathbf{I'} \cdot \boldsymbol{\omega} = \frac{1}{2}\left(I_1\,\omega_1^2 + I_2\,\omega_2^2 + I_3\,\omega_3^2\right). \tag{7.17}$$

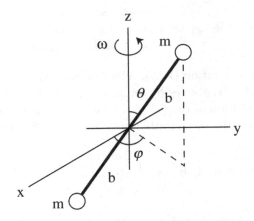

Fig. 7.2   Dumbbell with total mass $M = 2\,m$ and length $2\,b$.

Before proceeding further, we consider the example of a dumbbell composed of a massless rod of total length $2\,b$ (see Fig. 7.2), with equal point masses m placed at the ends of the rod, with positions

$$\mathbf{r}_{\pm} = \pm b\left[\sin\theta\left(\cos\varphi\,\widehat{\mathbf{x}} + \sin\varphi\,\widehat{\mathbf{y}}\right) + \cos\theta\,\widehat{\mathbf{z}}\,\right].$$

The center of mass is obviously located at the origin and the CM inertia tensor is $\mathbf{I} = 2\,m\,b^2\,\overline{\mathbf{I}}$, where

$$\overline{\mathbf{I}} = \begin{pmatrix} 1 - \cos^2\varphi\,\sin^2\theta & -\cos\varphi\,\sin\varphi\,\sin^2\theta & -\cos\varphi\,\cos\theta\,\sin\theta \\ -\cos\varphi\,\sin\varphi\,\sin^2\theta & 1 - \sin^2\varphi\,\sin^2\theta & -\sin\varphi\,\cos\theta\,\sin\theta \\ -\cos\varphi\,\cos\theta\,\sin\theta & -\sin\varphi\,\cos\theta\,\sin\theta & 1 - \cos^2\theta \end{pmatrix}.$$

(7.18)

After some tedious algebra, we find $\mathrm{Tr}(\mathbf{I}) = 4\,mb^2$, $\mathrm{Ad}(\mathbf{I}) = (2\,mb^2)^2$, and $\mathrm{Det}(\mathbf{I}) = 0$, and thus the cubic polynomial (7.11) has the single root $I_3 = 0$ and the double root $I_1 = I_2 = 2\,mb^2$, which makes the dumbbell a symmetric top (see Table 7.1).

The root $I_3 = 0$ clearly indicates that one of the three principal axes is the axis of symmetry of the dumbbell ($\widehat{\mathbf{e}}_3 = \widehat{r}$). The other two principal axes are located on the plane perpendicular to the symmetry axis (i.e., $\widehat{\mathbf{e}}_1 = \widehat{\theta}$ and $\widehat{\mathbf{e}}_2 = \widehat{\varphi}$). From these principal axes, we easily recover the rotation matrix

$$\mathsf{R} = \mathsf{R}_2(-\theta) \cdot \mathsf{R}_3(\varphi) = \begin{pmatrix} \cos\varphi\,\cos\theta & \sin\varphi\,\cos\theta & -\sin\theta \\ -\sin\varphi & \cos\varphi & 0 \\ \cos\varphi\,\sin\theta & \sin\varphi\,\sin\theta & \cos\theta \end{pmatrix},$$

where we used the notation (7.13). This two-step rotation describes, first, the rotation of the (x, y)-axes by an azimuthal angle φ about the z-axis, and, second, a rotation of the $(x', z' = z)$-axes by a polar angle $-\theta$ about the y'-axis. Hence, the new inertia tensor $\mathbf{I}' \equiv \mathsf{R} \cdot \mathbf{I} \cdot \mathsf{R}^\mathsf{T}$ becomes

$$\mathbf{I}' \equiv \mathsf{R} \cdot \mathbf{I} \cdot \mathsf{R}^\mathsf{T} = \begin{pmatrix} 2\,mb^2 & 0 & 0 \\ 0 & 2\,mb^2 & 0 \\ 0 & 0 & 0 \end{pmatrix}, \qquad (7.19)$$

and the principal axes of inertia for the dumbbell (Fig.7.2) are

$$\widehat{\mathbf{e}}_1 = \cos\theta\,(\cos\varphi\,\widehat{\mathbf{x}} + \sin\varphi\,\widehat{\mathbf{y}}) - \sin\theta\,\widehat{\mathbf{z}} = \widehat{\theta},$$
$$\widehat{\mathbf{e}}_2 = -\sin\varphi\,\widehat{\mathbf{x}} + \cos\varphi\,\widehat{\mathbf{y}} = \widehat{\varphi},$$
$$\widehat{\mathbf{e}}_3 = \sin\theta\,(\cos\varphi\,\widehat{\mathbf{x}} + \sin\varphi\,\widehat{\mathbf{y}}) + \cos\theta\,\widehat{\mathbf{z}} = \widehat{r}.$$

Indeed, the principal moment of inertia about the $\widehat{r}$-axis is zero, while the principal moments of inertia about the perpendicular $\widehat{\theta}$-axis and $\widehat{\varphi}$-axis are equally given as $2\,mb^2$.

7.2   Eulerian Rigid-Body Dynamics

Two representations exist for the description of rigid-body dynamics. In the Eulerian representation, the three components of the angular velocity $\boldsymbol{\omega}$ are treated as three independent dynamical variables (representing the three degrees of freedom associated with rotation dynamics). We note that Eulerian rigid-body dynamics does not represent a regular Lagrangian system (as discussed below), while its Hamiltonian formulation is given in terms of a noncanonical Poisson-bracket structure (see problem 8). Eulerian rigid-body dynamics describes rotations in the body frame of reference (i.e., the rigid-body's center-of-mass frame).

In the Lagrangian representation of rigid-body dynamics, the Lagrangian function is expressed in terms of three Eulerian angles, representing the configuration of the rotating right body in space (i.e., in the LAB frame of reference) and their velocities. Lagrangian rigid-body dynamics describes rotations in the space (LAB) frame of reference (this representation is discussed in Sec. 7.3).

7.2.1   *Euler Equations*

The time derivative of the angular momentum $\mathbf{L} = \mathbf{I} \cdot \boldsymbol{\omega}$ in the fixed (LAB) frame is given as

$$\left(\frac{d\mathbf{L}}{dt} \right)_f = \left(\frac{d\mathbf{L}}{dt} \right)_r + \boldsymbol{\omega} \times \mathbf{L} = \mathbf{N},$$

where $\mathbf{N}$ represents the external torque applied to the system (in the LAB frame) and $(d\mathbf{L}/dt)_r$ denotes the rate of change of $\mathbf{L}$ in the rotating frame. By choosing the body frame as the rotating frame, we find

$$\left(\frac{d\mathbf{L}}{dt} \right)_r = \mathbf{I} \cdot \dot{\boldsymbol{\omega}} = (I_1\,\dot{\omega}_1)\,\widehat{\mathbf{e}}_1 + (I_2\,\dot{\omega}_2)\,\widehat{\mathbf{e}}_2 + (I_3\,\dot{\omega}_3)\,\widehat{\mathbf{e}}_3, \qquad (7.20)$$

while

$$\boldsymbol{\omega} \times \mathbf{L} = -\,\widehat{\mathbf{e}}_1 \left[\omega_2\,\omega_3\,(I_2 - I_3) \right] - \widehat{\mathbf{e}}_2 \left[\omega_3\,\omega_1\,(I_3 - I_1) \right]$$
$$-\,\widehat{\mathbf{e}}_3 \left[\omega_1\,\omega_2\,(I_1 - I_2) \right]. \qquad (7.21)$$

Thus the time evolution of the angular momentum in the body frame of reference is described in terms of

$$\left. \begin{array}{l} I_1\,\dot{\omega}_1 - \omega_2\,\omega_3\,(I_2 - I_3) = N_1 \\ I_2\,\dot{\omega}_2 - \omega_3\,\omega_1\,(I_3 - I_1) = N_2 \\ I_3\,\dot{\omega}_3 - \omega_1\,\omega_2\,(I_1 - I_2) = N_3 \end{array} \right\}, \qquad (7.22)$$

which are known as the Euler equations for rigid-body motion. We note that the rate of change of the rotational kinetic energy (7.6) is expressed as

$$\frac{dK_{rot}}{dt} = \boldsymbol{\omega} \cdot \mathbf{I} \cdot \dot{\boldsymbol{\omega}} = \boldsymbol{\omega} \cdot (-\boldsymbol{\omega} \times \mathbf{L} + \mathbf{N}) = \mathbf{N} \cdot \boldsymbol{\omega}. \tag{7.23}$$

Hence, in the absence of external torque ($\mathbf{N} = 0$), not only is the kinetic energy conserved but so is the squared angular momentum $L^2 = \sum_{i=1}^{3} (I_i \omega_i)^2$, as can be verified from Eq. (7.22).

Lastly, in the absence of torque ($\mathbf{N} = 0$), the free-top Euler equations are

$$\left. \begin{array}{l} I_1 \dot{\omega}_1 = \omega_2 \omega_3 \, (I_2 - I_3) \\ I_2 \dot{\omega}_2 = \omega_3 \omega_1 \, (I_3 - I_1) \\ I_3 \dot{\omega}_3 = \omega_1 \omega_2 \, (I_1 - I_2) \end{array} \right\}. \tag{7.24}$$

While these equations possess the free-top Lagrangian

$$L = \frac{1}{2} \, \boldsymbol{\omega} \cdot \mathbf{I} \cdot \boldsymbol{\omega}, \tag{7.25}$$

the free-top Euler equations (7.24) cannot be obtained from an unconstrained variational principle $\delta \int L \, dt = 0$ with arbitrary variations $\delta\boldsymbol{\omega}$. Instead, by using the constrained variation[2]

$$\delta\boldsymbol{\omega} \equiv \dot{\boldsymbol{\xi}} + \boldsymbol{\omega} \times \boldsymbol{\xi}, \tag{7.26}$$

where $\boldsymbol{\xi}$ is an arbitrary vector that vanishes at the end points of the action integral (a dot refers to a time derivative in the rotating frame), we find

$$\delta L = \delta\boldsymbol{\omega} \cdot \mathbf{L} = \left(\dot{\boldsymbol{\xi}} + \boldsymbol{\omega} \times \boldsymbol{\xi} \right) \cdot \mathbf{L}$$

$$= \frac{d}{dt} \left(\boldsymbol{\xi} \cdot \mathbf{L} \right) - \boldsymbol{\xi} \cdot \left(\frac{d\mathbf{L}}{dt} + \boldsymbol{\omega} \times \mathbf{L} \right).$$

When this expression is now inserted in the variational principle $\delta \int L \, dt = 0$, we now readily obtain $d\mathbf{L}/dt + \boldsymbol{\omega} \times \mathbf{L} = 0$, from which Eqs. (7.24) are obtained.

7.2.2   *Euler Equations for a Torque-free Symmetric Top*

As an application of the Euler equations (7.22), we consider the case of the dynamics of a torque-free (or simply free) symmetric top, for which $\mathbf{N} = 0$ and $I_1 = I_2 \equiv I_\perp \neq I_3 \equiv I_\parallel$, i.e., $I_\perp$ denotes the moment of inertia

---

[2]D. D. Holm, J. E. Marsden, and T. S. Ratiu, *The EulerPoincaré Equations and Semidirect Products with Applications to Continuum Theories*, Advances in Mathematics **137**, 1-81 (1998).

associated with rotations that are perpendicular to the axis of symmetry (along the $\hat{e}_3$-axis) and $I_\parallel$ denotes the moment of inertia associated with rotations about the axis of symmetry of the top. Accordingly, the Euler equations (7.22) become

$$\left. \begin{array}{l} I_\perp \, \dot{\omega}_1 = \omega_2 \omega_3 \left(I_\perp - I_\parallel \right) \\ I_\perp \, \dot{\omega}_2 = \omega_3 \omega_1 \left(I_\parallel - I_\perp \right) \\ I_\parallel \, \dot{\omega}_3 = 0 \end{array} \right\} . \tag{7.27}$$

The last Euler equation states that if $I_\parallel \neq 0$, we have $\dot{\omega}_3 = 0$, i.e., ω_3 is a constant of motion.

7.2.2.1 *Body-frame Precession and Body Cone*

Next, using the constant component ω_3, we define the precession frequency

$$\omega_p = \omega_3 \left(\frac{I_\parallel}{I_\perp} - 1 \right), \tag{7.28}$$

which may be positive $(I_\parallel > I_\perp)$ or negative $(I_\parallel < I_\perp)$. The first two Euler equations in Eq. (7.27) become

$$\dot{\omega}_1(t) = - \omega_p \, \omega_2(t) \quad \text{and} \quad \dot{\omega}_2(t) = \omega_p \, \omega_1(t), \tag{7.29}$$

whose general solutions are

$$\omega_1(t) = \omega_\perp \cos(\omega_p t + \phi_0) \quad \text{and} \quad \omega_2(t) = \omega_\perp \sin(\omega_p t + \phi_0), \tag{7.30}$$

where $\omega_\perp$ is a constant amplitude of oscillation and ϕ_0 is an initial phase associated with initial conditions for $\omega_1(t)$ and $\omega_2(t)$.

Since ω_3 and $\omega_\perp^2 = \omega_1^2(t) + \omega_2^2(t)$ are constant, then the magnitude, $\omega = \sqrt{\omega_\perp^2 + \omega_3^2}$, of the angular velocity ω is also a constant. Thus, the angle between ω and $\hat{e}_3$:

$$\alpha \equiv \arccos \left(\frac{\omega_3}{\omega} \right) \tag{7.31}$$

is constant, with

$$\omega_3 = \omega \cos\alpha \quad \text{and} \quad \omega_\perp = \omega \sin\alpha. \tag{7.32}$$

The ω-dynamics simply involves a constant rotation with frequency ω_3 and a precession motion of ω about the $\hat{e}_3$-axis with a precession frequency ω_p. We, therefore, readily find the precession equation for ω in the body frame:

$$\frac{d\omega}{dt} = \omega_p \, \hat{e}_3 \times \omega. \tag{7.33}$$

As a result of this precession motion, the vector ω spans a *body* cone, with half-angle (7.31), and precesses with precession frequency $\omega_p > 0$ if $I_\parallel > I_\perp$ (for a pancake-shaped or oblate-spheroid symmetric top) or $\omega_p < 0$ if $I_\parallel < I_\perp$ (for a cigar-shaped or prolate-spheroid symmetric top).

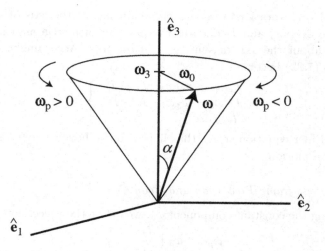

Fig. 7.3   Body cone in the Body frame of a free symmetric top, where the perpendicular angular velocity $\boldsymbol{\omega}_0 \equiv \omega_\perp \left[\cos(\omega_p t)\, \widehat{e}_1 + \sin(\omega_p t)\, \widehat{e}_2\right]$ precesses about the $\widehat{e}_3$-axis.

7.2.2.2   Examples: Oblate and Prolate Spheroids

We now consider two examples of a symmetric top. Our first example is Earth, which (to a good approximation) is an oblate spheroid (i.e., it is flattened at the poles), with moments of inertia (see problem 7)

$$I_\perp = \frac{1}{5} M \left(a^2 + c^2\right) \quad \text{and} \quad I_\parallel = \frac{2}{5} M a^2 > I_\perp, \qquad (7.34)$$

where $2\,c = 12,714$ km is the Pole-to-Pole distance and $2\,a = 12,756$ km is the equatorial diameter, so that

$$\frac{I_\parallel}{I_\perp} - 1 = \frac{a^2 - c^2}{a^2 + c^2} = 0.003298... = \epsilon.$$

The precession frequency (7.28) of the rotation axis of Earth is, therefore, $\omega_p = \epsilon\, \omega_3$, where $\omega_3 = 2\pi$ rad/day is the rotation frequency of the Earth. The precession motion repeats itself every ϵ^{-1} days or 303 days; the actual period is 430 days and the difference is partially due to the non-rigidity of Earth (i.e., it is not a solid of uniform density) and the fact that the Earth is not a pure oblate spheroid. A slower precession motion of approximately 26,000 years is introduced by the combined gravitational effect of the Sun and the Moon on one hand, and the fact that the Earth's rotation axis is at an angle 23.5^o to the Ecliptic plane.

Our second example is the American football, which can be approximated as a prolate spheroid, with the long axis $2c \simeq 28$ cm and the short axis $2a \simeq 17\,\text{cm}$. The moments of inertia (7.34) yield the ratio $I_\parallel / I_\perp = 1 + (17^2 - 28^2)/(17^2 + 28^2) \simeq 0.539$ for the case of a football that is a solid object with uniform mass density ρ. The experimental value[3] of the ratio $(I_\parallel / I_\perp)_{\text{exp}} \simeq 0.604$ shows the effect of mass distribution near the surface of the football (which is not a uniform solid) with thickness much less than the dimensions a and c.

7.2.2.3 *Space-frame Precession and Space Cone*

The fact that a symmetric top is torque-free implies that its rotational kinetic energy K is constant [see Eq. (7.23)] and, hence, $\mathbf{L} \cdot \boldsymbol{\omega} = 2K$ is constant. Since $\mathbf{L}$ itself is constant in magnitude and direction in the LAB (or fixed) frame, we may choose the $\hat{\mathbf{z}}$-axis to be along $\mathbf{L}$ (i.e., $\mathbf{L} = \ell \hat{\mathbf{z}}$). The vector $\boldsymbol{\omega}$, therefore, moves around z-axis along a *space* cone with half-angle β (i.e., $\mathbf{L} \cdot \boldsymbol{\omega} \equiv \ell \omega \cos \beta$), and obeys the precession equation in the space frame:

$$\frac{d\boldsymbol{\omega}}{dt} = \boldsymbol{\Omega} \times \boldsymbol{\omega}, \tag{7.35}$$

where the precession angular velocity is $\boldsymbol{\Omega} = \Omega \hat{\mathbf{z}}$, and the precession frequency Ω is defined below in Eq. (7.37).

In an analysis that will be carried out in Sec. 7.3.4, we will show that the space-cone half-angle is $\beta \equiv |\theta_0 - \alpha|$, where the body-cone half-angle α is defined in Eq. (7.31) and $\theta_0 \equiv \arccos(\hat{\mathbf{z}} \cdot \hat{\mathbf{e}}_3)$ is the angle between the $\hat{\mathbf{z}}$-axis (defined by $\mathbf{L}$) and the $\hat{\mathbf{e}}_3$-axis (about which the body-frame precession takes place; see Fig. 7.3). Since α and β are both constant angles, then so is the angle θ_0; it turns out that the unit vector $\hat{\mathbf{e}}_3$ precesses about the angular momentum $\mathbf{L}$ at the same frequency Ω as the angular velocity $\boldsymbol{\omega}$. Next, we will show that the relation between the body-cone angle α and the angle θ_0 is

$$\tan \theta_0 = \left(\frac{I_\perp}{I_\parallel} \right) \tan \alpha, \tag{7.36}$$

which shows that $\alpha < \theta_0$ for $I_\parallel < I_\perp$ (prolate spheroid) and $\alpha > \theta_0$ for $I_\parallel > I_\perp$ (oblate spheroid). Lastly, we will show that the space-cone precession frequency is $\Omega \equiv \dot{\varphi}$, where

$$\dot{\varphi} = \omega \sqrt{\sin^2 \alpha + (I_\parallel / I_\perp)^2 \cos^2 \alpha} = \omega \sqrt{1 + [(I_\parallel / I_\perp)^2 - 1] \cos^2 \alpha}. \tag{7.37}$$

We will also show in Sec. 7.3.4 that $\dot{\varphi} \equiv \ell / I_\perp$, where $\ell \equiv |\mathbf{L}| = \sqrt{I_\perp^2 (\omega_1^2 + \omega_2^2) + I_\parallel^2 \omega_3^2}$ denotes the magnitude of the angular momentum.

---

[3]See P. J. Brancazio, Am. J. Phys. **55**, 415 (1987).

7.2.3   *Euler Equations for a Free Asymmetric Top*

We now return to the general case of an asymmetric top ($I_1 > I_2 > I_3$) moving under torque-free conditions. Here, the ordering $I_1 > I_2 > I_3$ assumes that the principal axes corresponding to (I_1, I_2, I_3) have lengths (a, b, c) so that $a < b < c$. For example, for an asymmetric rectangular box (of mass M), we find

$$I_1 = \frac{M}{3} \left(c^2 + b^2\right) > I_2 = \frac{M}{3} \left(c^2 + a^2\right) > I_3 = \frac{M}{3} \left(b^2 + a^2\right).$$

Hence, rotation of an asymmetric top about its shortest axis (a) yields the largest moment of inertia (I_1), while rotation about its intermediate (b) and longest (c) axes yield the intermediate moment of inertia (I_2) and the smallest moment of inertia (I_3).

Taking into account this ordering of moments of inertia, Euler's equations (7.24) are written as

$$\left. \begin{array}{l} I_1\,\dot{\omega}_1 = \omega_2\,\omega_3\,(I_2 - I_3) \\ I_2\,\dot{\omega}_2 = -\,\omega_3\,\omega_1\,(I_1 - I_3) \\ I_3\,\dot{\omega}_3 = \omega_1\,\omega_2\,(I_1 - I_2) \end{array} \right\}. \qquad (7.38)$$

As previously mentioned, the torque-free Euler equations (7.38) have two constants of the motion: the conservation laws of kinetic energy

$$\kappa = \frac{1}{2}\left(I_1\,\omega_1^2 + I_2\,\omega_2^2 + I_3\,\omega_3^2\right) \equiv \frac{1}{2}\,I_0\,\Omega_0^2, \qquad (7.39)$$

and (squared) angular momentum

$$\ell^2 = I_1^2\,\omega_1^2 + I_2^2\,\omega_2^2 + I_3^2\,\omega_3^2 \equiv I_0^2\,\Omega_0^2, \qquad (7.40)$$

which are expressed in terms of the parameters

$$I_0 \equiv \ell^2/(2\,\kappa) \quad \text{and} \quad \Omega_0 \equiv 2\,\kappa/\ell. \qquad (7.41)$$

Figure 7.4 shows the numerical solution of the Euler equations (7.38) subject to the initial condition $(\omega_{10}, \omega_{20}, \omega_{30}) = (2, 0, 1)$ for different values of the ratio $I_1/I_3 > 1$ for a fixed ratio $I_2/I_3 > 1$. Note that in the limit $I_1 = I_2$ (corresponding to a symmetric top), the top evolves solely on the (ω_1, ω_2)-plane (top face in Fig. 7.4) at constant ω_3. As I_1 increases from I_2, the asymmetric top exhibits doubly-periodic behavior in the full three-dimensional $\boldsymbol{\omega}$-space until the motion becomes restricted again to the (ω_2, ω_3)-plane (right side face in Fig. 7.4) in the limit $I_1 \gg I_2$. In going from the case $I_1 = I_2$ to the case $I_1 \gg I_2$, it is clear that I_1 must cross

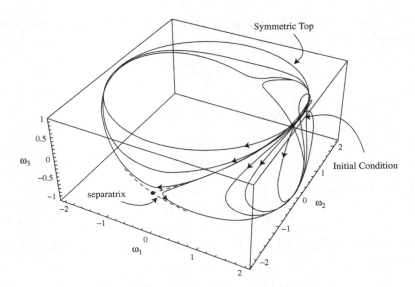

Fig. 7.4 Orbits of an asymmetric top with initial condition $(\omega_{10}, \omega_{20}, \omega_{30}) = (2, 0, 1)$ for different values of the ratio $I_1/I_3 > 1$ for a fixed ratio $I_2/I_3 > 1$.

a critical value that separates the two types of periodic (bounded) motion (see Fig. 7.4). The separatrix solution is defined by the critical value

$$I_{1c} = \frac{I_2}{2} + \sqrt{\frac{I_2^2}{4} + I_3(I_2 - I_3)\left(\frac{\omega_{30}}{\omega_{10}}\right)^2}, \qquad (7.42)$$

at constant I_2 and I_3 and given initial conditions $(\omega_{10}, \omega_{20}, \omega_{30})$. This critical value is obtained by substituting $I_0 = I_2$ in Eqs. (7.39)-(7.40), which become

$$I_2^2\left(\Omega_0^2 - \omega_{20}^2\right) = \begin{cases} I_1 I_2 \omega_{10}^2 + I_3 I_2 \omega_{30}^2 \\ \\ I_1^2 \omega_{10}^2 + I_3^2 \omega_{30}^2 \end{cases}$$

or $I_1^2 - I_1 I_2 + I_3(I_3 - I_2)(\omega_{30}/\omega_{10})^2 = 0$, whose positive solution for I_1 is given by Eq. (7.42).

We note that the existence of two constants of the motion, Eqs. (7.39) and (7.40), for the three Euler equations (7.38) means that we may express the Euler equations in terms of a single equation for ω_2:

$$\omega_2(\tau) \equiv \Omega_2(I_0)\, y(\tau), \qquad (7.43)$$

and introduce the following definitions

$$\omega_1(\tau) \equiv -\,\Omega_1(I_0)\, \sqrt{1 - y^2(\tau)}, \qquad (7.44)$$

$$\omega_3(\tau) \equiv \Omega_3(I_0)\, \sqrt{1 - m\, y^2(\tau)}, \qquad (7.45)$$

where $\tau(I_0) = [(I_1 - I_3)\,\Omega_1\,\Omega_3/(I_2\,\Omega_2)]\,t$ is the dimensionless time, the amplitudes are

$$\Omega_1(I_0) = \Omega_0 \sqrt{\frac{I_0\,(I_0 - I_3)}{I_1\,(I_1 - I_3)}}$$

$$\Omega_2(I_0) = \Omega_0 \sqrt{\frac{I_0\,(I_0 - I_3)}{I_2\,(I_2 - I_3)}}$$

$$\Omega_3(I_0) = \Omega_0 \sqrt{\frac{I_0\,(I_1 - I_0)}{I_3\,(I_1 - I_3)}}$$

and the modulus m in Eq. (7.45) is defined as

$$m(I_0) \equiv \frac{(I_0 - I_3)\,(I_1 - I_2)}{(I_2 - I_3)\,(I_1 - I_0)}. \tag{7.46}$$

By substituting Eqs. (7.43)-(7.45) into the equation for ω_2 in Eq. (7.38), we obtain

$$y'(\tau) = \sqrt{\left(1 - y^2(\tau)\right)\left(1 - m(I_0)\,y^2(\tau)\right)},$$

whose solution $y(\tau) = \text{sn}(\tau|m)$, subject to the initial condition $y(0) = 0$, is expressed in terms of the Jacobi elliptic function $\text{sn}(\tau|m)$ [see Eqs. (B.2)-(B.3)]. By requiring that the modulus (7.46) be positive, the parameter I_0 introduced in Eqs. (7.39)-(7.40) must satisfy $I_3 < I_0 < I_1$ and, hence, $0 \leq m(I_0) \leq 1$ for $I_3 \leq I_0 \leq I_2$ and $m(I_0) > 1$ for $I_2 < I_0 < I_1$ (with $m \to \infty$ as $I_0 \to I_1$).

The solutions for $\omega_1(\tau)$, $\omega_2(\tau)$ and $\omega_3(\tau)$, subject to the initial conditions $(\omega_{10}, \omega_{20}, \omega_{30}) = (-\Omega_1, 0, \Omega_3)$, can thus be expressed in terms of the Jacobi elliptic functions (sn, cn, dn) as [13]

$$(\omega_1,\ \omega_2,\ \omega_3) = (-\Omega_1\,\text{cn}\,\tau,\ \Omega_2\,\text{sn}\,\tau,\ \Omega_3\,\text{dn}\,\tau). \tag{7.47}$$

Lastly, we note that the separatrix (sp) solution of the free asymmetric top (see Fig. 7.4) corresponds to $I_0 = I_2$ (see problem 3), for Eq. (7.47) yields

$$\omega_1^{(\text{sp})}(t) = -\sqrt{\frac{I_2\,(I_2 - I_3)}{I_1\,(I_1 - I_3)}}\,\text{sech}(\Omega_0^*\,t), \tag{7.48}$$

$$\omega_2^{(\text{sp})}(t) = \tanh(\Omega_0^*\,t), \tag{7.49}$$

$$\omega_3^{(\text{sp})}(t) = \sqrt{\frac{I_2\,(I_1 - I_2)}{I_3\,(I_1 - I_3)}}\,\text{sech}(\Omega_0^*\,t), \tag{7.50}$$

where $\Omega_0^* = \Omega_0 \sqrt{(I_1 - I_2)\,(I_2 - I_3)/I_3\,I_1}$ and we used the limiting forms (B.10) of the Jacobi elliptic functions.

7.3 Lagrangian Rigid-Body Dynamics

In this Section, we develop a Lagrangian formulation of rigid-body dynamics represented in a three-dimensional configuration space based on three angles involving three separate rotations about a fixed point in the rigid body.

7.3.1 *Eulerian Angles as Generalized Coordinates*

The Lagrangian description of the physical state of a rotating object with principal moments of inertia (I_1, I_2, I_3) requires the definition of three Eulerian angles (φ, θ, ψ) in the body frame of reference (see Fig. 7.5).

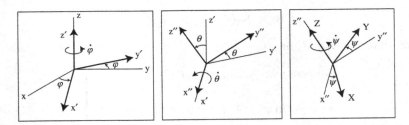

Fig. 7.5   Euler angles (φ, θ, ψ).

The first Eulerian angle φ (left figure in Fig. 7.5) is associated with the rotation of the fixed-frame unit vectors $(\widehat{x}, \widehat{y}, \widehat{z})$ about the z-axis. Through this rotation, we thus obtain the new unit vectors $(\widehat{x}', \widehat{y}', \widehat{z}')$ defined as

$$
\begin{pmatrix} \widehat{x}' \\ \widehat{y}' \\ \widehat{z}' \end{pmatrix} \equiv \overbrace{\begin{pmatrix} \cos\varphi & \sin\varphi & 0 \\ -\sin\varphi & \cos\varphi & 0 \\ 0 & 0 & 1 \end{pmatrix}}^{= \mathsf{R}_3(\varphi)} \cdot \begin{pmatrix} \widehat{x} \\ \widehat{y} \\ \widehat{z} \end{pmatrix} \tag{7.51}
$$

The second Eulerian angle θ (center figure in Fig. 7.5) is associated with the rotation of the unit vectors $(\widehat{x}', \widehat{y}', \widehat{z}')$ about the x'-axis. We thus obtain the new unit vectors $(\widehat{x}'', \widehat{y}'', \widehat{z}'')$ defined as

$$
\begin{pmatrix} \widehat{x}'' \\ \widehat{y}'' \\ \widehat{z}'' \end{pmatrix} \equiv \overbrace{\begin{pmatrix} 1 & 0 & 0 \\ 0 & \cos\theta & \sin\theta \\ 0 & -\sin\theta & \cos\theta \end{pmatrix}}^{= \mathsf{R}_1(\theta)} \cdot \begin{pmatrix} \widehat{x}' \\ \widehat{y}' \\ \widehat{z}' \end{pmatrix} \tag{7.52}
$$

The third Eulerian angle ψ (right figure in Fig. 7.5) is associated with the rotation of the unit vectors $(\widehat{x}'', \widehat{y}'', \widehat{z}'')$ about the z''-axis. We finally obtain the body-frame unit vectors $(\widehat{e}_1, \widehat{e}_2, \widehat{e}_3)$ defined as

$$
\begin{pmatrix} \widehat{e}_1 \\ \widehat{e}_2 \\ \widehat{e}_3 \end{pmatrix} \equiv \overbrace{\begin{pmatrix} \cos\psi & \sin\psi & 0 \\ -\sin\psi & \cos\psi & 0 \\ 0 & 0 & 1 \end{pmatrix}}^{= R_3(\psi)} \cdot \begin{pmatrix} \widehat{x}'' \\ \widehat{y}'' \\ \widehat{z}'' \end{pmatrix}. \tag{7.53}
$$

The resulting three unit vectors defined in Eq. (7.53) will correspond to the three principal axes of inertia defined in Sec. 7.1.4.

Hence, the relation between the fixed-frame unit vectors $\widehat{x}^j = (\widehat{x}, \widehat{y}, \widehat{z})$ and the body-frame unit vectors $\widehat{e}_i = (\widehat{e}_1, \widehat{e}_2, \widehat{e}_3)$ involves the matrix

$$
\mathsf{R} \equiv \mathsf{R}_3(\psi) \cdot \mathsf{R}_1(\theta) \cdot \mathsf{R}_3(\varphi),
$$

such that $\widehat{e}_i = R_{ij}\, \widehat{x}^j$, or

$$
\left. \begin{aligned}
\widehat{e}_1 &= \cos\psi\,\widehat{\perp} + \sin\psi\,(\cos\theta\,\widehat{\varphi} + \sin\theta\,\widehat{z}) \\[4pt]
\widehat{e}_2 &= -\sin\psi\,\widehat{\perp} + \cos\psi\,(\cos\theta\,\widehat{\varphi} + \sin\theta\,\widehat{z}) \\[4pt]
\widehat{e}_3 &= -\sin\theta\,\widehat{\varphi} + \cos\theta\,\widehat{z}
\end{aligned} \right\}, \tag{7.54}
$$

where $\widehat{\varphi} = -\sin\varphi\,\widehat{x} + \cos\varphi\,\widehat{y}$ and $\widehat{\perp} \equiv \widehat{\varphi} \times \widehat{z} = \cos\varphi\,\widehat{x} + \sin\varphi\,\widehat{y}$.

7.3.2 *Angular Velocity in Terms of Eulerian Angles*

According to Fig. 7.5, the angular velocity (7.16) is expressed in terms of the frequencies $(\dot{\varphi}, \dot{\theta}, \dot{\psi})$ as

$$
\boldsymbol{\omega} = \omega_1\widehat{e}_1 + \omega_2\widehat{e}_2 + \omega_3\widehat{e}_3 = \dot{\varphi}\widehat{z} + \dot{\theta}\widehat{x}' + \dot{\psi}\widehat{e}_3. \tag{7.55}
$$

The unit vectors $\widehat{z}$ and $\widehat{x}'$ are written in terms of the body-frame unit vectors $(\widehat{e}_1, \widehat{e}_2, \widehat{e}_3)$ as $\widehat{z} = \sin\theta\,(\sin\psi\,\widehat{e}_1 + \cos\psi\,\widehat{e}_2) + \cos\theta\,\widehat{e}_3$ and $\widehat{x}' = \widehat{x}'' = \cos\psi\,\widehat{e}_1 - \sin\psi\,\widehat{e}_2$. The angular velocity (7.55) can, therefore, be written exclusively in the body frame of reference in terms of the Euler basis vectors (7.54), where the body-frame angular frequencies are

$$
\left. \begin{aligned}
\omega_1 &= \dot{\varphi}\,\sin\theta\,\sin\psi + \dot{\theta}\,\cos\psi \\
\omega_2 &= \dot{\varphi}\,\sin\theta\,\cos\psi - \dot{\theta}\,\sin\psi \\
\omega_3 &= \dot{\psi} + \dot{\varphi}\,\cos\theta
\end{aligned} \right\}. \tag{7.56}
$$

Note that all three frequencies are independent of φ (i.e., $\partial\omega_i/\partial\varphi = 0$), while derivatives with respect to ψ and $\dot{\psi}$ are

$$\frac{\partial\omega_1}{\partial\psi} = \omega_2, \quad \frac{\partial\omega_2}{\partial\psi} = -\omega_1, \quad \text{and} \quad \frac{\partial\omega_3}{\partial\psi} = 0,$$

and

$$\frac{\partial\omega_1}{\partial\dot{\psi}} = 0 = \frac{\partial\omega_2}{\partial\dot{\psi}} \quad \text{and} \quad \frac{\partial\omega_3}{\partial\dot{\psi}} = 1.$$

Equations (7.56) allow us to relate Lagrangian rigid-body dynamics, expressed in terms of the angles (φ, θ, ψ) and their derivatives $(\dot{\varphi}, \dot{\theta}, \dot{\psi})$, to Eulerian rigid-body dynamics expressed in terms of the angular velocity (7.16).

Lastly, we note that the body-frame basis (7.54) rotates in space according to the equations of motion

$$\frac{d\hat{\mathbf{e}}_i}{dt} \equiv \boldsymbol{\omega} \times \hat{\mathbf{e}}_i \rightarrow \begin{cases} d\hat{\mathbf{e}}_1/dt = \dot{\psi}\hat{\mathbf{e}}_2 + \dot{\theta}\sin\psi\,\hat{\mathbf{e}}_3 + \dot{\varphi}\hat{\mathbf{z}} \times \hat{\mathbf{e}}_1 \\ \\ d\hat{\mathbf{e}}_2/dt = -\dot{\psi}\hat{\mathbf{e}}_1 + \dot{\theta}\cos\psi\,\hat{\mathbf{e}}_3 + \dot{\varphi}\hat{\mathbf{z}} \times \hat{\mathbf{e}}_2 \\ \\ d\hat{\mathbf{e}}_3/dt = -\dot{\theta}\left(\cos\theta\,\hat{\varphi} + \sin\theta\,\hat{\mathbf{z}}\right) + \dot{\varphi}\sin\theta\,\hat{\perp} \end{cases} \quad (7.57)$$

which are written in terms of the Eulerian angular velocities $(\dot{\varphi}, \dot{\theta}, \dot{\psi})$.

7.3.3 *Rotational Kinetic Energy of a Symmetric Top*

The rotational kinetic energy (7.6) for a symmetric top (with $I_1 = I_2 \equiv I_\perp \neq I_3 \equiv I_\parallel$) can be written as

$$K = \frac{1}{2}\left[I_\parallel\,\omega_3^2 + I_\perp\left(\omega_1^2 + \omega_2^2\right)\right],$$

or explicitly in terms of the Eulerian angles (φ, θ, ψ) and their time derivatives $(\dot{\varphi}, \dot{\theta}, \dot{\psi})$ as

$$K = \frac{1}{2}\left[I_\parallel\left(\dot{\psi} + \dot{\varphi}\cos\theta\right)^2 + I_\perp\left(\dot{\theta}^2 + \dot{\varphi}^2\sin^2\theta\right)\right]. \quad (7.58)$$

Hence, for the free symmetric top, the Lagrangian $L_0(\theta, \dot{\theta}; \dot{\varphi}, \dot{\psi})$ is simply given by Eq. (7.58).

7.3.3.1   *Lagrangian Dynamics of a Free Symmetric Top*

Since the Eulerian angles φ and ψ are ignorable coordinates, i.e., the torque-free Lagrangian (7.58) is independent of φ and ψ, their canonical angular momenta

$$p_\varphi = \frac{\partial L_0}{\partial \dot{\varphi}} = I_\parallel \, (\dot{\psi} + \dot{\varphi} \, \cos\theta) \, \cos\theta + I_\perp \, \sin^2\theta \, \dot{\varphi}, \qquad (7.59)$$

$$p_\psi = \frac{\partial L_0}{\partial \dot{\psi}} = I_\parallel \, (\dot{\psi} + \dot{\varphi} \, \cos\theta) \equiv I_\parallel \, \omega_3 \qquad (7.60)$$

are constants of motion. By inverting these relations, we obtain the equations of motion for the Eulerian angles (φ, ψ):

$$\dot{\varphi} = \frac{p_\varphi - p_\psi \, \cos\theta}{I_\perp \, \sin^2\theta} \quad \text{and} \quad \dot{\psi} = \omega_3 - \frac{(p_\varphi - p_\psi \, \cos\theta) \, \cos\theta}{I_\perp \, \sin^2\theta}, \qquad (7.61)$$

and the torque-free Lagrangian becomes

$$L_0 = \frac{1}{2} \left[I_\perp \, \dot{\theta}^2 + I_\parallel \, \omega_3^2 + \frac{(p_\varphi - p_\psi \, \cos\theta)^2}{I_\perp \, \sin^2\theta} \right]. \qquad (7.62)$$

By using the constants of motion (7.59)-(7.60), we now construct the Routhian for the free symmetric top:

$$
\begin{aligned}
R_0(\theta, \dot{\theta}; p_\varphi, p_\psi) &\equiv L_0 - p_\varphi \, \dot{\varphi} - p_\psi \, \dot{\psi} \\
&= \frac{I_\perp}{2} \, \dot{\theta}^2 - \frac{p_\psi^2}{2 \, I_\parallel} - \frac{(p_\varphi - p_\psi \, \cos\theta)^2}{2 \, I_\perp \, \sin^2\theta},
\end{aligned}
\qquad (7.63)
$$

where the third term represents the effective potential

$$V_0(\theta; p_\varphi, p_\psi) \equiv \frac{(p_\varphi - p_\psi \, \cos\theta)^2}{2 \, I_\perp \, \sin^2\theta} \geq 0 \qquad (7.64)$$

for the free symmetric top (see Fig. 7.6). Note that the term $p_\psi^2 / 2 I_\parallel$ in Eq. (7.63), which is a constant, can be removed from the Routhian R_0 without changing the equations of motion for θ.

The motion of a free symmetric top can now be described in terms of solutions of the Euler-Lagrange-Routh equation for the Eulerian angle θ:

$$
\begin{aligned}
\frac{d}{dt} \left(\frac{\partial R_0}{\partial \dot{\theta}} \right) = I_\perp \, \ddot{\theta} &= \frac{\partial R_0}{\partial \theta} \\
&= -\frac{(p_\varphi - p_\psi \, \cos\theta)}{I_\perp \, \sin\theta} \, \frac{(p_\psi - p_\varphi \, \cos\theta)}{\sin^2\theta}.
\end{aligned}
\qquad (7.65)
$$

For energies E such that

$$E - \frac{p_\psi^2}{2 \, I_\parallel} > V_0(\theta; p_\varphi, p_\psi), \qquad (7.66)$$

there exists two turning points $\theta_1 \leq \theta_2$ for θ so that the motion is periodic between $\theta_1 \leq \theta \leq \theta_2$. Once $\theta(t)$ is solved for given values of the principal moments of inertia $I_\perp$ and $I_\parallel$, and the invariant canonical angular momenta (7.59)-(7.60), the functions $\varphi(t)$ and $\psi(t)$ are determined from the time integration of Eqs. (7.61).

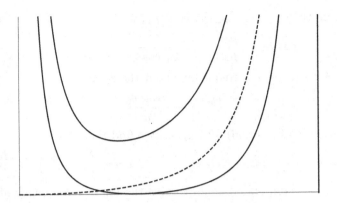

Fig. 7.6 Plots of $V_0(\theta; p_\varphi, p_\psi)$ versus $0 < \theta < \pi$ for the free symmetric top with various ratios of p_φ/p_ψ. The bottom solid curve (with $p_\varphi/p_\psi < 1$) has a minimum $V_{0\min} = 0$ at $\theta_0 = \arccos(p_\varphi/p_\psi)$; the dashed curve (with $p_\varphi = p_\psi$) has a minimum $V_{0\min} = 0$ at $\theta_0 = 0$; and the top solid curve (with $p_\varphi/p_\psi > 1$) has a minimum $V_{0\min} = (p_\varphi^2 - p_\psi^2)/2I_\perp$ at $\theta_0 = \arccos(p_\psi/p_\varphi)$.

7.3.3.2 *Relative Equilibria of the Free Symmetric Top*

Figure 7.6 shows plots of the effective potential (7.64) for various ratios of p_φ/p_ψ. When $p_\varphi/p_\psi < 1$ (bottom solid curve), the potential has a minimum and vanishes at $\theta_0 = \arccos(p_\varphi/p_\psi)$. When $p_\varphi = p_\psi$ (dashed curve), the potential (7.64) becomes $V_0(\theta; p_\varphi, p_\varphi) = (p_\varphi^2/2I_\perp)(1 - \cos\theta)/(1 + \cos\theta)$, which vanishes at $\theta_0 = 0$. When $p_\varphi/p_\psi > 1$ (top solid curve), the potential (7.64) does not vanish but has a minimum $V_{0\min} = (p_\varphi^2 - p_\psi^2)/2I_\perp$ at $\theta_0 = \arccos(p_\psi/p_\varphi)$.

Explicit solutions for the rotation of a free symmetric top can be obtained for the equilibrium case $\dot{\theta} = 0 = \ddot{\theta}$, i.e., when the angle $\theta = \theta_0$ is located at the minimum of the effective potential $V_0(\theta; p_\varphi, p_\psi)$, with $E = V_{0\min} + p_\psi^2/2I_\parallel$. This equilibrium is relative in the sense that the Eulerian angles (φ, ψ) do not have to be constants (i.e., $\dot{\varphi} \neq 0 \neq \dot{\psi}$).

First, in the case $p_\varphi = p_\psi \cos\theta_0 < p_\psi$, we find using Eqs. (7.59)-(7.60):

$$\cos\theta_0 = \frac{p_\varphi}{p_\psi} = \frac{I_\parallel(\dot{\psi} + \dot{\varphi}\cos\theta_0)\cos\theta_0 + I_\perp\sin^2\theta_0\,\dot{\varphi}}{I_\parallel(\dot{\psi} + \dot{\varphi}\cos\theta_0)}$$

$$= \cos\theta_0 + \left(\frac{I_\perp}{I_\parallel}\sin^2\theta_0\right)\frac{\dot{\varphi}}{\omega_3},$$

which yields $\dot{\varphi} = 0$ (for $\theta_0 \neq 0$) and $\dot{\psi} = \omega_3$, which is a constant of the motion for a free symmetric top. Hence, in this case, we find $p_\psi = I_\parallel\dot{\psi}$ and $E = I_\parallel\dot{\psi}^2/2$. For the case $p_\varphi = p_\psi$ (i.e., $\theta_0 = 0$), we find $\dot{\varphi} \equiv \omega_3 - \dot{\psi}$.

Next, in the case $p_\psi = p_\varphi \cos\theta_0 < p_\varphi$, we find using Eqs. (7.59)-(7.60):

$$\cos\theta_0 = \frac{p_\psi}{p_\varphi} = \frac{I_\parallel \,\omega_3}{I_\parallel \,\omega_3 \,\cos\theta_0 + I_\perp \,\sin^2\theta_0 \,\dot\varphi},$$

which yields the angular frequency about the $\widehat{z}$-axis:

$$\dot\varphi = \frac{I_\parallel \,\omega_3}{I_\perp \,\cos\theta_0} = \frac{p_\varphi - I_\perp \,\dot\varphi \,\sin^2\theta_0}{I_\perp \,\cos^2\theta_0} \;\rightarrow\; \dot\varphi = \frac{p_\varphi}{I_\perp}. \tag{7.67}$$

Note that, using Eq. (7.56) for ω_3, we also find the relation

$$\dot\varphi \,\cos\theta_0 = \frac{I_\parallel}{I_\perp} \left(\dot\psi + \dot\varphi \,\cos\theta_0 \right) \;\rightarrow\; \dot\varphi \,\cos\theta_0 \left(1 - \frac{I_\parallel}{I_\perp} \right) = \frac{I_\parallel}{I_\perp} \,\dot\psi,$$

which can be manipulated again to give the angular frequency about the $\widehat{e}_3$-axis:

$$\dot\psi = \left(1 - \frac{I_\parallel}{I_\perp} \right) \omega_3 \equiv -\,\omega_p, \tag{7.68}$$

where the body-cone precession frequency ω_p is defined in Eq. (7.28). Lastly, for the case $p_\psi = p_\varphi \cos\theta_0$, the total energy is $E = p_\psi^2/2I_\parallel + (p_\varphi^2 - p_\psi^2)/2I_\perp$.

7.3.4 *Space-frame Precession and Space-cone Solutions*

We now return to the problem of the space-frame precession of the angular velocity $\boldsymbol{\omega}$ (for a free symmetric top) about the constant angular-momentum axis (i.e., $\mathbf{L} \equiv \ell\widehat{z}$) along a space cone of half-angle β (i.e., $\mathbf{L} \cdot \boldsymbol{\omega} = \ell\omega \,\cos\beta$) discussed in Sec. 7.2.2.3 (see Fig. 7.7 below).

First, we use the fact that, since the angle θ_0 between the principal axis $\widehat{e}_3$ and the angular-momentum $\widehat{z}$-axis is constant, we find from Eq. (7.54):

$$\frac{d\widehat{e}_3}{dt} = \dot\varphi \,\sin\theta \,\widehat{\perp} \equiv \dot\varphi \,\widehat{z} \times \widehat{e}_3. \tag{7.69}$$

Since $\widehat{e}_3$ and $\boldsymbol{\omega}$ precess about the z-axis at the same rate, we find $\Omega \equiv \dot\varphi$ in Eq. (7.35). We note that, since the constant angle θ_0 must be at the minimum of the effective potential $V_0(\theta; p_\varphi, p_\psi)$ defined in Eq. (7.64), then $\dot\varphi$ must be either $\dot\varphi = 0$ (if $p_\varphi < p_\psi$) or $\dot\varphi = p_\varphi/I_\perp$ (if $p_\varphi \geq p_\psi$) [see Eq. (7.67)]. We, therefore, assume the latter since we want $\dot\varphi \neq 0$.

Second, at a time when $\psi = 0$, Eq. (7.56) yields $\omega_1 = 0$ (since $\dot\theta = 0$) and

$$\left. \begin{array}{l} \dot\varphi \,\sin\theta_0 = \omega_2 \equiv \omega \,\sin\alpha \\[2mm] \dot\varphi \,\cos\theta_0 = \omega_3 - \dot\psi \equiv \omega \,\cos\alpha + \omega_p = (I_\parallel/I_\perp)\,\omega \,\cos\alpha \end{array} \right\}, \tag{7.70}$$

where we used Eqs. (7.32) and (7.68). Hence, the ratio of these two equations yields the relation (7.36) between the body-cone half-angle α and θ_0.

Third, making use of the conservation laws of angular-momentum and energy (with $\omega_1 = 0$), we obtain

$$\ell \equiv \omega \sqrt{I_\parallel^2 \cos^2\alpha + I_\perp^2 \sin^2\alpha} = I_\parallel \, \omega \cos\alpha \sqrt{1 + \frac{I_\perp^2}{I_\parallel^2} \tan^2\alpha}$$

$$= I_\parallel \, \omega \cos\alpha \sqrt{1 + \tan^2\theta_0} = I_\parallel \, \omega \cos\alpha \sec\theta_0, \qquad (7.71)$$

and

$$\ell \cos\beta \equiv \mathbf{L} \cdot \hat{\boldsymbol{\omega}} = I_\parallel \, \omega \left(\cos^2\alpha + \frac{I_\perp}{I_\parallel} \sin^2\alpha \right)$$

$$= I_\parallel \, \omega \cos^2\alpha \left(1 + \tan\theta_0 \tan\alpha \right)$$

$$= I_\parallel \, \omega \cos\alpha \sec\theta_0 \left(\cos\alpha \cos\theta_0 + \sin\alpha \sin\theta_0 \right). \qquad (7.72)$$

By dividing Eq. (7.72) by Eq. (7.71), we obtain the relation

$$\cos\beta \equiv \cos(\theta_0 - \alpha) \;\rightarrow\; \beta = \begin{cases} \theta_0 - \alpha & (\text{if } I_\perp > I_\parallel) \\[2mm] \alpha - \theta_0 & (\text{if } I_\parallel > I_\perp) \end{cases} \qquad (7.73)$$

where we used Eq. (7.36).

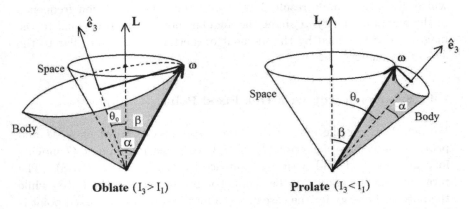

Oblate $(I_3 > I_1)$ **Prolate** $(I_3 < I_1)$

Fig. 7.7 Body (gray) and space cones for the cases of an oblate spheroid ($I_\parallel > I_\perp$) and a prolate spheroid ($I_\parallel < I_\perp$), where $I_\parallel \equiv I_3$ is the moment of inertia for rotations about the axis of symmetry and $I_\perp \equiv I_1 = I_2$ is the moment of inertia for rotations that are perpendicular to the axis of symmetry.

Figure 7.7 shows the body and space cones corresponding to the oblate-spheroid case ($I_\parallel > I_\perp$) and the prolate-spheroid case ($I_\parallel < I_\perp$). The angular momentum $\mathbf{L} = \ell\,\widehat{\mathbf{z}}$ is directed along the z-axis, while the angular velocity $\boldsymbol{\omega}$ lies on the edge of both cones. Both the angular velocity $\boldsymbol{\omega}$ and the unit vector $\widehat{\mathbf{e}}_3$ (also known as the symmetry axis) precess about the z-axis with precession frequency $\dot{\varphi}$ defined in Eq. (7.74); the symmetry cone traced by the precession of the symmetry axis is not shown. As $\boldsymbol{\omega}$ precesses about $\mathbf{L}$, the components $\boldsymbol{\omega}_\perp \equiv \boldsymbol{\omega} - \omega_3\,\widehat{\mathbf{e}}_3$ rotate about the $\widehat{\mathbf{e}}_3$-axis with precession frequency ω_p. Equation (7.21) shows that, for a symmetric top, we find $\boldsymbol{\omega} \times \mathbf{L} = \omega_3\,(I_\parallel - I_\perp)\,(\omega\,\widehat{\mathbf{e}}_1 - \omega_1\,\widehat{\mathbf{e}}_2)$ and, thus, $\widehat{\mathbf{e}}_3 \cdot (\boldsymbol{\omega} \times \mathbf{L}) = 0$. Hence, the three vectors $(\widehat{\mathbf{e}}_3, \boldsymbol{\omega}, \mathbf{L})$ lie on the same plane Π, and the plane Π precesses with angular velocity $\dot{\varphi}$.

Lastly, using Eq. (7.70), we find the space-frame precession frequency

$$\dot{\varphi} = \omega\,\sqrt{\sin^2\alpha + (I_\parallel/I_\perp)^2\,\sin^2\alpha} = \omega\sqrt{1 + [(I_\parallel/I_\perp)^2 - 1]\,\cos^2\alpha}$$

$$\equiv \sqrt{\omega^2 + [(I_\parallel/I_\perp)^2 - 1]\,\omega_3^2}.$$

We now note that Eq. (7.59) yields

$$p_\varphi = I_\perp\,\dot{\varphi} = \sqrt{I_\perp^2\,(\omega_\perp^2 + \omega_3^2) + \left(I_\parallel^2 - I_\perp^2\right)\omega_3^2}$$

$$= \sqrt{I_\perp^2\,\omega_\perp^2 + I_\parallel^2\,\omega_3^2} \equiv |\mathbf{L}|, \tag{7.74}$$

which gives the simple result $\dot{\varphi} \equiv |\mathbf{L}|/I_\perp$, i.e., the precession frequency of the angular velocity $\boldsymbol{\omega}$ about the angular momentum $\mathbf{L}$ is equal to the magnitude $|\mathbf{L}|$ divided by the moment of inertia $I_\perp$ perpendicular to the axis of symmetry.

7.4   Symmetric Top with One Fixed Point

We now consider the case of a spinning symmetric top of mass M and principal moments of inertia $(I_\perp \neq I_\parallel)$ with one fixed point O moving in a gravitational field with constant acceleration g (see Fig. 7.8). The rotational kinetic energy of the symmetric top is given by Eq. (7.58) while the potential energy for the case of a symmetric top with one fixed point is

$$U(\theta) = Mgh\,\cos\theta, \tag{7.75}$$

where h is the distance from the fixed point O to the center of mass (CM) of the symmetric top. It is immediately clear that, without the effects of rotation (i.e., with $\dot{\varphi} = 0 = \dot{\psi}$), this problem is analogous to the problem of an inverted pendulum with an unstable equilibrium at $\theta = 0$ and a stable

Fig. 7.8   Symmetric top with one fixed point.

equilibrium at $\theta = \pi$ (see problem 9). We shall now see that rotational effects allow a spinning symmetric top to remain standing with a range in motion $\theta_1 \le \theta \le \theta_2 < \pi$.

The Routhian for the symmetric top with one fixed point (also known as the *heavy* symmetric top) is

$$R(\theta, \dot{\theta}; p_\varphi, p_\psi) = \frac{1}{2} I_\perp \dot{\theta}^2 - V(\theta; p_\varphi, p_\psi), \qquad (7.76)$$

where the effective potential is

$$V(\theta; p_\varphi, p_\psi) = Mgh \cos\theta + \frac{(p_\varphi - p_\psi \cos\theta)^2}{2 I_\perp \sin^2\theta}. \qquad (7.77)$$

We see that rotational effects (with $p_\varphi \ne 0 \ne p_\psi$) prevent the symmetric top from reaching $\theta = \pi$. On the other hand, the case of $\theta = 0$ (the *sleeping* top) is considered in Sec. 7.4.3, where we show that, if the angular momentum p_ψ is large enough, the equilibrium point $\theta = 0$ is stable (see also Sec. 8.3.4).

The Euler-Lagrange-Routh equation of motion for θ is derived from the Routhian (7.76) as

$$I_\perp \ddot{\theta} = -V'(\theta) = Mgh \sin\theta - \frac{(p_\varphi - p_\psi \cos\theta)(p_\psi - p_\varphi \cos\theta)}{I_\perp \sin^3\theta},$$

where the first term represents the torque due to gravity while the second term is due to free-body rotational dynamics (7.65). The equations of motion for φ and ψ, on the other hand, are still given by Eqs. (7.61) since the gravitational potential (7.75) is independent of the Eulerian angles (φ, ψ).

We, henceforth, use the normalized Euler equations

$$\varphi' = \frac{(b - \cos\theta)}{\sin^2\theta} \quad \text{and} \quad \theta'' = a\sin\theta - \frac{(1 - b\cos\theta)(b - \cos\theta)}{\sin^3\theta}, \quad (7.78)$$

where time has been rescaled $d(\cdots)/d\tau \equiv (\cdots)' = \Omega^{-1} d(\cdots)/dt$, with $\Omega = p_\psi/I_\perp$ and the dimensionless parameters

$$\left.\begin{array}{l} a = Mgh/(p_\psi\,\Omega) \\ b = p_\varphi/p_\psi \\ \mathcal{E} = (E - I_\parallel\,\omega_3^2/2)/(p_\psi\Omega) \end{array}\right\}. \quad (7.79)$$

The dimensionless energy equation for the symmetric top with one fixed point is now expressed as

$$\mathcal{E} = \frac{1}{2}(\theta')^2 + a\cos\theta + \frac{(b - \cos\theta)^2}{2(1 - \cos^2\theta)} \equiv \frac{1}{2}(\theta')^2 + W(\cos\theta),$$

and the turning points $\theta_{1,2}$ are determined from the turning-point equation $\mathcal{E} = W(u)$, with $u = \cos\theta$. Since this turning-point equation yields the cubic equation $(a\,u - \mathcal{E})(1 - u^2) + (b - u)^2 = 0$ for u, we expect to find three roots (u_1, u_2, u_3). However, since the root $|u_3| > 1$ is not allowed for $u = \cos\theta \leq 1$, only two roots $(u_1 = \cos\theta_1 \geq u_2 = \cos\theta_2$, or $\theta_1 \leq \theta_2)$ remain.

Lastly, it can also been shown that the effective potential $W(u)$ has a single real minimum with $|u_0| < 1$ (see Fig. 7.9), which is determined by finding the roots of the quartic equation $a(1 - u^2)^2 - (b - u)(1 - ub) = 0$. We see from Fig. 7.9 that the minimum $u_0(a, b)$ is positive (i.e., $\theta_0 < \pi/2$) when $b > a$ and is negative (i.e., $\pi/2 < \theta_0 < \pi$) when $b < a$. The three remaining roots $(u_\alpha, u_\beta, u_\gamma)$ are a complex-conjugate pair $(u_\alpha, u_\beta = u_\alpha^*)$, where $u_\alpha = 1 = u_\beta$ at $b = 1$ and $a > 1/4$, and $u_\gamma < -1$ for all values of (a, b). The two turning points $u_1 \geq u_2$ satisfy the relation $u_1 \geq u_0 \geq u_2$ with the minimum u_0.

7.4.1 Nutation

We note that the azimuthal equation of motion φ' in Eq. (7.78) can change direction if $b - \cos\theta$ changes sign, which requires that $-1 < b < 1$. The normalized heavy-top equations (7.78) have been integrated for the fixed

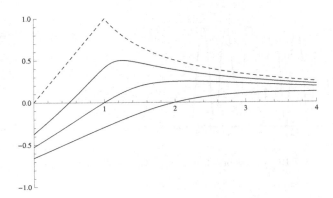

Fig. 7.9 Real root $u_0(a, b)$ corresponding to the single minimum (with $|u| < 1$) of the effective potential $W(u) = au + (b - u)/(1 - u^2)$. The plots of $u_0(a, b)$ versus b are shown from bottom to top with $a = (2, 1, 1/2)$ and $a = 0^+$ (dashed). Note that $u_0 \geq 0$ (i.e., $\theta_0 \leq \pi/2$) when $b \geq a$.

Table 7.2 Nutation in Azimuthal Rotation.

	$b = p_\varphi/p_\psi$	Sign of φ'
Case I	$b > \cos\theta_1 \geq \cos\theta$	$\varphi' > 0$
Case II	$b = \cos\theta_1 \geq \cos\theta$	$\varphi' \geq 0$; $\varphi' = 0$ at θ_1
Case III	$\cos\theta_1 > b = \cos\Theta \geq \cos\theta$	$\varphi' \geq 0$ $(\theta \geq \Theta)$ $\varphi' < 0$ $(\theta_1 \leq \theta < \Theta)$

value $a = 0.1$ and the initial conditions $\theta(0) = \theta_1 = 1$ (the lowest turning point) and $\varphi(0) = 0$, and three cases are shown in Fig. 7.10-7.12 (see also Table 7.2): the normalized heavy-top solutions in the (φ, θ)-plane ($\cos\theta$ increases downward) are shown on the left, and the spherical projection of the normalized heavy-top solutions $(\theta, \varphi) \to (\sin\theta\cos\varphi, \ \sin\theta\sin\varphi, \ \cos\theta)$ are shown on the right, where the initial condition is denoted by a dot ($\bullet$). Note that, since the initial condition is the turning point θ_1, the motion in θ takes place in the range $\theta_1 \leq \theta \leq \theta_2$, and three possible cases exist (see Table 7.2).

In Case I ($b > \cos\theta_1 \geq \cos\theta$), the azimuthal velocity φ' never changes sign and azimuthal precession occurs monotonically. (The case $b < \cos\theta_2 \leq \cos\theta$ also exhibits monotonic precession, with $\varphi' < 0$.) In Case II ($b = \cos\theta_1 \geq \cos\theta$), the azimuthal velocity φ' vanishes at $\theta = \theta_1$ (where θ' also vanishes since θ_1 is a turning point) and the heavy symmetric top exhibits

Fig. 7.10   Orbits of a heavy top – Case I ($b > \cos\theta_1 > \cos\theta$).

a *cusp* at $\theta = \theta_1$. In Case III ($\cos\theta_1 > b = \cos\Theta$), the azimuthal velocity

Fig. 7.11   Orbits of a heavy top – Case II ($b = \cos\theta_1 > \cos\theta$).

φ' vanishes for $\theta = \Theta$ and the heavy symmetric top exhibits a phase of *retrograde* motion (between $\theta_1 \leq \theta < \Theta$).

Fig. 7.12   Orbits of a heavy top – Case III ($\cos\theta_1 > b = \cos\Theta > \cos\theta$).

7.4.2 Slow and Fast Precession

We now investigate the motion of the symmetric top with a fixed point at the minimum angle $\theta_0 \equiv \arccos(u_0)$ for which $W'(u_0) = 0$ and $\mathcal{E} = W(u_0)$. For this case (see Fig. 7.9), when the dimensionless azimuthal frequency

$$\varphi'(u_0) = \frac{b - u_0}{1 - u_0^2} \equiv \Phi'$$

is inserted in

$$0 = W'(u_0) = a - \frac{(b - u_0)(1 - b\,u_0)}{(1 - u_0^2)^2} \equiv a - \Phi'\left(1 - u_0\,\Phi'\right),$$

we obtain the quadratic equation for Φ':

$$u_0\,(\Phi')^2 - \Phi' + a = 0,$$

which has the two solutions

$$\Phi'_\pm = \frac{1}{2\,u_0}\left(1 \pm \sqrt{1 - 4\,u_0\,a}\right). \tag{7.80}$$

These two solutions are real only if the radicand is positive:

$$u_0\,a = Mgh\,I_\perp\,\frac{\cos\theta_0}{p_\psi^2} < \frac{1}{4}. \tag{7.81}$$

This condition is obviously satisfied if $u_0 = \cos\theta_0 < 0$ (i.e., $\theta_0 > \pi/2$). Otherwise, it is satisfied if

$$p_\psi = I_\parallel\,\omega_3 > 2\,I_\perp\,\sqrt{\frac{Mgh}{I_\perp}}\,\cos\theta_0, \tag{7.82}$$

i.e., the spin frequency ω_3 must be large enough to satisfy the condition (7.82). Note that the two solutions (7.80) have the same sign if $0 < 4a\,u_0 < 1$, i.e., $\varphi'(u_0)$ has a fast component Φ'_+ and a slow component Φ'_-. If $u_0 < 0$ ($\theta_0 > \pi/2$), the spinning symmetric top lies below its fixed point, and the precession frequencies have opposite signs $\Phi'_+ < 0 < \Phi'_-$.

7.4.3 The Sleeping Top

As a last topic in our discussion of the problem of the symmetric top with one fixed, we now consider the case where a symmetric top is launched with initial conditions $\theta(0) = \theta_1 \equiv \arccos(b) \neq 0$, corresponding to a turning point, with $\dot{\theta}(0) = \dot{\varphi}(0) = 0$, with $\dot{\psi}(0) \neq 0$. In this case, the invariant canonical momenta are $p_\psi = I_\parallel\,\dot{\psi}(0)$ and $p_\varphi = p_\psi \cos\theta_1$ (i.e., the initial conditions correspond to Fig. 7.11). With these initial conditions, the dimensionless energy equation is

$$\mathcal{E} = \frac{1}{2}(\theta')^2 + a\cos\theta + \frac{(\cos\theta_1 - \cos\theta)^2}{2(1 - \cos^2\theta)} \equiv a\cos\theta_1,$$

which yields the equation

$$\frac{1}{2}(\theta')^2 = (\cos\theta_1 - \cos\theta)\left[a - \frac{(\cos\theta_1 - \cos\theta)}{2(1 - \cos^2\theta)}\right]. \tag{7.83}$$

Furthermore, if we consider the case of the *sleeping* top with the additional initial condition $\theta_1 = 0$, Eq. (7.83) becomes

$$(\theta')^2 = \left(\frac{1 - \cos\theta}{1 + \cos\theta}\right)\left[2a\cos\theta - (1 - 2a)\right],$$

or using the substitution $u = \cos\theta$:

$$(\Psi u')^2 = (1 - u)^2(u + 1 - \Psi^2), \tag{7.84}$$

where $\Psi^2 \equiv 1/2a = p_\psi^2/(2Mgh I_\perp)$. The solution of Eq. (7.84) for the special case $\Psi^2 = 2$ is $u(\tau) = 1 + 8\tau^{-2}$, which asymptotically approaches $u = 1$ as $\tau \to \infty$. Since this solution yields $u > 1$, it is discarded as unphysical because u must satisfy $u = \cos\theta \leq 1$. We look for solutions with $\Psi^2 \neq 2$ below.

The sleeping-top equation (7.84) has the following turning points (where $u' = 0$): $u_0 = 1$ (which is a double root) and $u_1 \equiv \Psi^2 - 1$. The equilibrium points for the sleeping top, on the other hand, are obtained from the acceleration equation

$$2\Psi^2 u'' = (1 - u)\left[(2\Psi^2 - 1) - 3u\right], \tag{7.85}$$

where the right side vanishes for $u_0 = 1$ and $u_2 = (2\Psi^2 - 1)/3$. Hence, $u_0 = 1$ is both a turning point and an equilibrium point of the sleeping top. For $\Psi^2 > 2$, the turning points (u_0, u_1) and equilibrium points (u_0, u_2) are ordered as $u_0 = 1 < u_2 < u_1$, while for $\Psi^2 < 2$, they are ordered as $u_1 < u_2 < u_0 = 1$. In Sec. 8.3.4, we investigate the stability of the sleeping top and show that the equilibrium point $u_0 = 1$ is stable if $\Psi^2 > 2$, which also automatically satisfies Eq. (7.82), while u_2 is stable for $\Psi^2 < 2$.

A sleeping top describes a symmetric top that is spinning about its axis of symmetry in the presence of a gravitational field (i.e., $\Psi \neq 0$). When the symmetric top is perfectly upright ($\theta = 0$), the gravitational torque vanishes ($Mgh\sin\theta = 0$) and the top remains upright. If the spin frequency is large enough (i.e., $\Psi^2 > 2$), a slight departure from $\theta = 0$ does not cause the top to fall but instead it returns to its upright position.

7.4.3.1 *Exact Solutions in Terms of the Weierstrass Function*

We now look for exact solutions of the sleeping-top equation (7.84) for arbitrary values of Ψ. Since the right side of Eq. (7.84) is a cubic polynomial in u, we may look for solutions expressed in terms of the Weierstrass elliptic function. First, we use

$$u(\tau) = w(\tau) + \frac{1}{3}\left(\Psi^2 + 1\right)$$

to transform Eq. (7.84) into the Weierstrass differential equation (B.19):

$$\left(2\Psi \frac{dw}{d\tau}\right)^2 = 4w^3 - g_2 w - g_3 \equiv f(w),$$

where $g_2 = 3B^2$, and $g_3 = B^3$, with $B \equiv \frac{2}{3}(\Psi^2 - 2)$.

Next, the roots of the cubic polynomial $f(w)$ are $w_1 = B$ (or $u_1 = \Psi^2 - 1$) and $w_{2,3} = -B/2$ (or $u_{2,3} = u_0 \equiv 1$). Because the discriminant $\Delta = g_2^3 - 27 g_3^2 = 0$ (due to the existence of a double root), the solution of the sleeping-top equation (7.85) corresponds to the singular Weierstrass case [8], which can be given in terms of elementary (trigonometric and hyperbolic-trigonomentric) functions as follows.

7.4.3.2 *Exact Solutions in Terms of Elementary Functions*

An exact solution of the sleeping-top equation (7.84) for $\Psi^2 > 2$ is expressed as

$$u = 1 + (\Psi^2 - 2) f^{-2}(z), \tag{7.86}$$

where $z = (\tau/2\Psi)\sqrt{\Psi^2 - 2}$ and the function $f(z)$ satisfies the differential equation

$$\left(\frac{df(z)}{dz}\right)^2 = \left(1 - f^2(z)\right), \tag{7.87}$$

which is obtained by combining the parts of Eq. (7.84) for $\Psi^2 > 2$:

$$(\Psi u')^2 = (\Psi^2 - 2)^3 (df/dz)^2 f^{-6}(z),$$
$$(1 - u)^2 = (\Psi^2 - 2)^2 f^{-4}(z),$$
$$(u + 1 - \Psi^2) = (\Psi^2 - 2)\left(1 - f^2(z)\right) f^{-2}(z).$$

Periodic solutions of Eq. (7.87) are $f(z) = \cos z$ or $\sin z$, but only $\cos z$ is finite at $z = 0$ (i.e., $\tau = 0$). Thus, the general solution for $\Psi^2 > 2$ is

$$u_+(\tau) = 1 + (\Psi^2 - 2) \sec^2\left(\frac{\tau}{2\Psi}\sqrt{\Psi^2 - 2}\right), \tag{7.88}$$

which satisfies the initial condition $u(0) = \Psi^2 - 1 = u_1$. Since this solution yields $u(\tau) > 1$, however, it must be rejected as unphysical because u must satisfy $u \equiv \cos\theta \le 1$. We will explore an approximate solution of Eq. (7.84) for $\Psi^2 > 2$ in Sec. 8.3.4, by studying the sleeping top as a two-degree-of-freedom dynamical problem.

Fig. 7.13   Plot of Eq. (7.89) between $u_-(-\infty) = 1$ (upper dashed line) and $u_-(0) = \Psi^2 - 1$ (lower dashed line).

An exact solution the sleeping-top equation (7.84) for $\Psi^2 < 2$ is obtained from Eq. (7.88) by writing $z = i\,(\tau/2\Psi)\,\sqrt{2 - \Psi^2}$ and using the identity $\sec(i\,\cdot) = \mathrm{sech}(\cdot)$:

$$u_-(\tau) = 1 + (\Psi^2 - 2)\,\mathrm{sech}^2\left(\frac{\tau}{2\Psi}\,\sqrt{2 - \Psi^2}\right), \qquad (7.89)$$

which satisfies the initial condition $u(0) = \Psi^2 - 1 < 1$. Figure 7.13 shows a plot of Eq. (7.89) for $\tau \le 0$, which asymptotically connects the two turning points $u_0 = 1 \equiv u_-(-\infty)$ and $u_1 = \Psi^2 - 1 \equiv u_-(0)$ of Eq. (7.84).

7.5   Summary

Chapter 7 investigated the rotation of a rigid body either in a three-dimensional body-frame space or a three-dimensional space-frame configuration space (or six-dimensional phase space). In the body frame, the rigid-body rotation was described in terms of the three components $(\omega_1, \omega_2, \omega_3)$ of the angular rotation frequency $\boldsymbol{\omega}$ defined in terms of the principal axes $(\widehat{e}_1, \widehat{e}_2, \widehat{e}_3)$ of the inertia tensor. In the space frame, the rotation of the rigid body was described in terms of the three Eulerian angles (φ, θ, ψ), and their

Table 7.3   Summary of Chapter 7: Rigid Body Motion.

Topic	Equation
Inertia Tensor for Discrete and Continuous Mass Distributions	(7.5) & (7.8)
Parallel-axes Theorem	(7.7)
Principal Axes and Moments of Inertia	(7.15)
Euler Equations with Torque	(7.22)
Euler Equations without Torque	(7.24)
Euler Equations for Torque-free Symmetric Top	(7.27)
Body Cone and Body-cone Angle	(7.31)
Torque-free Asymmetric Top	(7.43)-(7.45)
Eulerian Principal Axes	(7.54)
Body-frame Angular Frequencies	(7.56)
Routhian for Free Symmetric Top	(7.63)
Relative Equilibria for Free Symmetric Top	(7.67)-(7.68)
Space Cone and Space-cone Angle	(7.73)
Space-frame Precession	(7.74)
Euler-Lagrange-Routh Equation for Heavy Symmetric Top	(7.78)
Sleeping-top Equation and Solutions	(7.84) & (7.88)-(7.89)

associated velocities $(\dot{\varphi}, \dot{\theta}, \dot{\psi})$. Because the Lagrangian for a free symmetric top was independent of the two Eulerian angles (φ, ψ), their canonically-conjugate momenta p_φ and p_ψ could be used to reduce the rotation of the free symmetric top to the motion in θ. The precession motion of a free symmetric top could be analyzed in either the body frame or the space frame, in which the two axes $\widehat{\omega} \equiv \boldsymbol{\omega}/|\boldsymbol{\omega}|$ and $\widehat{e}_3$ were seen to precess about the angular momentum axes $\widehat{z} \equiv \mathbf{L}/|\mathbf{L}|$. The addition of a gravitational torque on the motion of a symmetric top with one fixed point did not break the invariance of (p_φ, p_ψ), but instead introduced conditions for the stability of a sleeping top that are further analyzed in the next Chapter. Table 7.3 presents a summary of the important topics of Chapter 7.

7.6   Problems

1. Consider a thin homogeneous rectangular plate of mass M and area ab that lies on the (x, y)-plane (with $0 \le x \le a$ and $0 \le y \le b$). Here, the infinitesimal mass dm is defined as $dm \equiv (M/ab)\, \delta(z)\, d^3x$, with

$$\int \delta(z)\, dz = 1 \quad \text{and} \quad \int z\, \delta(z)\, dz = 0 = \int z^2\, \delta(z)\, dz = 1.$$

(a) Show that the inertia tensor (calculated in the reference frame with its origin at one corner of the plate) takes the form

$$\mathbf{J} = \begin{pmatrix} A & -C & 0 \\ -C & B & 0 \\ 0 & 0 & A+B \end{pmatrix},$$

and find suitable expressions for A, B, and C in terms of M, a, and b.

(b) Show that by performing a rotation of the coordinate axes about the z-axis through an angle θ, the new inertia tensor is

$$\mathbf{J}'(\theta) = \mathsf{R}_3(\theta) \cdot \mathbf{J} \cdot \mathsf{R}_3^\top(\theta) = \begin{pmatrix} A' & -C' & 0 \\ -C' & B' & 0 \\ 0 & 0 & A'+B' \end{pmatrix},$$

where

$$A' = A \cos^2\theta + B \sin^2\theta - C \sin 2\theta$$
$$B' = A \sin^2\theta + B \cos^2\theta + C \sin 2\theta$$
$$C' = C \cos 2\theta - \frac{1}{2}(B-A)\sin 2\theta.$$

(c) Show that $A' + B' = A + B$ and
$$(A'+B')(A'B'-C'^2) = (A+B)(AB-C^2),$$
i.e., the trace and determinant of $\mathbf{J}$ are invariant.

(d) When

$$\theta = \frac{1}{2}\arctan\left(\frac{2C}{B-A}\right),$$

the off-diagonal component C' vanishes and the x'-axis and y'-axis become principal axes. Calculate expressions for A' and B' in terms of M, a, and b for this particular angle.

(e) Calculate the inertia tensor $\mathbf{I}$ in the CM frame by using the Parallel-Axis Theorem and show that

$$I_1 = \frac{M b^2}{12}, \quad I_2 = \frac{M a^2}{12}, \quad \text{and} \quad I_3 = \frac{M}{12}\left(b^2 + a^2\right).$$

2. Derive the moment of inertia (7.18).

3. (a) The Euler equation for an asymmetric top ($I_1 > I_2 > I_3$) with $L^2 = 2 I_2 K$ is $\dot{\omega}_2 = \alpha\left(\Omega^2 - \omega_2^2\right)$, where

$$\Omega^2 = \frac{2K}{I_2} \quad \text{and} \quad \alpha = \sqrt{\left(1 - \frac{I_2}{I_1}\right)\left(\frac{I_2}{I_3} - 1\right)}$$

Solve for $\omega_2(t)$ with the initial condition $\omega_2(0) = 0$.

(b) Use the solution $\omega_2(t)$ found in Part (a) to find the solutions $\omega_1(t)$ and $\omega_3(t)$ given by Eqs. (7.44) and (7.45) for $m = 1$.

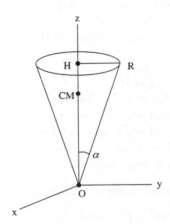

Fig. 7.14   Problem 4.

4. Consider a circular cone of height H and base radius $R = H \tan \alpha$ with uniform mass density $\rho = 3\, M/(\pi H R^2)$.

(a) Show that the non-vanishing components of the inertia tensor **I** calculated from the vertex O of the cone are

$$J_{xx} = J_{yy} = \frac{3}{5} M \left(H^2 + \frac{R^2}{4} \right) \quad \text{and} \quad J_{zz} = \frac{3}{10} M R^2$$

(b) Show that the principal moments of inertia calculated in the CM frame (located at a height $h = 3H/4$ on the symmetry axis) are

$$I_\perp = \frac{3}{20} M \left(R^2 + \frac{H^2}{4} \right) \quad \text{and} \quad I_\parallel = \frac{3}{10} M R^2$$

5. Show that the Euler basis vectors $(\hat{e}_1, \hat{e}_2, \hat{e}_3)$ are determined by Eq. (7.54).

6. (a) Use Euler's equations (7.22) to find the torque needed to rotate a rectangular plate of sides a and b about a diagonal with constant angular velocity $\boldsymbol{\omega}$.

(b) Show that this torque vanishes if the rectangular plate is a square (i.e., if $a = b$).

7. As a result of its daily rotation, the shape of Earth is approximated as an oblate spheroid with equatorial radius $a = 6,378$ km and polar radius $c = 6,357$ km. The gravitational potential is expressed in spherical coordinates as

$$\Phi(r,\theta) = -\frac{GM}{r} + \frac{G}{2\,r^3}\left(I_\| - I_\perp\right)\left(3\cos^2\theta - 1\right) - \frac{1}{2}\,\omega^2\,r^2\,\sin^2\theta,$$

where M denotes Earth's mass, ω denotes its rotation angular frequency, and G is the gravitational universal constant. The first term is the gravitational potential for a spherical non-rotating Earth, the second term represents the correction due to its non-spherical shape, and the third term represents the effects of Earth's rotation.

(a) Show that the principal moments of inertia are

$$I_\perp = \frac{M}{5}\left(a^2 + c^2\right) \quad \text{and} \quad I_\| = \frac{2\,M}{5}\,a^2 > I_\perp.$$

(b) Compute the gravitational acceleration $\mathbf{g} \equiv -\nabla\Phi$, and calculate its magnitude on the equator $(r,\theta) = (a, \pi/2)$ and at the north pole $(r,\theta) = (c, 0)$.

(c) Compare the directions and magnitudes of the corrections to the gravitational acceleration due to the centrifugal term and the non-spherical term.

8. In the absence of external torque, the Euler equations (7.22) can be written as

$$\frac{dL_i}{dt} = \{L_i,\, K\} = -\widehat{\mathsf{i}}\cdot\boldsymbol{\omega}\times\mathbf{L} = -\,\epsilon_{ijk}\,\omega_j\left(I_k\,\omega_k\right),$$

where the *Poisson bracket* $\{\ ,\ \}$ is defined in terms of two arbitrary functions $F(\mathbf{L})$ and $G(\mathbf{L})$ as

$$\{F,\, G\} = -\,\mathbf{L}\cdot\frac{\partial F}{\partial\mathbf{L}}\times\frac{\partial G}{\partial\mathbf{L}}.$$

Hence a general function $F(\mathbf{L})$ of angular momentum evolves according to the Hamilton's equation

$$\frac{dF}{dt} = \{F,\, K\} = -\,\frac{\partial F}{\partial\mathbf{L}}\cdot\boldsymbol{\omega}\times\mathbf{L}.$$

(a) Show that any function of $|\mathbf{L}|$ is a constant of the motion for rigid body dynamics.

(b) Show that the Poisson bracket satisfies the Jacobi identity

$$\{F, \{G, H\}\} + \{G, \{H, F\}\} + \{H, \{F, G\}\} = 0,$$

for three arbitrary functions F, G, and H.

9. In the absence of rotation ($\dot{\varphi} = 0 = \dot{\psi}$), the heavy-top equation of motion for the angle θ is

$$I_\perp \ddot{\theta} = Mgh \, \sin\theta, \tag{7.90}$$

which represents the equation for an inverted pendulum. If we choose $\theta = \pi$ (the stable equilibrium point) as the point of lowest potential energy, then the energy equation for this problem is expressed as

$$E = \frac{1}{2} I_\perp \, \dot{\theta}^2 + Mgh \, (1 + \cos\theta) = \frac{1}{2} I_\perp \, \dot{\theta}^2 + 2Mgh \left(1 - \sin^2(\theta/2)\right).$$

(a) Show that, for the "initial" conditions $\theta(0) = \pi$ and $\dot{\theta}(0) = -2\Omega$, where $\Omega^2 \equiv Mgh/I_\perp$ (i.e., the energy is $E = 2Mgh$), the solution for a "falling" top is

$$\theta(t) = 2 \arcsin\left[\, \text{sech}(\Omega t)\,\right], \tag{7.91}$$

with $\dot{\theta}(t) = -2\Omega \, \text{sech}(\Omega t)$. Note that the solution of Eq. (7.90) can only be found with initial conditions $\theta(0)$ and $\dot{\theta}(0)$ that are different from the unstable separatrix point $(\theta, \dot{\theta}) = (0, 0)$, since an unstable separatrix point is approached as an asymptotic limit $t \to \pm\infty$.

(b) Show that Eq. (7.91) is a solution of Eq. (7.90).

Chapter 8

Normal-Mode Analysis

8.1 Stability of Equilibrium Points

The nonlinear (one-dimensional) force equation $m\ddot{x} = -V'(x)$ has equilibrium points (labeled x_0) where $V'(x_0)$ vanishes. The stability of the equilibrium point x_0 is determined by the sign of $V''(x_0)$: the equilibrium point x_0 is stable if $V''(x_0) > 0$ or unstable if $V''(x_0) < 0$. Great care must be taken, of course, in finding the solution $x(t)$ for the unstable orbit, which must satisfy $\lim_{t\to\pm\infty} x(t) = x_0$ (i.e., one should not look for a solution in the vicinity of x_0 because it is a separatrix point; see Sec. 7.4.3.2 and problem 9 of Chap. 7). We now consider two examples from previous chapters.

8.1.1 *Bead on a Rotating Hoop*

As a first example, we return to the problem of a bead of mass m sliding freely on a hoop of radius R rotating with angular velocity Ω in a constant gravitational field with acceleration g (see Chap. 2). The Lagrangian for this system is

$$L(\theta, \dot{\theta}) = \frac{m}{2} R^2\dot{\theta}^2 + \left(\frac{m}{2} R^2\Omega^2 \sin^2\theta + mgR \cos\theta\right) = \frac{m}{2} R^2\dot{\theta}^2 - V(\theta),$$

where $V(\theta)$ denotes the effective potential, and the Euler-Lagrange equation (2.40) for θ is

$$mR^2\ddot{\theta} = -V'(\theta) = -mR^2\Omega^2 \sin\theta\,(\nu - \cos\theta), \tag{8.1}$$

where $\nu = g/(R\Omega^2)$.

The equilibrium points of Eq. (8.1) are $\theta = 0$ (for all values of ν) and $\theta = \arccos(\nu)$ if $\nu < 1$. The stability of the equilibrium point $\theta = \theta_0$ is determined by the sign of

$$V''(\theta_0) = mR^2\Omega^2 \left[\nu \cos\theta_0 - (2\cos^2\theta_0 - 1) \right].$$

Hence,

$$V''(0) = mR^2\Omega^2 (\nu - 1) \tag{8.2}$$

is positive (i.e., $\theta = 0$ is stable) if $\nu > 1$ or negative (i.e., $\theta = 0$ is unstable) if $\nu < 1$. In the latter case, when $\nu < 1$ and the second equilibrium point $\theta_0 = \arccos(\nu)$ is allowed, we find

$$V''(\theta_0) = mR^2\Omega^2 \left[\nu^2 - (2\nu^2 - 1) \right] = mR^2\Omega^2 (1 - \nu^2) > 0, \tag{8.3}$$

and thus the equilibrium point $\theta_0 = \arccos(\nu)$ is stable when $\nu < 1$.

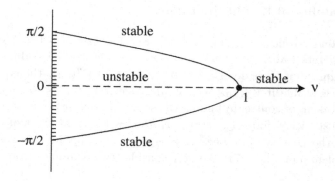

Fig. 8.1   Bifurcation diagram for the bead on a rotating-hoop problem.

Figure 8.1 shows the bifurcation diagram for the problem of a bead on a rotating hoop. Here, we see that for $\nu > 1$, a single stable equilibrium exists at $\theta = 0$. For $\nu < 1$, however, the equilibrium point $\theta = 0$ is unstable and new stable equilibrium points appear at $\theta = \pm \arccos(\nu)$. The critical value $\nu \equiv g/(R\Omega^2) = 1$ is the bifurcation point for the bead on a rotating hoop.

8.1.2   *Circular Orbits in Central-Force Fields*

As our second example, we consider the radial force equation

$$\mu\ddot{r} = \frac{\ell^2}{\mu r^3} - k r^{n-1} = -V'(r),$$

studied in Chap. 4 for a central-force field $F(r) = -k r^{n-1}$ (here, μ is the reduced mass of the system, the azimuthal angular momentum ℓ is a constant of the motion, and the constant $k > 0$ for an attractive force). The equilibrium point at $r = \rho$ is defined by the relation $V'(\rho) = 0$:

$$\rho^{n+2} = \frac{\ell^2}{\mu k}. \tag{8.4}$$

The second derivative of the effective potential is

$$V''(\rho) = \frac{\ell^2}{\mu \rho^4} \left(3 + (n-1) \frac{k \mu}{\ell^2} \rho^{n+2} \right) = \frac{\ell^2}{\mu \rho^4} (2+n). \tag{8.5}$$

Hence, $V''(\rho)$ is positive if $n > -2$, and, thus, circular orbits are stable in central-force fields $F(r) = -k r^{n-1}$ if $n > -2$.

8.2 Small Oscillations about Stable Equilibria

Once an equilibrium point x_0 is shown to be stable, i.e., $V''(x_0) > 0$, we may expand $x = x_0 + \delta x$ about the equilibrium point x_0 (with $|\delta x| \ll |x_0|$) to find the *linearized* force equation

$$m \, \delta \ddot{x} = -V''(x_0) \, \delta x, \tag{8.6}$$

which has oscillatory behavior with frequency

$$\omega(x_0) = \sqrt{\frac{V''(x_0)}{m}}. \tag{8.7}$$

We first look at the problem of a bead on a rotating hoop, where the frequency of small oscillations $\omega(\theta_0)$ is either given in Eq. (8.2) as

$$\omega(0) = \sqrt{\frac{V''(0)}{mR^2}} = \Omega \sqrt{\nu - 1}$$

for $\theta_0 = 0$ and $\nu > 1$, or is given in Eq. (8.3) as

$$\omega(\theta_0) = \sqrt{\frac{V''(\theta_0)}{mR^2}} = \Omega \sqrt{1 - \nu^2}$$

for $\theta_0 = \arccos(\nu)$ and $\nu < 1$. We note that $\omega = 0$ at $\nu = 1$, which means that the period of oscillation is infinite.

Next, we look at the frequency of small oscillations about the stable circular orbit in a central-force field $F(r) = -k r^{n-1}$ (with $n > -2$). Here, from Eq. (8.5), we find

$$\omega = \sqrt{\frac{V''(\rho)}{\mu}} = \sqrt{\frac{k (2+n)}{\mu \rho^{2-n}}},$$

where $\ell^2 = \mu k \, \rho^{2+n}$ was used. We note that for the Kepler problem ($n = -1$), the period of small oscillations $T = 2\pi/\omega$ is expressed as

$$T^2 = \frac{(2\pi)^2 \, \mu}{k} \, \rho^3,$$

which is precisely the statement of Kepler's Third Law for circular orbits [see Eq. (4.43)]. Hence, a small perturbation of a stable Keplerian circular orbit does not change its orbital period.

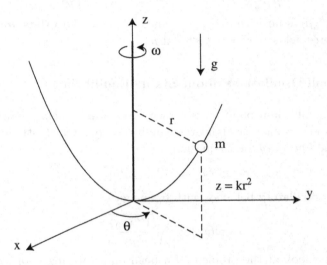

Fig. 8.2   Bead on a rotating parabolic wire.

As a last example of linear stability, we consider the case of a time-dependent equilibrium. A rigid parabolic wire having equation $z = k r^2$ is fastened to a vertical shaft rotating at constant angular velocity $\dot\theta = \omega$. A bead of mass m is free to slide along the wire in the presence of a constant gravitational field with potential $U(z) = mg \, z$ (see Fig. 8.2). The Lagrangian for this mechanical system is given as

$$L(r, \dot r) = \frac{m}{2} \left(1 + 4 k^2 r^2\right) \dot r^2 + m \left(\frac{\omega^2}{2} - g \, k\right) r^2,$$

and the Euler-Lagrange equation of motion is easily obtained as

$$\left(1 + 4 k^2 r^2\right) \ddot r + 4 k^2 r \, \dot r^2 = \left(\omega^2 - 2 g k\right) r. \qquad (8.8)$$

Note that when $\omega^2 < 2 \, gk$, we see that the bead moves in an effective potential represented by an isotropic simple harmonic oscillator with spring constant $\sqrt{m\,(2\,gk - \omega^2)}$ (i.e., the radial position of the bead is bounded), while when $\omega^2 > 2\,gk$, the bead appears to move on the surface of an inverted paraboloid and, thus, the radial position of the bead in this case is unbounded.

We now investigate the stability of the linearized motion $r(t) = r_0 + \delta r(t)$ about an initial radial position r_0. The linearized equation for $\delta r(t)$ is

$$\delta\ddot{r} = \left(\frac{\omega^2 - 2\,gk}{1 + 4\,k^2 r_0^2}\right) \delta r,$$

so that the radial position $r = r_0$ is stable if $\omega^2 < 2\,gk$ and unstable if $\omega^2 > 2\,gk$. In the stable case ($\omega^2 < 2\,gk$), the bead oscillates back and forth with $0 \leq r(t) \leq r_0$, although the motion can be rather complex (see problem 1). For the special case $\omega^2 = 2\,gk$, the linearized equation $\delta\ddot{r} = 0$ implies that the radial dynamics $r(t) = r_0$ is marginally stable. In the unstable case ($\omega^2 > 2\,gk$), the radial position of the bead increases exponentially as it spirals outward away from the initial radial position r_0.

8.3   Normal-Mode Analysis of Coupled Oscillations

Coupled oscillators can exchange energy periodically as a result of the coupling mechanism. In the problem of the Foucault pendulum (see Fig. 6.6), for example, if the pendulum motion is started in the East-West plane, the Coriolis force (the coupling mechanism) allows energy to be transfered to the pendulum motion in the North-South plane and this transfer continues until motion in the East-West plane has disappeared. This transfer process between oscillations in the East-West and North-South planes generates the standard precession motion of the Foucault pendulum.

The normal-mode analysis enables us to determine the characteristic oscillation frequencies exhibited by coupled linear oscillators. For nonlinear coupled oscillators, the nonlinear equations of motion must be linearized first before obtaining the characteristic frequencies of small oscillations.

8.3.1   *Normal-Mode Analysis*

8.3.1.1   *One-degree-of-freedom Analysis*

As a way of introducing the normal-mode analysis for coupled linear equations obtained from a set of Euler-Lagrange equations, we begin with the normal-mode analysis of the one-degree-of-freedom equation

$$a\,\ddot{x}(t) + 2b\,\dot{x}(t) + c\,x(t) = 0, \tag{8.9}$$

with initial conditions $x(0) = A$ and $\dot{x}(0) = 0$, where (a,b,c) are positive constants. First, we introduce the normal-mode representation $x(t) \equiv \overline{x} \exp(-i\,\omega t)$, where $\overline{x}$ denotes a constant amplitude and ω is the normal-mode frequency. Next, we insert the normal-mode representation for $x(t)$ into Eq. (8.9) and obtain the quadratic equation for ω: $(-a\,\omega^2 - 2i\,b\,\omega + c)\,\overline{x} = 0$, whose solutions (since $\overline{x} \neq 0$) are

$$\omega_\pm = -i\,\frac{b}{a} \pm \sqrt{\frac{c}{a} - \frac{b^2}{a^2}} \equiv -i\,\nu \pm \Omega. \tag{8.10}$$

Hence, the general normal-mode solution of Eq. (8.9) is

$$x(t) = e^{-\nu t}\left(\overline{x}_+ \, e^{-i\,\Omega t} + \overline{x}_- \, e^{i\,\Omega t}\right), \tag{8.11}$$

where the constant amplitudes are now determined from the initial conditions. When this general solution is matched with the initial conditions $x(0) = A$ and $\dot{x}(0) = 0$, we obtain

$$\left.\begin{array}{r} \overline{x}_+ + \overline{x}_- = A \\[2mm] i\,\Omega\,(\overline{x}_- - \overline{x}_+) - \nu\,(\overline{x}_+ + \overline{x}_-) = 0 \end{array}\right\} \;\rightarrow\; \overline{x}_\pm = A\left(\frac{1}{2} \pm i\,\frac{\nu}{2\Omega}\right),$$

from which we find the solution

$$x(t) = A\,e^{-\nu t}\left(\cos\Omega t + \frac{\nu}{\Omega}\,\sin\Omega t\right).$$

8.3.1.2   *Two-degree-of-freedom Analysis*

We now turn our attention to a general set of coupled linear second-order differential equations

$$\left.\begin{array}{l} \ddot{x} = -a\,x + b\,y \\[2mm] \ddot{y} = b\,x - d\,y \end{array}\right\}, \tag{8.12}$$

where (a,b,d) are positive constants. These equations could be derived as Euler-Lagrange equations from the Lagrangian

$$L = \frac{1}{2}\left(\dot{x}^2 + \dot{y}^2\right) - \frac{1}{2}\left(a\,x^2 + d\,y^2 - 2b\,x\,y\right).$$

We first introduce the normal-mode representations $x(t) = \overline{x}\exp(-i\,\omega t)$ and $y(t) = \overline{y}\exp(-i\,\omega t)$, where $(\overline{x},\overline{y})$ are arbitrary constant amplitudes while ω is the common normal-mode frequency. In the second step, we construct the symmetric matrix equation

$$\begin{pmatrix} \omega^2 - a & b \\ b & \omega^2 - d \end{pmatrix} \cdot \begin{pmatrix} \overline{x} \\ \overline{y} \end{pmatrix} = 0, \tag{8.13}$$

which is expressed in terms of the constants $(\overline{x}, \overline{y})$. Since we can find non-trivial solutions $(\overline{x}, \overline{y}) \neq (0, 0)$ only if the determinant of the matrix in Eq. (8.13) vanishes, we obtain the relation

$$\left(\omega^2 - a\right)\left(\omega^2 - d\right) = b^2,$$

which yields the normal-mode frequencies $(\pm\omega_+, \pm\omega_-)$:

$$\omega_\pm^2 = \lambda \pm \sqrt{\Delta^2 + b^2}, \tag{8.14}$$

where $\lambda \equiv (a + d)/2$ and $\Delta \equiv (a - d)/2$. In the third step, we determine the normal-mode amplitudes $(\overline{x}_\pm, \overline{y}_\pm)$, which are constructed as solutions of the normal-mode equation

$$\begin{aligned}
0 &= \begin{pmatrix} \omega_\pm^2 - a & b \\ b & \omega_\pm^2 - d \end{pmatrix} \cdot \begin{pmatrix} \overline{x}_\pm \\ \overline{y}_\pm \end{pmatrix} \\
&= \begin{pmatrix} -\Delta \pm \sqrt{\Delta^2 + b^2} & b \\ b & \Delta \pm \sqrt{\Delta^2 + b^2} \end{pmatrix} \cdot \begin{pmatrix} \overline{x}_\pm \\ \overline{y}_\pm \end{pmatrix},
\end{aligned}$$

which yields the solution

$$\overline{y}_\pm = \frac{b\,\overline{x}_\pm}{\Delta \mp \sqrt{\Delta^2 + b^2}} = \left(-\frac{\Delta}{b} \mp \sqrt{1 + \frac{\Delta^2}{b^2}}\right)\overline{x}_\pm. \tag{8.15}$$

If we write

$$b \equiv c\,\sin(2\mu) \quad \text{and} \quad \Delta \equiv c\,\cos(2\mu), \tag{8.16}$$

then Eq. (8.15) yields

$$\overline{y}_\pm = \frac{\sin(2\mu)\,\overline{x}_\pm}{\cos(2\mu) \mp 1} \rightarrow \begin{cases} \overline{y}_+ = -(\cot\mu)\,\overline{x}_+ \\ \overline{y}_- = (\tan\mu)\,\overline{x}_- \end{cases}$$

and, thus, the general solutions for $(x(t), y(t))$ are

$$\left. \begin{aligned} x(t) &= A_+ \cos(\omega_+ t) + A_- \cos(\omega_- t) \\ y(t) &= -(A_+ \tan\mu)\cos(\omega_+ t) + (A_- \cot\mu)\cos(\omega_- t) \end{aligned} \right\}, \tag{8.17}$$

where the constants $A_\pm$ are determined from initial conditions for $(x(t), y(t))$, with $(\dot{x}, \dot{y}) = (0, 0)$ used for simplicity. With the solutions Eq. (8.17), for example, we easily verify that

$$
\begin{aligned}
\ddot{x} + a\,x &= A_+ \left(-\omega_+^2 + a \right) \cos(\omega_+ t) + A_- \left(-\omega_-^2 + a \right) \cos(\omega_- t) \\
&= A_+ \left(-b \tan\mu \right) \cos(\omega_+ t) + A_- \left(b \cot\mu \right) \cos(\omega_- t) \\
&\equiv b\, y(t).
\end{aligned}
$$

Lastly, we can construct two independent normal-mode solutions

$$
\eta_+(t) \equiv \frac{\cot\mu\, x(t) - y(t)}{\tan\mu + \cot\mu} = A_+ \cos(\omega_+ t), \tag{8.18}
$$

$$
\eta_-(t) \equiv \frac{\tan\mu\, x(t) + y(t)}{\tan\mu + \cot\mu} = A_- \cos(\omega_- t), \tag{8.19}
$$

which oscillate at their respective normal-mode frequencies (i.e., $\ddot{\eta}_\pm = -\omega_\pm^2\, \eta_\pm$). The amplitude A_+ vanishes when $y(t) = x(t) \cot\mu$, or the amplitude A_- vanishes when $y(t) = -x(t) \tan\mu$.

8.3.2   *Coupled Simple Harmonic Oscillators*

Fig. 8.3   Coupled identical masses and springs.

We begin our study of linearly-coupled oscillators by considering the following coupled system comprised of two block-and-spring systems, with identical mass m and identical spring constant k, coupled by means of a spring of constant K (see Fig. 8.3). The coupled equations

$$
m\ddot{x} = -(k + K)\,x + K\,y \quad \text{and} \quad m\ddot{y} = -(k + K)\,y + K\,x \tag{8.20}
$$

are derived as Euler-Lagrange equations from the Lagrangian

$$
L = \frac{m}{2} \left(\dot{x}^2 + \dot{y}^2 \right) - \frac{k}{2} \left(x^2 + y^2 \right) - \frac{K}{2} \left(x - y \right)^2.
$$

The solutions for $x(t)$ and $y(t)$ are obtained by normal-mode analysis as follows.

First, we write $x(t)$ and $y(t)$ in the normal-mode representation $x(t) = \bar{x}\,\exp(-i\omega t)$ and $y(t) = \bar{y}\,\exp(-i\omega t)$, where $\bar{x}$ and $\bar{y}$ are constant oscillation amplitudes and the normal-mode frequency ω is to be solved in terms of the system parameters (m, k, K). Next, substituting the normal-mode representation into Eq. (8.20), we obtain the following normal-mode matrix equation

$$\begin{pmatrix} \omega^2\,m - (k+K) & K \\ K & \omega^2\,m - (k+K) \end{pmatrix} \begin{pmatrix} \bar{x} \\ \bar{y} \end{pmatrix} = 0, \qquad (8.21)$$

which couples the amplitudes $\bar{x}$ and $\bar{y}$. Comparison with Eq. (8.13) yields

$$\left. \begin{aligned} a &= d = (k+K)/m \\ b &= K/m \\ \lambda &= (a+d)/2 = (k+K)/m \\ \Delta &= (a-d)/2 = 0 \end{aligned} \right\}, \qquad (8.22)$$

which implies that $\cos(2\mu) = 0 \rightarrow \mu = \pi/4$ according to Eq. (8.16), so that $c = K/m$ and $\tan\mu = 1 = \cot\mu$ in Eq. (8.17), where the normal-mode frequencies are

$$\omega_\pm^2 = \lambda \pm b = \frac{(k+K)}{m} \pm \frac{K}{m} = \begin{cases} \omega_+^2 = (k+2\,K)/m \\ \\ \omega_-^2 = k/m \end{cases} \qquad (8.23)$$

Figure 8.4 shows the normalized solution of the coupled equations (8.20), where time is normalized as $t \rightarrow \sqrt{k/m}\,t$ for the weak coupling $(K < k)$ case. Note that the two eigenfrequencies are said to be *commensurate* if the ratio $\omega_+/\omega_- = \sqrt{1 + 2K/k}$ is expressed as a rational number ρ for values of the ratio $K/k = (\rho^2 - 1)/2$ and that for commensurate eigenfrequencies, the graph of the solutions on the (x, y)-plane generates the so-called Lissajous figures. For non-commensurate eigenfrequencies, however, the graph of the solutions on the (x, y)-plane shows more complex behavior.

Lastly, we define the normal-mode coordinates

$$\eta_\pm(t) \equiv x(t) \mp y(t). \qquad (8.24)$$

Figure 8.5 shows the graphs of the normal coordinates $\eta_\pm(t) = x(t) \mp y(t)$, which clearly displays the single-frequency behavior predicted by the present normal-mode analysis. The solutions $\eta_\pm(t)$ are of the form

$$\eta_\pm = A_\pm \cos(\omega_\pm t + \varphi_\pm),$$

where $A_\pm$ and $\varphi_\pm$ are constants (determined from initial conditions). The general solution of Eqs. (8.20) can, therefore, be written explicitly in terms of the normal coordinates $\eta_\pm$ as

$$\begin{pmatrix} x(t) \\ y(t) \end{pmatrix} = \frac{A_-}{2} \cos(\omega_- t + \varphi_-) \pm \frac{A_+}{2} \cos(\omega_+ t + \varphi_+).$$

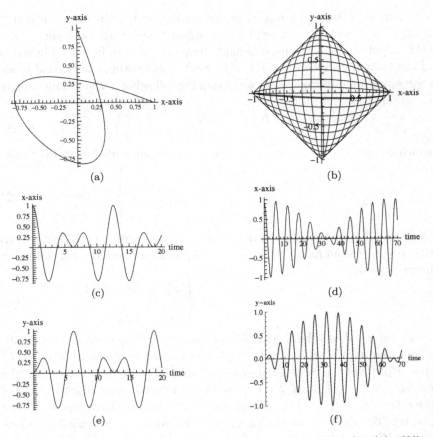

Fig. 8.4   Weak-coupling normalized solutions of the coupled equations (8.20) for $K/k =$ 5/8 (plots a, c, e) and $K/k = 0.1$ (plots b, d, f). Plots (a)-(b) show the parametric plots of $y(t)$ versus $x(t)$; plots (c)-(d) show $x(t)$ versus time t; and plots (e)-(f) show $y(t)$ versus time t.

8.3.3   Coupled Nonlinear Oscillators

Our next example considers coupled nonlinear oscillators, represented by the following system composed of two pendula of identical length ℓ but different masses m_1 and m_2 coupled by means of a spring of constant k in the presence of a gravitational field of constant acceleration g (see Fig. 8.6). Here, the distance D between the two points of attach of the pendula is equal to the length of the spring in its relaxed state and we assume, for simplicity, that the masses always stay on the same horizontal line.

Using the generalized coordinates (θ_1, θ_2) defined in Fig. 8.6, the

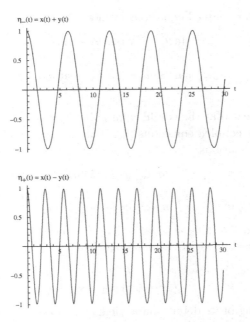

Fig. 8.5 Normal coordinates $\eta_-(t)$ and $\eta_+(t)$ as a function of normalized time for the case $K/k = 2$ with normalized frequencies 1 and $\sqrt{5}$, respectively.

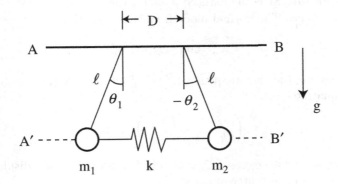

Fig. 8.6 Coupled pendula.

Lagrangian for this system is

$$L = \frac{\ell^2}{2}\left(m_1\,\dot{\theta}_1^2 + m_2\,\dot{\theta}_2^2\right) - g\ell\,[\,m_1\,(1 - \cos\theta_1) + m_2\,(1 - \cos\theta_2)\,]$$
$$- \frac{k\ell^2}{2}\,(\sin\theta_1 - \sin\theta_2)^2,$$

and the nonlinear coupled equations of motion are

$$m_1 \ddot{\theta}_1 = - m_1 \omega_g^2 \sin\theta_1 - k (\sin\theta_1 - \sin\theta_2) \cos\theta_1 \left.\right\}$$
$$m_2 \ddot{\theta}_2 = - m_2 \omega_g^2 \sin\theta_2 + k (\sin\theta_1 - \sin\theta_2) \cos\theta_2 \left.\right\}, \qquad (8.25)$$

where $\omega_g^2 = g/\ell$.

It is quite clear that the equilibrium point $\theta_1 = 0 = \theta_2$ is stable and the expansion of the coupled equations (8.25) about this equilibrium yields the coupled linear equations

$$m_1 \ddot{q}_1 = - m_1 \omega_g^2 q_1 - k (q_1 - q_2) \left.\right\}$$
$$m_2 \ddot{q}_2 = - m_2 \omega_g^2 q_2 + k (q_1 - q_2), \left.\right\} \qquad (8.26)$$

where $\theta_1 = q_1 \ll 1$ and $\theta_2 = q_2 \ll 1$. The normal-mode matrix associated with these coupled linear equations (8.26) is

$$\begin{pmatrix} (\omega^2 - \omega_g^2) m_1 - k & k \\ k & (\omega^2 - \omega_g^2) m_2 - k \end{pmatrix} \begin{pmatrix} \overline{q}_1 \\ \overline{q}_2 \end{pmatrix} = 0, \qquad (8.27)$$

and the vanishing of its determinant yields the relation

$$\left[(\omega^2 - \omega_g^2) \mu - k \right] (\omega^2 - \omega_g^2) = 0,$$

where $\mu = m_1 m_2 / M$ is the reduced mass for the system and $M = m_1 + m_2$ is the total mass. The normal-mode frequencies are thus

$$\omega_-^2 = \omega_g^2 \quad \text{and} \quad \omega_+^2 = \omega_g^2 + \frac{k}{\mu}.$$

By inserting the eigenfrequency $\omega_-^2 = \omega_g^2$ into Eq. (8.27), we obtain the matrix equation

$$\begin{pmatrix} -k & k \\ k & -k \end{pmatrix} \begin{pmatrix} \overline{q}_{1-} \\ \overline{q}_{2-} \end{pmatrix} = 0,$$

whose solution yields $\overline{q}_{2-} = \overline{q}_{1-} \equiv \overline{Q}$, which represents a net displacement of the motion of the center of mass:

$$\overline{Q} = \frac{m_1}{M} \overline{q}_{1-} + \frac{m_2}{M} \overline{q}_{2-}.$$

By inserting the eigenfrequency $\omega_+^2 = \omega_g + k/\mu$ into Eq. (8.27), on the other hand, we obtain the matrix equation

$$\begin{pmatrix} k (m_1/\mu - 1) & k \\ k & k (m_2/\mu - 1) \end{pmatrix} \begin{pmatrix} \overline{q}_{1+} \\ \overline{q}_{2+} \end{pmatrix} = 0,$$

whose solution yields

$$\overline{q}_{2+} = -(m_1/\mu - 1)\,\overline{q}_{1+} = -(M/m_2 - 1)\,\overline{q}_{1+} = -(m_1/m_2)\,\overline{q}_{1+}.$$

In this case, the center of mass does not move since $m_1\,\overline{q}_{1+} + m_2\,\overline{q}_{2+} \equiv 0$.
Lastly, we may solve for q_1 and q_2 as

$$q_1 = \eta_- + \frac{m_2}{M}\,\eta_+ \quad \text{and} \quad q_2 = \eta_- - \frac{m_1}{M}\,\eta_+, \qquad (8.28)$$

where $\eta_\pm = A_\pm\,\cos(\omega_\pm\,t + \varphi_\pm)$ are general solutions of the normal-mode
equations $\ddot{\eta}_\pm = -\omega_\pm^2\,\eta_\pm$. Here, η_- denotes the displacement of the center
of mass

$$\frac{m_1}{M}\,q_1 + \frac{m_2}{M}\,q_2 = \eta_- + \left(\frac{\mu}{M} - \frac{\mu}{M}\right)\,\eta_+ = \eta_-,$$

while η_+ denotes the separation $\eta_+ \equiv q_1 - q_2$.

8.3.4 Stability of the Sleeping Top

8.3.4.1 Normal-mode Analysis of the Sleeping Top

Lastly, we return to the problem of the sleeping top discussed in Sec. 7.4.3.
Here, the equilibrium points of Eq. (7.85) are $u_0 = 1$ and $u_2 = (2\,\Psi^2 - 1)/3$,
where the first derivative of the potential $V(u)$ vanishes, with

$$V'(u) = (u - 1)\left[(2\,\Psi^2 - 1) - 3\,u\right].$$

The second derivative of the potential $V(u)$ is $V''(u) = 2\left[(\Psi^2 + 1) - 3\,u\right]$,
which yields

$$\left.\begin{array}{l} V''(u_0) = 2\,(\Psi^2 - 2) \\[2mm] V''(u_2) = -2\,(\Psi^2 - 2) \end{array}\right\}, \qquad (8.29)$$

and, thus, the equilibrium points $u_0 = 1$ and $u_2 = (2\,\Psi^2 - 1)/3$ are stable
and unstable, respectively, when $\Psi^2 > 2$. Since $u_2 = (2\,\Psi^2 - 1)/3 > 1$ when
$\Psi^2 > 2$, however, this unstable point is physically irrelevant since u must
satisfy $u = \cos\theta \le 1$.

When $\Psi^2 < 2$, on the other hand, $u_0 = 1$ becomes unstable, while
$u_2 = (2\,\Psi^2 - 1)/3$ becomes stable. Since $-\frac{1}{3} < u_2 < 1$ for $0 < \Psi^2 < 2$, the
stable equilibrium point u_2 is physically relevant.

8.3.4.2 *Two-dimensional Analysis of the Sleeping Top*

Following a treatment presented by Whittaker [20], we return to the problem of the symmetric top with one fixed point expressed in terms of the Eulerian angles (φ, θ, ψ), where the Lagrangian is expressed as

$$L = \frac{I_\perp}{2}\left(\dot{\theta}^2 + \dot{\varphi}^2 \sin^2\theta\right) + \frac{I_\parallel}{2}\left(\dot{\psi} + \dot{\varphi}\cos\theta\right)^2 - Mgh\cos\theta.$$

In Sec. 7.4, we eliminated the ignorable angles (ψ, φ) and considered the reduced Lagrangian dynamics for the angle θ.

If we are interested in investigating the two-dimensional rotational dynamics of the symmetric top with one fixed point, however, we can construct the Routhian $R_\perp \equiv L - p_\psi \dot{\psi}$:

$$R_\perp = \frac{I_\perp}{2}\left(\dot{\theta}^2 + \dot{\varphi}^2 \sin^2\theta\right) + Mgh\left(1 - \cos\theta\right) - p_\psi \dot{\varphi}\left(1 - \cos\theta\right), \quad (8.30)$$

where the canonical angular momentum $p_\psi \equiv \partial L/\partial\dot{\psi}$ is a constant of motion and we have omitted the terms $p_\psi \dot{\varphi} - (Mgh + p_\psi^2/2I_\parallel) \equiv d\chi/dt$ since they can be represented as an exact time derivative.

We now study the two-dimensional rotation dynamics in the vicinity of $\theta = 0$ using Eq. (8.30). For this purpose, we introduce the Cartesian coordinates coordinates $x = \sin\theta\cos\varphi$ and $y = \sin\theta\sin\varphi$, which represent the projection of a point on the unit sphere onto the (x,y)-plane. With these coordinates, we readily obtain

$$\left. \begin{array}{c} x^2 + y^2 = \sin^2\theta = (1 - \cos\theta)(1 + \cos\theta) \\[2mm] x\dot{y} - y\dot{x} = \dot{\varphi}\sin^2\theta = \dot{\varphi}(1 - \cos\theta)(1 + \cos\theta) \\[2mm] \dot{x}^2 + \dot{y}^2 = \dot{\theta}^2\cos^2\theta + \dot{\varphi}^2\sin^2\theta \end{array} \right\}. \quad (8.31)$$

If we now use the approximation that both x and y are small in magnitude (i.e., θ is small but $\dot{\varphi}$ is finite and may even be large), we obtain

$$\left. \begin{array}{c} \dot{\theta}^2 + \dot{\varphi}^2 \sin^2\theta \simeq \dot{x}^2 + \dot{y}^2 \\[2mm] 1 - \cos\theta \simeq \frac{1}{2}(x^2 + y^2) \\[2mm] \dot{\varphi}(1 - \cos\theta) \simeq \frac{1}{2}(x\dot{y} - y\dot{x}) \end{array} \right\}, \quad (8.32)$$

which, when substituted into the Routhian (8.30), yields the reduced sleeping-top Lagrangian

$$L(x, y; \dot{x}, \dot{y}) = \frac{I_\perp}{2}\left(\dot{x}^2 + \dot{y}^2\right) + \frac{Mgh}{2}\left(x^2 + y^2\right)$$
$$- \frac{p_\psi}{2}\left(x\dot{y} - y\dot{x}\right). \quad (8.33)$$

The first term in Eq. (8.33) represents the rotational kinetic energy, the second term is a destabilizing potential energy (analogous to an inverted pendulum; see also problem 9 of Chap. 7), and the third term represents the stabilizing non-inertial (Coriolis) energy $-p_\psi\,\hat{\mathbf{z}}\cdot(\mathbf{r}\times\dot{\mathbf{r}})$, where $p_\psi\hat{\mathbf{z}}$ plays a role similar to the angular velocity $2\,\omega$ involved in the discussion of the Coriolis acceleration observed in non-inertial frames in Sec. 6.3.

The coupled equations of motion for this problem are the Euler-Lagrange equations

$$\left.\begin{array}{c} I_\perp\,\ddot{x} \;-\; Mgh\,x = -\,p_\psi\,\dot{y} \\[2mm] I_\perp\,\ddot{y} \;-\; Mgh\,y = p_\psi\,\dot{x} \end{array}\right\}, \qquad (8.34)$$

which are identical in form with the Coriolis-corrected equations (6.51) for the Foucault pendulum. By introducing the definitions $\Omega^2 \equiv 2\,Mgh/I_\perp$ and $p_\psi/I_\perp \equiv \Psi\,\Omega$ (see Eq. (7.84)), the normal-mode analysis of Eq. (8.34), with $(x,y) = (\overline{x},\overline{y})\,\exp(-i\,\omega t)$, yields the matrix equation

$$\begin{pmatrix} \omega^2 + \Omega^2/2 & i\,\Psi\,\Omega\,\omega \\[2mm] -i\,\Psi\,\Omega\,\omega & \omega^2 + \Omega^2/2 \end{pmatrix}\cdot\begin{pmatrix} \overline{x} \\[2mm] \overline{y} \end{pmatrix} = 0. \qquad (8.35)$$

The nontrivial solution $(\overline{x},\overline{y}) \neq (0,0)$ requires ω to be a root of the quartic polynomial

$$\omega^4 \;+\; \Omega^2\,(1-\Psi^2)\,\omega^2 \;+\; \Omega^4/4 = 0.$$

We thus easily find

$$\omega_\pm^2 \;=\; \frac{\Omega^2}{2}\left[(\Psi^2-1)\pm\Psi\,\sqrt{\Psi^2-2}\right] \equiv \left(\frac{\Omega}{2}\left[\Psi\pm\sqrt{\Psi^2-2}\right]\right)^2,$$

which yield four normal-mode frequencies $\pm\,(\Omega/2)\,\nu_+$ and $\pm\,(\Omega/2)\,\nu_-$, where $\nu_\pm = \Psi\pm\sqrt{\Psi^2-2}$. These frequencies are real if $\Psi^2 > 2$ (and, therefore, the motion is stable) or complex-valued (and, therefore, potentially unstable) if $\Psi^2 < 2$ (i.e., for $\Psi = 1$, we find $\nu_\pm = 1\pm i$).

We conclude our normal-mode analysis of the sleeping-top problem by constructing the normal-mode coordinates $(\overline{x}_\pm,\overline{y}_\pm)$:

$$\begin{pmatrix} \omega_\pm^2 + \Omega^2/2 & i\,\Psi\,\Omega\,\omega_\pm \\[2mm] -i\,\Psi\,\Omega\,\omega_\pm & \omega_\pm^2 + \Omega^2/2 \end{pmatrix}\cdot\begin{pmatrix} \overline{x}_\pm \\[2mm] \overline{y}_\pm \end{pmatrix} = 0, \qquad (8.36)$$

where

$$\omega_\pm^2 + \frac{\Omega^2}{2} \;=\; \Omega\,\Psi\left(\frac{\Omega}{2}\left[\Psi\pm\sqrt{\Psi^2-2}\right]\right) \equiv \frac{\Omega^2}{2}\,\Psi\,\nu_\pm,$$

and, thus, we find $\overline{x}_\pm \pm i\,\overline{y}_\pm = 0$. The normal-mode solutions for $x(t)$ and $y(t)$ are, therefore, expressed as

$$x(t) = \frac{A_+}{2}\cos\left(\frac{\Omega}{2}\nu_+ t\right) + \frac{A_-}{2}\cos\left(\frac{\Omega}{2}\nu_- t\right), \tag{8.37}$$

$$y(t) = \frac{A_+}{2}\sin\left(\frac{\Omega}{2}\nu_+ t\right) + \frac{A_-}{2}\sin\left(\frac{\Omega}{2}\nu_- t\right), \tag{8.38}$$

which satisfy the coupled equations (8.34) for arbitrary constant coefficients $A_\pm$. These solutions can be rearranged, with $x = \sin\theta\,\cos\varphi$ and $y = \sin\theta\,\sin\varphi$, to obtain an expression for $x + i\,y \equiv \sin\theta\,\exp(i\,\varphi) \simeq \theta\,\exp(i\,\varphi)$:

$$\theta(t)\,e^{i\,\varphi(t)} = \frac{A_+}{2}\exp\left(i\,\frac{\Omega}{2}\nu_+ t\right) + \frac{A_-}{2}\exp\left(i\,\frac{\Omega}{2}\nu_- t\right)$$

$$= A\sin\left(\frac{\Omega}{2}\sqrt{\Psi^2 - 2}\,t\right)\exp\left(i\,\frac{\Omega}{2}\Psi\,t\right), \tag{8.39}$$

where the last expression satisfies the initial condition $\theta(0) = 0$.

Fig. 8.7   Orbits $(x(t), y(t))$ of a sleeping top for $\Psi^2 = 4$ (solid) and $\Psi^2 = 1$ (dashed), with initial conditions $(x_0, y_0) = (0, 0)$.

Figure 8.7 shows the sleeping-top orbits on the (x, y)-plane for $\Psi^2 = 4$ (solid) and $\Psi^2 = 1$ (dashed). The effect of the Coriolis-like deflection is

Table 8.1   Summary of Chapter 8: Normal-Mode Analysis.

Topic	Equation
Small Oscillation about a Stable Equilibrium	(8.6)-(8.7)
Normal-mode Analysis for Two Degrees of Motion	(8.12)-(8.19)
Coupled Simple Harmonic Oscillators	(8.20)-(8.24)
Coupled Nonlinear Oscillators	(8.25)-(8.28)
Stability Analysis of the Sleeping Top	(8.30)-(8.39)

quite obvious when the sleeping top is stable ($\Psi^2 > 2$). As the top begins to fall (because of the gravitational torque) from a slight departure from $\theta = 0$, the Coriolis-like deflection associated by p_ψ is strong enough to bring the top back to $\theta = 0$. Note that this Coriolis-like deflection is stronger as the top speeds up in its fall (i.e., as it moves radially outward, the radius of curvature increases). When $\Psi^2 < 2$ (dashed curve in Fig. 8.7), however, the Coriolis-like deflection is not strong enough and the gravitational torque causes the top to continue its fall (see problem 9 in Chapter 7).

8.4   Summary

Chapter 8 studied the stability of equilibria of Euler-Lagrange equations. When an equilibrium point was stable, one could calculate a normal-mode frequency of oscillation based on the normal-mode analysis, which also yielded the normal modes of oscillation. Linear and nonlinear coupled oscillators were considered and the stability of the sleeping top was discussed in detail. Table 8.1 presents a summary of the important topics of Chapter 8.

8.5   Problems

1. This problem deals with the numerical integration of the radial equation of motion (8.8), which is expressed in dimensionless form as

$$(1 + 4\rho^2)\, \rho'' + 4\rho\, (\rho')^2 = (\Omega - 2)\rho, \qquad (8.40)$$

where Ω is a dimensionless parameter and $\rho' = d\rho/d\tau$ is defined in terms of the dimensionless time τ.

(a) Find expressions for ρ, τ, and Ω.

(b) Integrate Eq. (8.40) (with initial conditions $\rho_0 = 1$ and $\rho_0' = 0$) for (I) $\Omega < 2$, (II) $\Omega = 2$, and (III) $\Omega > 2$.

(c) Compare the orbits obtained in Part (b) with the stability analysis of this problem found in Sec. 8.2.

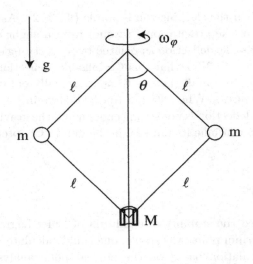

Fig. 8.8   Problem 2.

2. The following compound pendulum is composed of two identical masses m attached by massless rods of identical length ℓ to a ring of mass M, which is allowed to slide up and down along a vertical axis in a gravitational field with constant g (see Fig. 8.8). The entire system rotates about the vertical axis with an azimuthal angular frequency ω_φ.

(a) Show that the Lagrangian for the system can be written as

$$L(\theta, \dot{\theta}) = \ell^2 \dot{\theta}^2 \left(m + 2M \sin^2 \theta \right) + m\ell^2 \omega_\varphi^2 \sin^2 \theta + 2 \left(m + M \right) g\ell \, \cos\theta$$

(b) Identify the equilibrium points for the system and investigate their stability.

(c) Determine the frequency of small oscillations about each stable equilibrium point found in Part (b).

3. Consider the same problem as in Sec. (8.3.2) but now with different masses $m_1 \neq m_2$ (see Fig. 8.9). Calculate the eigenfrequencies and eigenvectors (normal coordinates) for this system.

Fig. 8.9   Problem 3.

4. Find the eigenfrequencies associated with small oscillations of the system shown in Fig. 8.10.

Fig. 8.10   Problem 4.

5. Two blocks of identical mass m are attached by massless springs (with identical spring constant k) as shown in Fig. 8.11. The Lagrangian for this system is

$$L(x, \dot{x}; y, \dot{y}) = \frac{m}{2} \left(\dot{x}^2 + \dot{y}^2 \right) - \frac{k}{2} \left[x^2 + (y - x)^2 \right],$$

where x and y denote departures from equilibrium.

Fig. 8.11   Problem 5.

(a) Derive the Euler-Lagrange equations for x and y.

(b) Show that the eigenfrequencies for small oscillations for this system are

$$\omega_{\pm}^2 = \frac{\omega_k^2}{2}\left(3 \pm \sqrt{5}\right),$$

where $\omega_k^2 = k/m$.

(c) Show that the eigenvectors associated with the eigenfrequencies $\omega_{\pm}$ are represented by the relations

$$\overline{y}_{\pm} = \frac{1}{2}\left(1 \mp \sqrt{5}\right)\overline{x}_{\pm}$$

where $(\overline{x}_{\pm}, \overline{y}_{\pm})$ represent the normal-mode amplitudes.

6. An infinite sheet with surface mass density σ has a hole of radius R cut into it. A particle of mass m sits (in equilibrium) at the center of the circle. Assuming that the sheet lies on the (x, y)-plane (with the hole centered at the origin) and that the particle is displaced by a small amount $z \ll R$ along the z-axis, calculate the frequency of small oscillations.

7. Two identical masses are connected by two identical massless springs and are constrained to move on a circle (see Fig. 8.12). Of course, the two masses are in equilibrium when they are diametrically opposite points on the circle. Solve for the normal modes of the system.

8. Consider a pendulum of mass m attached at a point O with the help of a massless rigid rod on length ℓ. Here, point O is located at a distance $R > \ell$ from a axis of rotation and is rotating at an angular velocity Ω about the axis of rotation (see Fig. 8.13).

(a) Show that there are two equilibrium configurations to this problem,

Fig. 8.12   Problem 7.

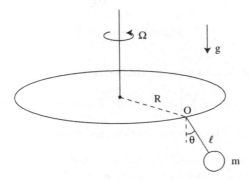

Fig. 8.13   Problem 8.

which are obtained from finding the roots to the transcendental equation

$$(R - \ell \sin\theta)\, \Omega^2 \cos\theta \,=\, g \sin\theta.$$

(b) Show that one equilibrium configuration is stable while the other is unstable.

9. Two particles of identical masses are connected to each other by a spring (with constant k) and are allowed to move without friction on a hoop of radius R (see Fig. 8.14). The angles θ_1 and θ_2 are expressed in terms of the generalized coordinates θ and φ as $\theta_1 = \theta$ and $\theta_2 = \theta + \varphi - \Theta$, where θ is the angular displacement of the right mass from the vertical, Θ is the angular separation between the two masses when the spring is at equilibrium, and φ is the angular displacement of the spring away from equilibrium.

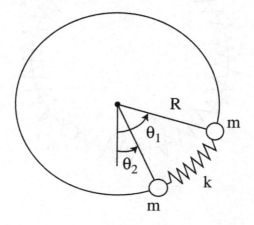

Fig. 8.14   Problem 9.

(a) Show that the Lagrangian for this system is

$$L = \frac{m}{2} R^2 \left[\dot{\theta}^2 + \left(\dot{\theta} + \dot{\varphi} \right)^2 \right] - \frac{k}{2} (R\,\varphi)^2 + mgR \left[\cos\theta + \cos\left(\theta + \varphi - \Theta\right) \right],$$

and derive the Euler-Lagrange equations for θ and φ.

(b) Show that the potential $U(\theta, \varphi)$ has a global minimum at θ_0 and φ_0, which satisfy the transcendental equation

$$\theta_0 = \frac{1}{2} \left(\Theta - \varphi_0 \right) \equiv \frac{1}{2} \left(\Theta - \Omega^2 \sin\theta_0 \right),$$

where $\Omega^2 \equiv \omega_g^2 / \omega_k^2$ is the ratio of the squared pendulum frequency $\omega_g^2 = g/R$ and the squared spring frequency $\omega_k^2 = k/m$.

(c) Find the eigenfrequencies and eigenvectors for the normal modes of small oscillations about the equilibrium defined by θ_0 and φ_0.

Chapter 9

Continuous Lagrangian Systems

This last Chapter, in fact, represents the beginning for some of the most important applications of Lagrangian methods in physics, namely those that apply to classical mechanics, special and general relativisty, or classical and quantum field theories. So far in this textbook, Lagrangian methods have been applied to derive equations of motion for particles or rigid bodies. The Noether method has also been applied to obtain the conservation laws of energy and momentum for these systems whenever symmetries existed for the corresponding Lagrangians.

While a systematic presentation of the Lagrangian formulation of field equations cannot be undertaken at this level, a few examples have nonetheless been selected that give a flavor of the power of the Lagrangian method for continuous systems.

9.1 Waves on a Stretched String

9.1.1 *Wave Equation*

The equation describing transverse waves propagating on a stretched string of constant linear mass density ρ under constant tension T is

$$\rho \frac{\partial^2 u(x,t)}{\partial t^2} = T \frac{\partial^2 u(x,t)}{\partial x^2}, \tag{9.1}$$

where $u(x,t)$ denotes the amplitude of the wave at position x along the string at time t. General solutions to this linear wave equation involve arbitrary functions $g(x \pm v\,t)$, where $v = \sqrt{T/\rho}$ represents the speed of waves propagating on the string. Indeed, we find

$$\rho\, \partial_t^2 g(x \pm v\,t) = \rho v^2\, g'' = T\, g'' = T\, \partial_x^2 g(x \pm v\,t).$$

The interpretation of the two different signs is that $g(x - vt)$ represents a wave propagating to the right while $g(x + vt)$ represents a wave propagating to the left. The general solution of the wave equation (9.1) is

$$u(x, t) = A_- g(x - vt) + A_+ g(x + vt),$$

where $A_\pm$ are arbitrary constants determined from initial conditions

9.1.2 Lagrangian Formulation

The question we now ask is whether the wave equation (9.1) can be derived from a variational principle

$$\delta \int \mathcal{L}(u, \partial_t u, \partial_x u) \, dx \, dt = 0, \qquad (9.2)$$

where the Lagrangian *density* $\mathcal{L}(u, \partial_t u, \partial_x u)$ is a function of the dynamical variable $u(x, t)$ and its space-time derivatives. Here, the variation of the Lagrangian density $\mathcal{L}$ in Eq. (9.2) is expressed as

$$\delta \mathcal{L} = \delta u \, \frac{\partial \mathcal{L}}{\partial u} + \frac{\partial \delta u}{\partial t} \, \frac{\partial \mathcal{L}}{\partial(\partial_t u)} + \frac{\partial \delta u}{\partial x} \, \frac{\partial \mathcal{L}}{\partial(\partial_x u)},$$

where $\delta u(x, t)$ is a general variation of $u(x, t)$ subject to the condition that it vanishes at the integration boundaries in Eq. (9.2), and we used the substitutions $\delta(\partial u/\partial t) = \partial \delta u/\partial t$ and $\delta(\partial u/\partial x) = \partial \delta u/\partial x$. By re-arranging terms, the variation of $\mathcal{L}$ can be written as

$$\delta \mathcal{L} = \delta u \left\{ \frac{\partial \mathcal{L}}{\partial u} - \frac{\partial}{\partial t} \left(\frac{\partial \mathcal{L}}{\partial(\partial_t u)} \right) - \frac{\partial}{\partial x} \left(\frac{\partial \mathcal{L}}{\partial(\partial_x u)} \right) \right\}$$
$$+ \frac{\partial}{\partial t} \left(\delta u \, \frac{\partial \mathcal{L}}{\partial(\partial_t u)} \right) + \frac{\partial}{\partial x} \left(\delta u \, \frac{\partial \mathcal{L}}{\partial(\partial_x u)} \right). \qquad (9.3)$$

When we insert this expression for $\delta \mathcal{L}$ into the variational principle (9.2), we obtain

$$\int dx \, dt \, \delta u \left\{ \frac{\partial \mathcal{L}}{\partial u} - \frac{\partial}{\partial t} \left(\frac{\partial \mathcal{L}}{\partial(\partial_t u)} \right) - \frac{\partial}{\partial x} \left(\frac{\partial \mathcal{L}}{\partial(\partial_x u)} \right) \right\} = 0, \qquad (9.4)$$

where the last two terms in Eq. (9.3) cancel out because δu vanishes on the integration boundaries. Since the variational principle (9.4) is true for general variations δu, we obtain the Euler-Lagrange equation for the dynamical field $u(x, t)$:

$$\frac{\partial}{\partial t} \left(\frac{\partial \mathcal{L}}{\partial(\partial_t u)} \right) + \frac{\partial}{\partial x} \left(\frac{\partial \mathcal{L}}{\partial(\partial_x u)} \right) = \frac{\partial \mathcal{L}}{\partial u}. \qquad (9.5)$$

The question we posed earlier now focuses on deciding what form the Lagrangian density must take. Here, the answer is surprisingly simple: the kinetic energy density of the wave is $\rho\,(\partial_t u)^2/2$, while the potential energy density is $T\,(\partial_x u)^2/2$, and thus the Lagrangian density for waves on a stretched string is

$$\mathcal{L}(u, \partial_t u, \partial_x u;\; x, t) \;=\; \frac{\rho}{2}\left(\frac{\partial u}{\partial t}\right)^2 - \frac{T}{2}\left(\frac{\partial u}{\partial x}\right)^2, \qquad (9.6)$$

where it is assumed (for simplicity) that the mass density ρ and tension T are uniform and time-independent. Since $\partial \mathcal{L}/\partial u = 0$, we find

$$\frac{\partial}{\partial t}\left(\frac{\partial \mathcal{L}}{\partial(\partial_t u)}\right) = \frac{\partial}{\partial t}\left(\rho\,\frac{\partial u}{\partial t}\right) = \rho\,\frac{\partial^2 u}{\partial t^2},$$

$$\frac{\partial}{\partial x}\left(\frac{\partial \mathcal{L}}{\partial(\partial_x u)}\right) = \frac{\partial}{\partial x}\left(-\,T\,\frac{\partial u}{\partial x}\right) = -\,T\,\frac{\partial^2 u}{\partial x^2},$$

and Eq. (9.1) is indeed represented as an Euler-Lagrange equation (9.5) in terms of the Lagrangian density (9.6).

The energy density $\mathcal{E}$ of a stretched string can also be calculated by using the Legendre transformation:

$$\mathcal{E} \;\equiv\; \frac{\partial u}{\partial t}\,\frac{\partial \mathcal{L}}{\partial(\partial_t u)} - \mathcal{L} \;=\; \frac{\rho}{2}\left(\frac{\partial u}{\partial t}\right)^2 + \frac{T}{2}\left(\frac{\partial u}{\partial x}\right)^2.$$

By using the wave equation (9.1), we readily find that the time derivative of the energy density

$$\frac{\partial \mathcal{E}}{\partial t} \;=\; \frac{\partial}{\partial x}\left(T\,\frac{\partial u}{\partial t}\,\frac{\partial u}{\partial x}\right)$$

can be expressed as an energy conservation law $\partial_t \mathcal{E} + \partial \mathcal{S} = 0$, where the energy-density flux is defined as $\mathcal{S} \equiv -\,T\,\partial_x u\,\partial_t u$. The next Section will present the general variational formulation of classical field theory, which enables us to show that the wave equation (9.1) also satisfies the momentum conservation law $\partial_t \mathcal{P} + \partial_x \Pi = 0$, where the momentum density is $\mathcal{P} \equiv \mathcal{S}/v^2$ and the momentum-density flux is $\Pi \equiv \mathcal{E}$.

9.2 Variational Principle for Field Theory*

The simple example of transverse waves on a stretched string allows us to view the Euler-Lagrange equation (9.5) as a generalization of the Euler-Lagrange equations

$$\frac{d}{dt}\left(\frac{\partial L}{\partial \dot{q}^i}\right) \;=\; \frac{\partial L}{\partial q^i},$$

in terms of the generalized coordinates q^i. We now spend some time investigating the Lagrangian description of continuous systems, in which the dynamical variable is the single-component field $\psi(\mathbf{x}, t)$, which can easily be generalized to fields with multiple components.

9.2.1 *Lagrangian Formulation*

Classical and quantum field theories rely on variational principles based on the existence of action functionals. The typical action functional is of the form

$$\mathcal{A}[\psi] = \int d^4x \, \mathcal{L}(\psi, \, \partial_\mu \psi), \tag{9.7}$$

where the wave function $\psi(\mathbf{x}, t)$ represents the state of the system at position $\mathbf{x}$ (in n-dimensional space) and time t, while the entire physical content of the theory is carried by the Lagrangian density $\mathcal{L}$. We, henceforth, use the convenient four-vector notation $\partial_\mu = (c^{-1}\partial_t, \, \nabla)$ in Eq. (9.7) and we use the space-like metric tensor[1] $g^{\mu\nu} = \text{diag}(-1, +1, +1, +1)$.

The variational principle is based on the stationarity of the action functional (9.7):

$$0 = \delta\mathcal{A}[\psi] = \frac{d}{d\epsilon}\left(\mathcal{A}[\psi + \epsilon\,\delta\psi]\right)_{\epsilon=0} = \int \delta\mathcal{L}(\psi, \, \partial_\mu\psi) \, d^4x. \tag{9.8}$$

Here, the functional variation of the Lagrangian density is

$$\delta\mathcal{L} = \frac{\partial\mathcal{L}}{\partial\psi}\,\delta\psi + \frac{\partial\mathcal{L}}{\partial(\partial_\mu\psi)}\,\frac{\partial\delta\psi}{\partial x^\mu}$$

$$\equiv \delta\psi\left[\frac{\partial\mathcal{L}}{\partial\psi} - \frac{\partial}{\partial x^\mu}\left(\frac{\partial\mathcal{L}}{\partial(\partial_\mu\psi)}\right)\right] + \frac{\partial\Lambda^\mu}{\partial x^\mu}, \tag{9.9}$$

where

$$\frac{\partial\mathcal{L}}{\partial(\partial_\mu\psi)}\,\frac{\partial\delta\psi}{\partial x^\mu} = \frac{\partial\mathcal{L}}{\partial(\nabla\psi)}\cdot\nabla\delta\psi + \frac{\partial\mathcal{L}}{\partial(\partial_t\psi)}\,\frac{\partial\delta\psi}{\partial t},$$

and the exact space-time divergence $\partial_\mu\Lambda^\mu$ in Eq. (9.9) is obtained by rearranging terms, with

$$\Lambda^\mu = \delta\psi\,\frac{\partial\mathcal{L}}{\partial(\partial_\mu\psi)} \quad \text{and} \quad \frac{\partial\Lambda^\mu}{\partial x^\mu} = \frac{\partial}{\partial t}\left(\delta\psi\,\frac{\partial\mathcal{L}}{\partial(\partial_t\psi)}\right) + \nabla\cdot\left(\delta\psi\,\frac{\partial\mathcal{L}}{\partial(\nabla\psi)}\right).$$

The variational principle (9.8) then yields

$$0 = \int d^4x \, \delta\psi\left[\frac{\partial\mathcal{L}}{\partial\psi} - \frac{\partial}{\partial x^\mu}\left(\frac{\partial\mathcal{L}}{\partial(\partial_\mu\psi)}\right)\right],$$

---

[1]For two four-vectors $A^\mu = (A^0, \mathbf{A})$ and $B^\mu = (B^0, \mathbf{B})$, we have $A \cdot B = A_\mu B^\mu = \mathbf{A}\cdot\mathbf{B} - A^0\,B^0$, where $A_0 = -A^0$.

where the exact divergence $\partial_\mu \Lambda^\mu$ drops out under the assumption that the variation $\delta\psi$ vanish on the integration boundaries. Following the standard rules of Calculus of Variations, the Euler-Lagrange equation for the field ψ is

$$\frac{\partial}{\partial x^\mu}\left(\frac{\partial \mathcal{L}}{\partial(\partial_\mu \psi)}\right) = \frac{\partial}{\partial t}\left(\frac{\partial \mathcal{L}}{\partial(\partial_t \psi)}\right) + \nabla\cdot\left(\frac{\partial \mathcal{L}}{\partial(\nabla\psi)}\right) = \frac{\partial \mathcal{L}}{\partial\psi}. \qquad (9.10)$$

The generalization to a multiple-component field simply involves replacing the single field ψ with the field component ψ^a (where the component index $a \geq 2$).

9.2.2 *Noether Method and Conservation Laws*

Since the Euler-Lagrange equation (9.10) holds true for arbitrary field variations $\delta\psi$, the variation of the Lagrangian density $\mathcal{L}$ is now expressed as the Noether equation

$$\delta\mathcal{L} \equiv \frac{\partial \Lambda^\mu}{\partial x^\mu} = \frac{\partial}{\partial x^\mu}\left[\delta\psi\,\frac{\partial \mathcal{L}}{\partial(\partial_\mu \psi)}\right], \qquad (9.11)$$

which associates symmetries with conservation laws $\partial_\mu \mathcal{J}^\mu = 0$.

9.2.2.1 *Energy-Momentum Conservation Law*

The conservation of energy-momentum (a four-vector quantity) involves a symmetry of the Lagrangian with respect to constant *space-time* translations $x^\nu \to \overline{x}^\mu \equiv x^\nu + \delta x^\nu$, where $\delta x^\mu = (c\,\delta t, \delta\mathbf{x})$. The variation $\delta\psi$ is no longer arbitrary in Eq. (9.9) but is required to be of the form

$$\delta\psi = -\delta x^\nu\,\partial_\nu \psi, \qquad (9.12)$$

which follows from the scalar-invariance property $\overline{\psi}(\overline{x}) \equiv \psi(x)$, with $\overline{\psi} \equiv \psi + \delta\psi$. The variation $\delta\mathcal{L}$, on the other hand, is expressed as

$$\delta\mathcal{L} = -\delta x^\nu\left[\partial_\nu \mathcal{L} - (\partial_\nu \mathcal{L})_\psi\right], \qquad (9.13)$$

where $(\partial_\nu \mathcal{L})_\psi$ denotes the explicit derivative of $\mathcal{L}$ at constant ψ.

The Noether equation (9.11) can now be written as[2]

$$\frac{\partial}{\partial x^\mu}\left(\mathcal{L}\,g^\mu{}_\nu - \frac{\partial \mathcal{L}}{\partial(\partial_\mu \psi)}\,\partial_\nu \psi\right) = \left(\frac{\partial \mathcal{L}}{\partial x^\nu}\right)_\psi. \qquad (9.14)$$

---

[2]Compare with Eq. (2.48): $d/dt\,(\delta t\,L - \delta\mathbf{q}\cdot\partial L/\partial\dot{\mathbf{q}}) = \delta t\,(\partial L/\partial t)_{\mathbf{q},\dot{\mathbf{q}}}$.

If the Lagrangian is explicitly independent of the space-time coordinates, i.e., $(\partial_\nu \mathcal{L})_\psi = 0$, the energy-momentum conservation law $\partial_\mu T^{\mu\nu} = 0$ is written in terms of the energy-momentum tensor

$$T^{\mu\nu} \equiv \mathcal{L}\, g^{\mu\nu} - \frac{\partial \mathcal{L}}{\partial(\partial_\mu \psi)}\, \frac{\partial \psi}{\partial x_\nu}. \tag{9.15}$$

We note that the derivation of the energy-momentum conservation law is the same for classical and quantum fields. A similar procedure would lead to the conservation of angular momentum by considering symmetries of the Lagrangian density under arbitrary rotations.

We briefly return to the problem of waves on a stretched string (Sec. 9.1) and derive the components of the energy-momentum tensor (9.15) for the Lagrangian density (9.6):

$$T^{00} \equiv -\mathcal{L} + \frac{\partial \mathcal{L}}{\partial(\partial_t u)}\, \frac{\partial u}{\partial t} = \frac{\rho}{2}\left(\frac{\partial u}{\partial t}\right)^2 + \frac{T}{2}\left(\frac{\partial u}{\partial x}\right)^2 \equiv \mathcal{E}, \tag{9.16}$$

$$T^{x0} \equiv \frac{1}{v}\frac{\partial \mathcal{L}}{\partial(\partial_x u)}\, \frac{\partial u}{\partial t} = -\frac{T}{v}\frac{\partial u}{\partial x}\frac{\partial u}{\partial t} \equiv \frac{\mathcal{S}}{v}, \tag{9.17}$$

$$T^{0x} \equiv -v\frac{\partial \mathcal{L}}{\partial(\partial_t u)}\, \frac{\partial u}{\partial x} = -\rho v \frac{\partial u}{\partial t}\frac{\partial u}{\partial x} \equiv \mathcal{P}\, v, \tag{9.18}$$

$$T^{xx} \equiv \mathcal{L} - \frac{\partial \mathcal{L}}{\partial(\partial_x u)}\, \frac{\partial u}{\partial x} = \frac{\rho}{2}\left(\frac{\partial u}{\partial t}\right)^2 + \frac{T}{2}\left(\frac{\partial u}{\partial x}\right)^2 \equiv \Pi. \tag{9.19}$$

We have thus shown that the energy-momentum conservation laws $\partial_t \mathcal{E} + \partial_x \mathcal{S} = 0$ and $\partial_t \mathcal{P} + \partial_x \Pi = 0$ can be expressed as $\partial_\mu T^{\mu\nu} = 0$, where $\partial_\mu = (v^{-1}\partial_t, \partial_x)$ with $v^2 = T/\rho$. If we compare Eqs. (9.17)-(9.18), we immediately conclude that the tensor $T^{\mu\nu}$ is symmetric: $T^{0x} = T^{x0}$, since $T/v \equiv \rho\, v$.

9.2.2.2 Wave-Action Conservation Law

Waves are known to exist on a great variety of media. When waves are supported by a spatially nonuniform or time-dependent medium, the conservation law of energy or momentum no longer apply and instead energy or momentum is transfered between the medium and the waves. There is, however, one conservation law which still applies and the quantity being conserved is known as the *wave action*.

The derivation of a wave-action conservation law differs for classical fields and quantum fields. The difference is related to the fact that, whereas classical fields are generally represented by real-valued wave functions (i.e.,

$\psi^* = \psi$), the wave functions of quantum field theories are complex-valued (i.e., $\psi^* \neq \psi$).

The first step in deriving a wave-action conservation law in classical field theory involves transforming the real-valued wave function ψ into a complex-valued wave function ψ. Next, variations of ψ and its complex conjugate $\psi^* \neq \psi$ are of the form

$$\delta\psi \equiv i\delta\varphi\,\psi \quad \text{and} \quad \delta\psi^* \equiv -i\delta\varphi\,\psi^*, \tag{9.20}$$

obtained by introducing an infinitesimal phase shift $\delta\varphi$ in $\psi \to \overline{\psi} \equiv \psi \exp(i\,\delta\varphi)$ and $\psi^* \to \overline{\psi}^* \equiv \psi^* \exp(-i\,\delta\varphi)$. Lastly, we transform the classical Lagrangian density $\mathcal{L}$ into a real-valued Lagrangian density $\mathcal{L}_R(\psi, \psi^*)$ such that $\delta\mathcal{L}_R \equiv 0$ (i.e., $\mathcal{L}_R$ is explicitly independent of the phase of the wave function ψ). The wave-action conservation law is, therefore, expressed in the form $\partial_\mu\,\mathcal{J}^\mu = 0$, where the wave-action four-density is

$$\mathcal{J}^\mu \equiv 2\,\text{Im}\left[\psi\,\frac{\partial\mathcal{L}_R}{\partial(\partial_\mu\psi)}\right], \tag{9.21}$$

where $\text{Im}[\cdots]$ denotes the imaginary part [i.e., $\text{Im}(a^*b) = (a^*b - ab^*)/2i$].

The standard method in deriving the wave-action four-density (9.21) makes use of the *eikonal* representation for the real-valued wave field $\psi(\mathbf{x}, t)$:

$$\psi(\mathbf{x}, t) = \widetilde{\psi}(\epsilon\mathbf{x}, \epsilon t)\,e^{i\,\Theta(\epsilon\mathbf{x},\epsilon t)/\epsilon} + \widetilde{\psi}^*(\epsilon\mathbf{x}, \epsilon t)\,e^{-i\,\Theta(\epsilon\mathbf{x},\epsilon t)/\epsilon}, \tag{9.22}$$

where $\widetilde{\psi}$ denotes the complex-valued eikonal amplitude and Θ denotes the eikonal phase. The small parameter $\epsilon \ll 1$ indicates that the space-time gradient

$$\partial_\mu\psi = \left(i\,k_\mu\,\widetilde{\psi} + \epsilon\,\widetilde{\psi}_{,\mu}\right)e^{i\Theta/\epsilon} + \left(-i\,k_\mu\,\widetilde{\psi}^* + \epsilon\,\widetilde{\psi}^*_{,\mu}\right)e^{-i\Theta/\epsilon}$$

is expressed (to lowest order in ϵ) in terms of the wave four-vector

$$k_\mu = \epsilon^{-1}\,\partial_\mu\Theta \equiv \Theta_{,\mu} = (-\omega/c,\,\mathbf{k}), \tag{9.23}$$

where we used the eikonal relations (3.15). Hence, to lowest order in ϵ, the Lagrangian density $\mathcal{L}(\psi, \partial_\mu\psi)$ for a real-valued wave field ψ becomes the real-valued eikonal-averaged Lagrangian density $\overline{\mathcal{L}}_R(\widetilde{\psi}; k_\mu)$ for the complex-valued wave field $\widetilde{\psi}$. The wave-action density (9.21) now simply becomes

$$\mathcal{J}^\mu \equiv \epsilon\,\frac{\partial\overline{\mathcal{L}}_R}{\partial(\partial_\mu\Theta)} = \frac{\partial\overline{\mathcal{L}}_R}{\partial k_\mu}, \tag{9.24}$$

and the wave-action conservation law becomes the Euler-Lagrange equation

$$\frac{\partial}{\partial x^\mu}\left(\frac{\partial\overline{\mathcal{L}}_R}{\partial(\partial_\mu\Theta)}\right) = \frac{\partial\overline{\mathcal{L}}_R}{\partial\Theta} \equiv 0,$$

which follows from the fact that $\mathcal{L}_R$ is independent of the eikonal phase Θ but not its space-time derivatives $\partial_\mu\Theta$.

9.3 Schroedinger's Equation

A simple yet important example for a quantum field theory is provided by the Schroedinger equation for a spinless particle of mass m subjected to a real-valued potential energy function $U(\mathbf{x}, t)$. The Lagrangian density for the Schroedinger equation is given as

$$\mathcal{L}_R = -\frac{\hbar^2}{2m}\, \nabla\psi^* \cdot \nabla\psi + \frac{i\hbar}{2}\left(\psi^* \frac{\partial\psi}{\partial t} - \psi \frac{\partial\psi^*}{\partial t}\right) - U\,|\psi|^2. \quad (9.25)$$

The Schroedinger equation for ψ is derived as an Euler-Lagrange equation (9.10) in terms of ψ^*, where

$$\frac{\partial\mathcal{L}_R}{\partial(\partial_t\psi^*)} = -\frac{i\hbar}{2}\,\psi \;\rightarrow\; \frac{\partial}{\partial t}\left(\frac{\partial\mathcal{L}_R}{\partial(\partial_t\psi^*)}\right) = -\frac{i\hbar}{2}\frac{\partial\psi}{\partial t},$$

$$\frac{\partial\mathcal{L}_R}{\partial(\nabla\psi^*)} = -\frac{\hbar^2}{2m}\,\nabla\psi \;\rightarrow\; \nabla\cdot\left(\frac{\partial\mathcal{L}_R}{\partial(\nabla\psi^*)}\right) = -\frac{\hbar^2}{2m}\,\nabla^2\psi,$$

$$\frac{\partial\mathcal{L}_R}{\partial\psi^*} = \frac{i\hbar}{2}\frac{\partial\psi}{\partial t} - U\,\psi.$$

By combining these derivatives, the Euler-Lagrange equation (9.10) for the Schroedinger Lagrangian (9.25) becomes

$$i\hbar\frac{\partial\psi}{\partial t} = -\frac{\hbar^2}{2m}\,\nabla^2\psi + U\,\psi, \quad (9.26)$$

while the Schroedinger equation for ψ^* is as an Euler-Lagrange equation (9.10) in terms of ψ:

$$-i\hbar\frac{\partial\psi^*}{\partial t} = -\frac{\hbar^2}{2m}\,\nabla^2\psi^* + U\,\psi^*, \quad (9.27)$$

which is simply the complex-conjugate equation of Eq. (9.26).

The energy-momentum conservation law for the Schroedinger equation (9.26) is now derived by Noether method. Because the potential $U(\mathbf{x}, t)$ is in general spatially nonuniform and time dependent, the energy-momentum contained in the wave function is not conserved and energy-momentum is exchanged between the wave function and the potential U. For example, the energy *transfer* equation is

$$\frac{\partial\mathcal{E}}{\partial t} + \nabla\cdot\mathbf{S} = |\psi|^2\frac{\partial U}{\partial t}, \quad (9.28)$$

where the energy density $\mathcal{E}$ and energy density flux $\mathbf{S}$ are given explicitly as

$$\mathcal{E} = -\mathcal{L}_R + \frac{i\hbar}{2}\left(\psi^* \frac{\partial\psi}{\partial t} - \psi \frac{\partial\psi^*}{\partial t}\right)$$

$$\mathbf{S} = -\frac{\hbar^2}{2m}\left(\frac{\partial\psi}{\partial t}\nabla\psi^* + \frac{\partial\psi^*}{\partial t}\nabla\psi\right).$$

The momentum transfer equation, on the other hand, is

$$\frac{\partial \mathbf{P}}{\partial t} + \nabla \cdot \mathsf{T} = - |\psi|^2 \, \nabla U, \qquad (9.29)$$

where the momentum density $\mathbf{P}$ and momentum density tensor T are given explicitly as

$$\mathbf{P} = \frac{i\hbar}{2} \, (\psi \, \nabla \psi^* - \psi^* \, \nabla \psi)$$

$$\mathsf{T} = \mathcal{L}_R \, \mathsf{I} + \frac{\hbar^2}{2m} \, (\nabla \psi^* \, \nabla \psi + \nabla \psi \, \nabla \psi^*).$$

Note that Eqs. (9.28) and (9.29) are both exact equations for any time-dependent, nonuniform potential $U(\mathbf{x}, t)$.

Whereas energy-momentum is transfered between the wave function ψ and the potential V, the amount of wave-action contained in the wave function is conserved. Indeed, the wave-action conservation law is

$$\frac{\partial \mathcal{J}}{\partial t} + \nabla \cdot \mathbf{J} = 0, \qquad (9.30)$$

where, according to Eq. (9.21), the wave-action density $\mathcal{J}$ and wave-action density flux $\mathbf{J}$ are

$$\mathcal{J} = \hbar \, |\psi|^2 \quad \text{and} \quad \mathbf{J} = \frac{\hbar^2}{m} \, \mathrm{Im} \, (\psi^* \, \nabla \psi). \qquad (9.31)$$

Thus wave-action conservation law is none other than the law of conservation of probability associated with the normalization condition

$$\int |\psi|^2 \, d^3 x = 1$$

for bounds states or the conservation of the number of quanta in a scattering problem.

Lastly, by substituting the *ansatz*

$$\psi \equiv \sqrt{\rho} \, \exp(i\mathcal{S}/\hbar) \qquad (9.32)$$

in the Schroedinger Lagrangian density (9.25), where $\rho > 0$ and $\mathcal{S}$ are real-valued functions, we can easily obtain the "classical" Lagrangian density

$$\mathcal{L}_C = - \rho \left(\frac{\partial \mathcal{S}}{\partial t} + \frac{|\nabla \mathcal{S}|^2}{2m} + U \right)$$

in the classical limit $\hbar \to 0$. The variational principle

$$\delta \int \mathcal{L}_C(\rho; \mathcal{S}, \partial_t \mathcal{S}, \nabla \mathcal{S}) \, dt = 0$$

with respect to variations $\delta\rho$ then yields the Hamilton-Jacobi equation (3.8)

$$\frac{\partial S}{\partial t} + \frac{|\nabla S|^2}{2m} + U \equiv \frac{\partial S}{\partial t} + H(\mathbf{x}, \nabla S; t) = 0,$$

where H is the Hamiltonian function with momentum $\mathbf{p} \equiv \nabla S$ defined in terms of S (see Table 3.1). We thus see the explicit connection between the Hamilton-Jacobi equation for classical particle dynamics and the Schroedinger equation (9.26) for quantum mechanics. The Euler-Lagrange equation for ρ (corresponding to variations δS), on the other hand, yields the conservation law

$$\frac{\partial \rho}{\partial t} + \nabla \cdot \left(\rho \, \frac{\nabla S}{m} \right) = 0,$$

which is identical to the wave-action conservation law (9.30), with $\mathcal{J} \equiv \hbar \rho$ and $\mathbf{J} \equiv \mathcal{J} \nabla S / m$.

9.4 Euler Equations for a Perfect Fluid

Given Euler's role in the development of the Calculus of Variations in Chap. 1, it is extremely fitting to end this textbook on Lagrangian Mechanics by considering the Euler equations of motion for a perfect fluid

$$\frac{\partial \rho}{\partial t} + \nabla \cdot (\rho \, \mathbf{u}) = 0, \tag{9.33}$$

$$\rho \left(\frac{\partial}{\partial t} + \mathbf{u} \cdot \nabla \right) \mathbf{u} = - \nabla p, \tag{9.34}$$

$$\frac{\partial S}{\partial t} + \mathbf{u} \cdot \nabla S = 0. \tag{9.35}$$

Equation (9.33) represents the particle (mass) conservation law, where $\rho(\mathbf{x}, t)$ denotes the mass density of the fluid and $\mathbf{u}(\mathbf{x}, t)$ denotes the fluid velocity. Equation (9.34) represents Newton's Second Law for the perfect fluid, which states that the fluid moves under the influence of a pressure gradient force $- \nabla p(\mathbf{x}, t)$. Equation (9.35) represents the conservation of entropy $S(\mathbf{x}, t)$, which states that entropy is *advected* with the fluid.

According to the First Law of Thermodynamics, the mass density ρ and the entropy S (per unit mass) of the fluid can be used as independent variables so that a change $\delta\varepsilon(\rho, S)$ in the internal energy (per unit mass) of the fluid can be expressed as

$$\delta\varepsilon = T \, \delta S - p \, \delta\rho^{-1}, \tag{9.36}$$

where T and p denote the temperature and pressure of the fluid.

9.4.1   *Lagrangian Formulation*

The Euler-Lagrange formulation of the Euler fluid equations (9.33)-(9.35) is based on the Lagrangian density

$$\mathcal{L} = \frac{1}{2}\rho\,|\mathbf{u}|^2 - \rho\varepsilon + \phi\left(\frac{\partial\rho}{\partial t} + \nabla\cdot\rho\,\mathbf{u}\right) - \rho\lambda\,\frac{dS}{dt}, \qquad (9.37)$$

where ϕ and λ are Lagrange multipliers used to enforce the particle conservation law (9.33) and the constraint (9.35) that entropy is advected by the fluid. Here, the variational fields are represented by the seven-component field $\psi^a = (\rho, \mathbf{u}, S; \phi, \lambda)$ and the Euler-Lagrange equation for each component ψ^a is

$$\frac{\partial}{\partial t}\left(\frac{\partial\mathcal{L}}{\partial(\partial_t\psi^a)}\right) + \nabla\cdot\left(\frac{\partial\mathcal{L}}{\partial(\nabla\psi^a)}\right) = \frac{\partial\mathcal{L}}{\partial\psi^a}. \qquad (9.38)$$

We note that the Lagrangian density (9.37) does not have space-time derivatives for all fields. For example, it is immediately obvious that we recover the constraint equations (9.33) and (9.35) from variations of the Lagrangian density with respect to the Lagrange multipliers ϕ and λ (i.e., $\partial\mathcal{L}/\partial\phi = 0 = \delta\mathcal{L}/\partial\lambda$).

The Euler-Lagrange equation for the mass density ρ is

$$\frac{d\phi}{dt} \equiv \frac{\partial\phi}{\partial t} + \mathbf{u}\cdot\nabla\phi = \frac{1}{2}\,|\mathbf{u}|^2 - h, \qquad (9.39)$$

where $h \equiv \partial(\rho\varepsilon)/\partial\rho = \varepsilon + p/\rho$ is the *enthalpy* of the fluid. The Euler-Lagrange equation for the entropy S is

$$\frac{d\lambda}{dt} = T, \qquad (9.40)$$

where we made use of Eqs. (9.33) and (9.36). Lastly, the Euler-Lagrange equation for the fluid velocity $\mathbf{u}$ yields

$$\mathbf{u} = \nabla\phi + \lambda\,\nabla S, \qquad (9.41)$$

which introduces a decomposition of the fluid velocity in terms of a curl-free term (i.e., $\nabla\times\nabla\phi = 0$) and a term that is proportional to the entropy gradient.

We now show that the equation of motion (9.34) is contained in Eqs. (9.39)-(9.41) as follows. First, we write the partial time derivative of Eq. (9.41)

$$\frac{\partial\mathbf{u}}{\partial t} = \nabla\frac{\partial\phi}{\partial t} + \frac{\partial\lambda}{\partial t}\,\nabla S + \lambda\,\nabla\frac{\partial S}{\partial t}$$

$$= \nabla\left(\frac{1}{2}\,|\mathbf{u}|^2 - h - \mathbf{u}\cdot\nabla\phi\right) + (T - \mathbf{u}\cdot\nabla\lambda)\,\nabla S - \lambda\,\nabla(\mathbf{u}\cdot\nabla S).$$

By rearranging terms, we find

$$\frac{\partial \mathbf{u}}{\partial t} = \nabla \mathbf{u} \cdot (\mathbf{u} - \nabla \phi - \lambda \nabla S) - \mathbf{u} \cdot \nabla (\nabla \phi + \lambda \nabla S)$$
$$- \nabla h + T \nabla S.$$

Lastly, by using Eq. (9.41) and the identity $T \nabla S - \nabla h \equiv - \rho^{-1} \nabla p$, we recover Eq. (9.34).

9.4.2 *Energy-Momentum Conservation Laws*

We now derive the energy-momentum conservation laws for the Euler fluid equations (9.33)-(9.35). The Noether equation for the Lagrangian density (9.37) is expressed as

$$\delta \mathcal{L} = \frac{\partial}{\partial t} \left(\phi \, \delta \rho - \rho \lambda \, \delta S \right) + \nabla \cdot \left[(\delta \rho \, \mathbf{u} + \rho \, \delta \mathbf{u}) \phi - \rho \mathbf{u} \, \lambda \, \delta S \right]. \quad (9.42)$$

Here, the variations $(\delta \rho, \delta \mathbf{u}, \delta S)$ are expressed in terms of space-time translations $\boldsymbol{\xi} \equiv \delta \mathbf{x} - \mathbf{u} \, \delta t$ as

$$\delta \rho = - \nabla \cdot (\boldsymbol{\xi} \, \rho), \quad (9.43)$$

$$\delta \mathbf{u} = \frac{\partial \boldsymbol{\xi}}{\partial t} + \mathbf{u} \cdot \nabla \boldsymbol{\xi} - \boldsymbol{\xi} \cdot \nabla \mathbf{u}, \quad (9.44)$$

$$\delta S = - \boldsymbol{\xi} \cdot \nabla S, \quad (9.45)$$

which satisfy the constraint equations

$$\frac{\partial \delta \rho}{\partial t} = - \nabla \cdot (\delta \rho \, \mathbf{u} + \rho \, \delta \mathbf{u}), \quad (9.46)$$

$$\frac{\partial \delta S}{\partial t} = - \delta \mathbf{u} \cdot \nabla S - \mathbf{u} \cdot \nabla \delta S. \quad (9.47)$$

By substituting Eqs. (9.43)-(9.45) on the right side of the Noether equation (9.42), we obtain

$$\delta \mathcal{L} = \frac{\partial}{\partial t} (\rho \, \boldsymbol{\xi} \cdot \mathbf{u}) + \nabla \cdot \left[\mathbf{u} (\rho \, \boldsymbol{\xi} \cdot \mathbf{u}) - \rho \boldsymbol{\xi} \left(\frac{1}{2} |\mathbf{u}|^2 - h \right) \right], \quad (9.48)$$

after carrying out several cancellations as well as using the identity

$$\nabla \cdot [\rho \phi (\mathbf{u} \boldsymbol{\xi} - \boldsymbol{\xi} \mathbf{u})] = \nabla \times (\rho \phi \, \boldsymbol{\xi} \times \mathbf{u}),$$

and using Eq. (9.39) for $d\phi/dt$.

First, the energy conservation law

$$\frac{\partial \mathcal{E}}{\partial t} + \nabla \cdot \mathbf{S} = 0 \quad (9.49)$$

Table 9.1   Summary of Chapter 9: Continuous Lagrangian Systems.

Topic	Equation
Lagrangian Density for a Wave on a Stretched String	(9.6)
General Euler-Lagrange Field Equation	(9.10)
Noether Equation and Conservation Laws	(9.11)-(9.15)
Lagrangian Formulation of the Schroedinger Equation	(9.25)-(9.31)
Lagrangian Formulation of the Euler Equations for a Perfect Fluid	(9.37)-(9.53)

is associated with time-translation symmetry for which $\delta\mathcal{L} = -\delta t \, \partial\mathcal{L}/\partial t$, where the energy density and energy-density flux are

$$\mathcal{E} = \rho \, |\mathbf{u}|^2 - \mathcal{L} = \rho \left(\frac{1}{2} |\mathbf{u}|^2 + \varepsilon \right), \tag{9.50}$$

$$\mathbf{S} = \rho \, \mathbf{u} \left(\frac{1}{2} |\mathbf{u}|^2 + \varepsilon \right) + p \, \mathbf{u} \equiv (\mathcal{E} + p) \, \mathbf{u}. \tag{9.51}$$

Second, the momentum conservation

$$\frac{\partial \mathbf{P}}{\partial t} + \nabla \cdot \mathsf{T} = 0 \tag{9.52}$$

is associated with space-translation symmetry for which $\delta\mathcal{L} = -\delta\mathbf{x} \cdot \nabla\mathcal{L}$, where the momentum density and stress tensor are

$$\mathbf{P} = \rho \, \mathbf{u} \quad \text{and} \quad \mathsf{T} = \rho \, \mathbf{u}\mathbf{u} + p \, \mathbf{I}. \tag{9.53}$$

The Euler fluid equations also possess a wave-action conservation law, which requires us to introduce a fluid reference state on which waves propagate.

We note in closing that the Euler fluid equations (9.33)-(9.35) possess a different Lagrangian formulation (see problem 5) that makes use of constrained variations for the fluid fields $(\rho, \mathbf{u}, S)$ without the use of Lagrange multipliers (ϕ, λ).

9.5   Summary

Chapter 9 presented a brief introduction to the Lagrangian formulation of classical and quantum field equations. The Noether method was also presented and applied to the derivation of the energy-momentum and action conservation laws. Table 9.1 presents a summary of the important topics of Chapter 9.

9.6 Problems

1. Verify Eqs. (9.28)-(9.29) for an arbitrary potential $U(\mathbf{x}, t)$ in the Schroedinger equation (9.26).

2. When we insert the ansatz (9.32) into the Schroedinger equation (9.26), we obtain two equations defined as the real and imaginary parts of the resulting Schroedinger equation. Derive these two equations and give their interpretations in the classical limit $\hbar \to 0$.

3. Show that Eqs. (9.39)-(9.41) have the Euler-Lagrange form (9.38).

4. Show that Eqs. (9.43)-(9.45) satisfy the constraint equations (9.46)-(9.47).

5. Consider a perfect fluid under the influence of an external force (e.g., gravity) and subject to the equation of motion

$$\rho \left(\frac{\partial}{\partial t} + \mathbf{u} \cdot \nabla \right) \mathbf{u} = - \nabla p - \rho \nabla \Phi,$$

where $\Phi(\mathbf{x}, t)$ denotes the scalar potential (per unit mass) associated with the external force. Show that the new equation of motion can be derived from the new Lagrangian density

$$\mathcal{L}' = \frac{1}{2} \rho |\mathbf{u}|^2 - \rho (\varepsilon + \Phi) + \phi \left(\frac{\partial \rho}{\partial t} + \nabla \cdot \rho \mathbf{u} \right) - \rho \lambda \frac{dS}{dt}.$$

6. * Show that the Euler fluid equations (9.33)-(9.35) can be formulated in terms of a constrained variational principle, with the Lagrangian density

$$\mathcal{L} = \frac{1}{2} \rho |\mathbf{u}|^2 - \rho \, \varepsilon(\rho, S),$$

where the constrained variations

$$\delta \rho = - \nabla \cdot (\rho \boldsymbol{\xi}),$$
$$\delta \mathbf{u} = \frac{\partial \boldsymbol{\xi}}{\partial t} + \mathbf{u} \cdot \nabla \boldsymbol{\xi} - \boldsymbol{\xi} \cdot \nabla \mathbf{u},$$
$$\delta S = - \boldsymbol{\xi} \cdot \nabla S,$$

are expressed in terms of the virtual fluid displacement $\boldsymbol{\xi}$.

Appendix A

Basic Mathematical Methods

Appendix A introduces, first, a simple solution for finding the roots of a general cubic polynomial, which is a problem that is encountered often enough throughout the textbook. Second, we present a compendium of definite integrals that are evaluated by either the trigonometric-substitution method or the hyperbolic-trigonometric-substitution method. Third, an explicit derivation of the Frenet-Serret formulas is presented for an arbitrary curve in three-dimensional space used in Chapters 1-2 and 6. Fourth, some basic concepts in linear algebra that a student may have acquired before taking this course are summarized. Hopefully, this material will assist the student in following the presentation in Chapters 7 and 8. Lastly, some general comments are made concerning the numerical analysis of the nonlinear (and coupled) differential equations presented in this textbook.

A.1 Roots of a General Cubic Polynomial

The problem of finding the roots of a cubic polynomial arises often enough in physics, that it is worthwhile to present a simple solution based on a standard trigonometric identity

$$\cos\phi \;=\; \cos\left(3 \cdot \frac{\phi}{3}\right) \;=\; 4\cot^3(\phi/3) \,-\, 3\cos(\phi/3), \qquad \text{(A.1)}$$

where ϕ may be real or complex-valued.

We begin with the problem of finding the roots of a general cubic polynomial

$$f(x) \;=\; 4\,x^3 \,+\, a\,x^2 \,+\, b\,x \,+\, c, \qquad \text{(A.2)}$$

where the constants (a, b, c) are arbitrary, and the coefficient 4 appears in analogy of Eq. (A.1). In order to continue this analogy, we need to find a transformation $x = z + \alpha$ that will eliminate the z^2-term with a proper choice of α. First, we introduce the Taylor expansion

$$f(z + \alpha) = f(\alpha) + f'(\alpha) z + \frac{1}{2} f''(\alpha) z^2 + 4 z^3,$$

and request that $f''(\alpha)$ vanish, which yields an expression for α in terms of the coefficient a:

$$f''(\alpha) = 24 \alpha + 2 a = 0 \quad \to \quad \alpha = -\frac{a}{12}.$$

Hence, with this choice for α, the cubic polynomial (A.2) becomes

$$P(z) = 4 z^3 + f'(\alpha) z + f(\alpha) \equiv 4 z^3 - g_2 z - g_3, \tag{A.3}$$

where the new coefficients are $g_2 \equiv -f'(\alpha)$ and $g_3 \equiv -f(\alpha)$. The notation used here is taken from the Weierstrass equation (B.19).

Secondly, we define two parameters β and ϕ (which may or may not be real):

$$g_2 \equiv 3 \beta^2 \quad \text{and} \quad g_3 \equiv \beta^3 \cos \phi, \tag{A.4}$$

so that the first root of Eq. (A.3) is

$$z_1 \equiv \beta \cos(\phi/3). \tag{A.5}$$

Note that we easily verify that

$$4 z^3 - g_2 z - g_3 = \beta^3 \left[4 \cos^3(\phi/3) - 3 \cos(\phi/3) - \cos \phi \right] = 0,$$

which follows from the trigonometric identity (A.1).

Thirdly, if we divide Eq. (A.3) by $(z - z_1)$, we obtain the quadratic polynomial

$$4 z^2 + 4 \beta \cos(\phi/3) z + \beta^2 \left[4 \cos^2(\phi/3) - 3 \right],$$

whose roots are

$$z_{2,3} = \beta \left(-\frac{1}{2} \cos(\phi/3) \pm \frac{\sqrt{3}}{2} \sin(\phi/3) \right) \equiv -\beta \cos\left(\frac{\pi \pm \phi}{3} \right). \tag{A.6}$$

Hence, the three roots of the cubic polynomial (A.2) are

$$x_1 = \alpha + \beta \cos(\phi/3) \quad \text{and} \quad x_{2,3} = \alpha - \beta \cos[(\pi \pm \phi)/3],$$

where

$$\alpha = -\frac{a}{12}, \quad \beta = \sqrt{-\frac{1}{3} f'(\alpha)}, \quad \text{and} \quad \phi = \arccos\left[\frac{-f(\alpha)}{(-f'(\alpha)/3)^{3/2}} \right].$$

Lastly, we note that the roots of Eq. (A.2) are multiple if the discriminant $\Delta \equiv g_2^3 - 27 g_3^2$ vanishes, where

$$\Delta = 27 \beta^6 (1 - \cos^2 \phi) = 27 \beta^6 \sin^2 \phi.$$

Hence, two roots are equal when $\phi = 0$: $x_1 = \alpha + \beta$ and $x_{2,3} = \alpha - \beta/2$ or $\phi = \pi$: $x_{1,2} = \alpha + \beta/2$ and $x_3 = \alpha - \beta$. When the constants (a, b, c) are real in Eq. (A.2), then either all roots are real or one root is real and the other two are complex-conjugate of each other (see Fig. B.5).

A.2 Integration by Trigonometric Substitution

The method of trigonometric substitution has a special place in physics. It is used in most chapters of this textbook. In this Section, we present a compendium of integrals that are solved either by the trigonometric-substitution method or the hyperbolic-trigonometric-substitution method. The goal of this method is to convert an integral of the form $\int f(\sqrt{1 \pm x^2})\, dx$ into a trigonometric-function integral that can be readily evaluated. The reader may consult Gradshteyn and Ryzhik [8] for additional useful formulas.

A.2.1 *Trigonometric Functions*

The trigonometric-substitution method relies on the following trigonometric identities

$$\cos^2 \theta + \sin^2 \theta = 1 \quad \text{and} \quad 1 + \tan^2 \theta = \sec^2 \theta,$$

and the indefinite integral

$$\int \sec \theta \, d\theta = \ln(\sec \theta + \tan \theta),$$

which leads to the indefinite integrals

$$\int^{\arccos x} \sec \theta \, d\theta = \ln \left(\frac{1}{x} + \frac{1}{x} \sqrt{1 - x^2} \right), \tag{A.7}$$

$$\int^{\arcsin x} \sec \theta \, d\theta = \ln \left(\frac{1 + x}{\sqrt{1 - x^2}} \right) = \frac{1}{2} \ln \left(\frac{1 + x}{1 - x} \right), \tag{A.8}$$

$$\int^{\arctan x} \sec \theta \, d\theta = \ln \left(\sqrt{x^2 + 1} + x \right), \tag{A.9}$$

$$\int^{\text{arcsec} \, x} \sec \theta \, d\theta = \ln \left(x + \sqrt{x^2 - 1} \right). \tag{A.10}$$

1. Sine substitution ($t = \sin \theta \;\rightarrow\; dt/\sqrt{1 - t^2} = d\theta$)

$$\theta = \int_0^{\sin \theta} \frac{dt}{\sqrt{1 - t^2}} \;\rightarrow\; \arcsin x = \int_0^x \frac{dt}{\sqrt{1 - t^2}} \tag{A.11}$$

2. Cosine substitution ($t = \cos \theta \;\rightarrow\; dt/\sqrt{1 - t^2} = -\, d\theta$)

$$\theta = \int_{\cos \theta}^1 \frac{dt}{\sqrt{1 - t^2}} \;\rightarrow\; \arccos x = \int_x^1 \frac{dt}{\sqrt{1 - t^2}} \tag{A.12}$$

3. Tangent substitution I $(t = \tan\theta \;\to\; dt/(t^2 + 1) = d\theta)$

$$\theta = \int_0^{\tan\theta} \frac{dt}{t^2 + 1} \;\to\; \arctan x = \int_0^x \frac{dt}{t^2 + 1} \qquad \text{(A.13)}$$

Tangent substitution II $(y = \tan\theta \;\to\; dy/\sqrt{y^2 + 1} = \sec\theta\, d\theta)$

$$\int_0^x \frac{dy}{\sqrt{y^2 + 1}} = \int_0^{\arctan x} \sec\theta\, d\theta = \ln\left(x + \sqrt{x^2 + 1}\right) \qquad \text{(A.14)}$$

4. Secant substitution $(y = \sec\theta \;\to\; dy/\sqrt{y^2 - 1} = \sec\theta\, d\theta)$

$$\int_1^x \frac{dy}{\sqrt{y^2 - 1}} = \int_0^{\operatorname{arcsec} x} \sec\theta\, d\theta = \ln\left(x + \sqrt{x^2 - 1}\right) \qquad \text{(A.15)}$$

A.2.2 *Hyperbolic-Trigonometric Functions*

The hyperbolic-trigonometric-substitution method relies on the following hyperbolic-trigonometric identities

$$\cosh^2 z - \sinh^2 z = 1 \quad \text{and} \quad 1 - \tanh^2 z = \operatorname{sech}^2 z,$$

where

$$\cosh z = \cos(i\,z) = \frac{1}{2}\left(e^z + e^{-z}\right) \text{ and } \sinh z = -i\,\sin(i\,z) = \frac{1}{2}\left(e^z - e^{-z}\right).$$

1. Hyperbolic-sine substitution $(y = \sinh z \;\to\; dy/\sqrt{y^2 + 1} = dz)$

$$z = \int_0^{\sinh z} \frac{dy}{\sqrt{y^2 + 1}} \;\to\; \operatorname{arcsinh} z = \int_0^z \frac{dy}{\sqrt{y^2 + 1}} \qquad \text{(A.16)}$$

Tangent substitution $(y = \tan t \;\to\; dy/\sqrt{y^2 + 1} = \sec t\, dt)$

$$\operatorname{arcsinh} z = \int_0^{\arctan z} \sec t\, dt = \ln\left(z + \sqrt{z^2 + 1}\right) \qquad \text{(A.17)}$$

2. Hyperbolic-cosine substitution $(y = \cosh z \;\to\; dy/\sqrt{y^2 - 1} = dz)$

$$z = \int_1^{\cosh z} \frac{dy}{\sqrt{y^2 - 1}} \;\to\; \operatorname{arccosh} z = \int_1^z \frac{dy}{\sqrt{y^2 - 1}} \qquad \text{(A.18)}$$

Secant substitution $(y = \sec t \;\to\; dy/\sqrt{y^2 - 1} = \sec t\, dt)$

$$\operatorname{arccosh} z = \int_0^{\operatorname{arcsec} z} \sec t\, dt = \ln\left(z + \sqrt{z^2 - 1}\right) \qquad \text{(A.19)}$$

3. Hyperbolic-tangent substitution $(y = \tanh z \;\to\; dy/(1 - y^2) = dz)$

$$z = \int_0^{\tanh z} \frac{dy}{1 - y^2} \;\to\; \text{arctanh}\, z = \int_0^z \frac{dy}{1 - y^2} \qquad (A.20)$$

Sine substitution $(y = \sin t \;\to\; dy/(1 - y^2) = \sec t\, dt)$

$$\text{arctanh}\, z = \int_0^{\arcsin z} \sec t\, dt = \frac{1}{2} \ln\left(\frac{1 + z}{1 - z}\right) \qquad (A.21)$$

A.3 Frenet-Serret Formulas

The geometric properties of spatial curves expressed in terms of the Frenet-Serret formulas (Jean Frédéric Frenet, 1816-1900; Jospeh Alfred Serret, 1819-1895) are discussed in many places in the textbook, e.g., in the context of light rays (Chap. 1), the Maupertuis Principle (Chap. 2), the scattering of a particle by a hard surface (Chap. 5), and the Coriolis acceleration (Chap. 6). Here, we review the derivation of the Frenet-Serret formulas and discuss other applications.

A.3.1 *Frenet Frame*

Consider a curve

$$\mathbf{r}(t) = x(t)\widehat{\mathbf{x}} + y(t)\widehat{\mathbf{y}} + z(t)\widehat{\mathbf{z}} \qquad (A.22)$$

in three-dimensional space parameterized by time t. The infinitesimal length element along the curve $ds(t) = v(t)\, dt$ is also parameterized by time t, with $v(t) \equiv |\dot{\mathbf{r}}|$ denoting the speed along the curve at time t.

The Frenet-Serret formulas associated with the curvature κ and torsion τ of the curve (A.22) are defined in terms of the right-handed set of unit vectors $(\widehat{\mathbf{t}}, \widehat{\mathbf{n}}, \widehat{\mathbf{b}})$, where $\widehat{\mathbf{t}}$ denotes the *tangent* unit vector, $\widehat{\mathbf{n}}$ denotes the *normal* unit vector, and $\widehat{\mathbf{b}}$ denotes the *binormal* unit vector.

First, by definition, the tangent unit vector is

$$\widehat{\mathbf{t}} \equiv \frac{d\mathbf{r}}{ds} = \frac{\dot{\mathbf{r}}(t)}{v(t)}. \qquad (A.23)$$

The definitions of the curvature κ and the normal unit vector $\widehat{\mathbf{n}}$ are

$$\frac{d\widehat{\mathbf{t}}}{ds} = \frac{1}{v}\frac{d}{dt}\left(\frac{\dot{\mathbf{r}}}{v}\right) = \frac{\ddot{\mathbf{r}}}{v^2} - \frac{\dot{v}}{v^2}\widehat{\mathbf{t}} = \frac{\widehat{\mathbf{t}}}{v^2} \times (\ddot{\mathbf{r}} \times \widehat{\mathbf{t}})$$

$$= \left(\frac{\dot{\mathbf{r}} \times \ddot{\mathbf{r}}}{v^3}\right) \times \widehat{\mathbf{t}} = \kappa\left(\widehat{\mathbf{b}} \times \widehat{\mathbf{t}}\right) \equiv \kappa\widehat{\mathbf{n}}, \qquad (A.24)$$

where $\dot{v} \equiv d|\dot{\mathbf{r}}|/dt = \widehat{\mathbf{t}} \cdot \ddot{\mathbf{r}}$, so that the curvature is defined as

$$\kappa \equiv \frac{|\dot{\mathbf{r}} \times \ddot{\mathbf{r}}|}{v^3}, \tag{A.25}$$

while the normal and binormal unit vectors are defined as

$$\widehat{\mathbf{n}} \equiv \frac{\dot{\mathbf{r}} \times (\ddot{\mathbf{r}} \times \dot{\mathbf{r}})}{\kappa \, v^4} = \widehat{\mathbf{t}} \times \left(\frac{\ddot{\mathbf{r}} \times \dot{\mathbf{r}}}{|\ddot{\mathbf{r}} \times \dot{\mathbf{r}}|} \right), \tag{A.26}$$

and

$$\widehat{\mathbf{b}} \equiv \frac{\dot{\mathbf{r}} \times \ddot{\mathbf{r}}}{|\dot{\mathbf{r}} \times \ddot{\mathbf{r}}|}. \tag{A.27}$$

Hence, the curve (A.22) exhibits curvature if its velocity $\dot{\mathbf{r}}$ and acceleration $\ddot{\mathbf{r}}$ are not collinear. We note that for a two-dimensional curve $\mathbf{r} = x(s)\,\widehat{\mathbf{x}} + y(s)\,\widehat{\mathbf{y}}$, we find

$$\widehat{\mathbf{t}} = \frac{d\mathbf{r}}{ds} = x'\,\widehat{\mathbf{x}} + y'\,\widehat{\mathbf{y}} \equiv \cos\phi\,\widehat{\mathbf{x}} + \sin\phi\,\widehat{\mathbf{y}},$$

where $\phi(s)$ denotes the tangential angle. With this definition, we readily show that the curvature is defined as

$$\kappa \equiv \left| \frac{d\widehat{\mathbf{t}}}{ds} \right| = \frac{d\phi}{ds}.$$

Second, we obtain the following expression for the derivative of the normal unit vector (A.26):

$$\frac{d\widehat{\mathbf{n}}}{ds} = \frac{d\widehat{\mathbf{t}}}{ds} \times \left(\frac{\ddot{\mathbf{r}} \times \dot{\mathbf{r}}}{|\ddot{\mathbf{r}} \times \dot{\mathbf{r}}|} \right) + \widehat{\mathbf{t}} \times \frac{d}{ds} \left(\frac{\ddot{\mathbf{r}} \times \dot{\mathbf{r}}}{|\ddot{\mathbf{r}} \times \dot{\mathbf{r}}|} \right)$$

$$= \left[\widehat{\mathbf{t}} \times \left(\frac{\ddot{\mathbf{r}} \times \dot{\mathbf{r}}}{v^3} \right) \right] \times \left(\frac{\ddot{\mathbf{r}} \times \dot{\mathbf{r}}}{|\ddot{\mathbf{r}} \times \dot{\mathbf{r}}|} \right) + \frac{\widehat{\mathbf{t}}}{v} \times \left[\widehat{\mathbf{b}} \times \left(\frac{(\ddot{\mathbf{r}} \times \dot{\mathbf{r}}) \times \widehat{\mathbf{b}}}{|\ddot{\mathbf{r}} \times \dot{\mathbf{r}}|} \right) \right]$$

$$= \kappa \left(\widehat{\mathbf{t}} \times \widehat{\mathbf{b}} \right) \times \widehat{\mathbf{b}} + \left(\frac{\widehat{\mathbf{t}} \cdot \left[(\dddot{\mathbf{r}} \times \dot{\mathbf{r}}) \times \widehat{\mathbf{b}} \right]}{v\,|\ddot{\mathbf{r}} \times \dot{\mathbf{r}}|} \right) \widehat{\mathbf{b}} \equiv -\kappa\widehat{\mathbf{t}} + \tau\widehat{\mathbf{b}}, \tag{A.28}$$

where the torsion

$$\tau \equiv \frac{\widehat{\mathbf{t}} \cdot \left[(\dddot{\mathbf{r}} \times \dot{\mathbf{r}}) \times \widehat{\mathbf{b}} \right]}{v\,|\ddot{\mathbf{r}} \times \dot{\mathbf{r}}|} = \frac{\widehat{\mathbf{n}} \cdot (\dddot{\mathbf{r}} \times \dot{\mathbf{r}})}{\kappa\,v^4} = \frac{\dot{\mathbf{r}} \cdot (\ddot{\mathbf{r}} \times \dddot{\mathbf{r}})}{\kappa^2\,v^6} \tag{A.29}$$

is defined in terms of the triple product $\dot{\mathbf{r}} \cdot (\ddot{\mathbf{r}} \times \dddot{\mathbf{r}})$. Hence, the torsion requires that the rate of change of acceleration $\dddot{\mathbf{r}} = d\ddot{\mathbf{r}}/dt$ (known as *jerk*) along the curve have a nonvanishing component perpendicular to the plane constructed by the velocity $\dot{\mathbf{r}}$ and the acceleration $\ddot{\mathbf{r}}$. We readily find that a curve that is restricted to a plane possesses zero torsion.

Third, we obtain the expression for the derivative of the binormal unit vector (A.27):

$$\frac{d\widehat{\mathbf{b}}}{ds} = \frac{\widehat{\mathbf{b}} \times \left[\left(\dot{\mathbf{r}} \times \ddot{\mathbf{r}}\right) \times \widehat{\mathbf{b}}\right]}{v\,|\dot{\mathbf{r}} \times \ddot{\mathbf{r}}|} = \alpha\widehat{\mathbf{t}} + \beta\widehat{\mathbf{n}} \equiv -\tau\,\widehat{\mathbf{n}}, \qquad (A.30)$$

where

$$\alpha \equiv \widehat{\mathbf{t}} \cdot \frac{d\widehat{\mathbf{b}}}{ds} = \frac{\widehat{\mathbf{t}} \cdot \left(\dot{\mathbf{r}} \times \ddot{\mathbf{r}}\right)}{\kappa\,v^4} = \frac{\dot{\mathbf{r}} \cdot \left(\dot{\mathbf{r}} \times \ddot{\mathbf{r}}\right)}{\kappa\,v^5} \equiv 0,$$

and

$$\beta \equiv \widehat{\mathbf{n}} \cdot \frac{d\widehat{\mathbf{b}}}{ds} = \frac{\widehat{\mathbf{n}} \cdot \left(\dot{\mathbf{r}} \times \dddot{\mathbf{r}}\right)}{\kappa\,v^4} = \left[\frac{\dot{\mathbf{r}} \times (\ddot{\mathbf{r}} \times \dot{\mathbf{r}})}{\kappa^2\,v^8}\right] \cdot \left(\dot{\mathbf{r}} \times \dddot{\mathbf{r}}\right)$$

$$= \frac{\dddot{\mathbf{r}} \cdot \left(\dot{\mathbf{r}} \times \ddot{\mathbf{r}}\right)}{\kappa^2\,v^6} = -\tau.$$

The equations (A.24), (A.28), and (A.30) are refered to as the Frenet-Serret formulas, which describes the evolution of the unit vectors $(\widehat{\mathbf{t}}, \widehat{\mathbf{n}}, \widehat{\mathbf{b}})$ along the curve (A.22) in terms of the curvature (A.25) and the torsion (A.29).

Lastly, we note that by introducing the *Darboux* vector (Gaston Darboux, 1842-1917)

$$\boldsymbol{\omega} \equiv \tau\widehat{\mathbf{t}} + \kappa\widehat{\mathbf{b}}, \qquad (A.31)$$

the Frenet-Serret equations (A.24), (A.28), and (A.30) may be written as

$$\frac{d\widehat{\mathbf{e}}_i}{ds} \equiv \boldsymbol{\omega} \times \widehat{\mathbf{e}}_i, \qquad (A.32)$$

where $\widehat{\mathbf{e}}_i = (\widehat{\mathbf{t}}, \widehat{\mathbf{n}}, \widehat{\mathbf{b}})$ denotes a component of the so-called Frenet *frame*. Hence, curvature is a measure of the rotation of the Frenet frame about the binormal unit vector $\widehat{\mathbf{b}}$, while torsion is the measure of the rotation of the Frenet frame about the tangent unit vector $\widehat{\mathbf{t}}$.

A.3.2 Darboux Frame

If the curve $\mathbf{r}(t)$ lies on a surface $\mathcal{S}(\mathbf{r})$, then a Darboux *frame* $\widehat{\mathsf{E}}_i \equiv (\widehat{\mathsf{T}}, \widehat{\mathsf{N}}, \widehat{\mathsf{B}})$ can be constructed as follows. First, the tangent unit vector $\widehat{\mathsf{T}} \equiv \widehat{\mathbf{t}}$ is identical to the tangent unit vector $\widehat{\mathbf{t}} \equiv \dot{\mathbf{r}}/|\dot{\mathbf{r}}|$, which is also tangent to the surface $\mathcal{S}$, i.e., $\widehat{\mathsf{T}} \cdot \nabla\mathcal{S} = 0$. Next, the normal unit vector $\widehat{\mathsf{N}}$ is naturally chosen to be perpendicular to the surface $\mathcal{S}$, i.e., $\widehat{N} \equiv \pm \nabla \mathcal{S}/|\nabla \mathcal{S}|$, while the binormal unit vector $\widehat{B}$ satisfies the identity $\widehat{B} \equiv \widehat{T} \times \widehat{N}$.

The relation between the Frenet frame and the Darboux frame involves a rotation about the $\widehat{t}$-axis:

$$\begin{pmatrix} \widehat{T} \\ \widehat{N} \\ \widehat{B} \end{pmatrix} = \begin{pmatrix} 1 & 0 & 0 \\ 0 & \cos\theta & \sin\theta \\ 0 & -\sin\theta & \cos\theta \end{pmatrix} \cdot \begin{pmatrix} \widehat{t} \\ \widehat{n} \\ \widehat{b} \end{pmatrix}, \qquad (A.33)$$

where the rotation angle $\theta(s)$ may depend on the position s along the curve. The corresponding Frenet-Serret formulas for the Darboux frame are

$$\left. \begin{aligned} d\widehat{T}/ds &= \kappa \cos\theta\, \widehat{N} - \kappa \sin\theta\, \widehat{B} \\[4pt] d\widehat{N}/ds &= -\kappa \cos\theta\, \widehat{T} + (\tau + d\theta/ds)\widehat{B} \\[4pt] d\widehat{B}/ds &= \kappa \sin\theta\, \widehat{T} - (\tau + d\theta/ds)\widehat{N} \end{aligned} \right\} \;\to\; \frac{d\widehat{E}_i}{ds} = \boldsymbol{\Omega} \times \widehat{E}_i, \quad (A.34)$$

where

$$\boldsymbol{\Omega} \equiv \kappa \left(\cos\theta\, \widehat{B} + \sin\theta\, \widehat{N} \right) + \left(\tau + \frac{d\theta}{ds} \right) \widehat{T}, \qquad (A.35)$$

where $\kappa_g \equiv \kappa \cos\theta$ is called the *geodesic* curvature, $\kappa_n \equiv -\kappa \sin\theta$ is called the *normal* curvature, and $\tau_r \equiv \tau + d\theta/ds$ is called the *relative* torsion.

A.3.3 Example: Seiffert Spiral on the Unit Sphere

For example, consider the Seiffert spiral curve on the unit sphere:[1]

$$\mathbf{r}(s; k) = \operatorname{sn}(s|k^2) \left[\cos(ks)\widehat{x} + \sin(ks)\widehat{y} \right] + \operatorname{cn}(s|k^2)\widehat{z}, \qquad (A.36)$$

where s measures the length of the curve from the North Pole: $\mathbf{r}(0) = \widehat{z}$, the functions $(\operatorname{cn}, \operatorname{sn})$ are Jacobi elliptic functions (see App. B), and the modulus $k^2 < 1$. Because Eq. (A.36) describes a curve on the unit sphere (i.e., $|\mathbf{r}| = 1$), we write $\mathbf{r} = \widehat{r} \equiv \operatorname{sn}\widehat{\rho} + \operatorname{cn}\widehat{z}$ (we omit displaying the arguments of the Jacobi functions for convenience). When the value $k = 0$ is inserted into Eq. (A.36), we obtain a circle $\mathbf{r}(s; 0) = \sin s\, \widehat{x} + \cos s\, \widehat{z}$ on the (x, z)-plane. When the value $k = 1$ is inserted into Eq. (A.36), we obtain $\mathbf{r}(s; 1) = \tanh s\, (\cos s\, \widehat{x} + \sin s\, \widehat{y}) + \operatorname{sech} s\, \widehat{z}$, which quickly settles into a circle on the (x, y)-plane.

---

[1] See papers by P. Erdös, Am. J. phys. **68**, 888 (2000) and A. J. Brizard, Eur. J. Phys. **30**, 729-750 (2009).

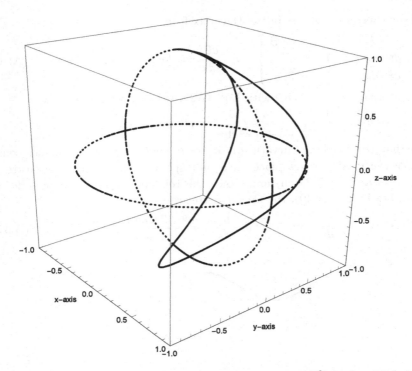

Fig. A.1 Plots of Seiffert spiral (A.36) from $s = 0$ to $s = 4\,\mathsf{K}(k^2)$ for $k = 0.5$ (solid). Circles (dotted) on the (x, y)-plane and (x, z)-plane are shown as guides.

Using properties of the Jacobi elliptic functions discussed in App. B, we readily find the tangent unit vector

$$\widehat{t} = \frac{d\widehat{r}}{ds} = \mathrm{dn}\,\left(\mathrm{cn}\,\widehat{\rho} - \mathrm{sn}\,\widehat{z}\right) + k\,\mathrm{sn}\,\widehat{\varphi} \equiv k\,\mathrm{sn}\,\widehat{\varphi} - \mathrm{dn}\,(\widehat{r} \times \widehat{\varphi}), \quad \text{(A.37)}$$

where $d\widehat{\rho}/ds \equiv k\,\widehat{\varphi}$. Next, we find

$$\frac{d\widehat{t}}{ds} = -\widehat{r} + 2\,k\,\mathrm{cn}\,\left[\mathrm{dn}\,\widehat{\varphi} + k\,\mathrm{sn}\,(\widehat{r} \times \widehat{\varphi})\right] \equiv -\widehat{r} + 2\,k\,\mathrm{cn}\,\widehat{\perp}, \quad \text{(A.38)}$$

so that the Frenet-Serret curvature is

$$\kappa(s) \equiv |d\widehat{t}/ds| = \sqrt{1 + 4\,k^2\,\mathrm{cn}^2(s|k^2)} \quad \text{(A.39)}$$

and the normal unit vector $\widehat{n}$ in the Frenet frame is $\widehat{n} \equiv \kappa^{-1}\,d\widehat{t}/ds \equiv \widehat{b} \times \widehat{t}$.

We now note that the unit vectors $(\widehat{t}, \widehat{\perp}, \widehat{r})$ satisfy the relation $\widehat{t} = \widehat{\perp} \times \widehat{r}$, where $\widehat{r}$ is perpendicular to the surface of the unit sphere. We may, therefore, construct the Darboux frame $(\widehat{T}, \widehat{N}, \widehat{B}) \equiv (\widehat{t}, -\widehat{r}, \widehat{\perp})$, which yields the

Frenet-Serret equations in the Darboux frame:

$$\left.\begin{array}{l} d\widehat{\mathsf{T}}/ds = \widehat{\mathsf{N}} + 2\,k\,\mathrm{cn}\,\widehat{\mathsf{B}} \\ d\widehat{\mathsf{N}}/ds = -\,\widehat{\mathsf{T}} \\ d\widehat{\mathsf{B}}/ds = -\,2\,k\,\mathrm{cn}\,\widehat{\mathsf{T}} \end{array}\right\} \rightarrow \frac{d\widehat{\mathsf{E}}_i}{ds} = \left(\widehat{\mathsf{B}} - 2\,k\,\mathrm{cn}\,\widehat{\mathsf{N}}\right) \times \widehat{\mathsf{E}}_i. \quad (A.40)$$

By comparing Eq. (A.40) with Eqs. (A.34)-(A.35), we readily find

$$\cos\theta(s) \equiv \frac{1}{\kappa} = \frac{1}{\sqrt{1 + 4\,k^2\,\mathrm{cn}^2}} \quad \text{and} \quad \sin\theta(s) = \frac{-\,2\,k\,\mathrm{cn}}{\sqrt{1 + 4\,k^2\,\mathrm{cn}^2}},$$

so that the geodesic curvature is $\kappa_g = \kappa\,\cos\theta = 1$ (which makes sense since the motion takes place on a unit sphere), the normal curvature is $\kappa_n = -\,\kappa\,\sin\theta = 2\,k\,\mathrm{cn}$, and, since the relative torsion $\tau_r = \tau + d\theta/ds$ vanishes in Eq. (A.40), the Frenet-Serret torsion is

$$\tau(s) = -\frac{d\theta(s)}{ds} = -\cot\theta\,\frac{d\ln\kappa}{ds} = \frac{-\,2\,k\,\mathrm{sn}(s|k^2)\,\mathrm{dn}(s|k^2)}{1 + 4\,k^2\,\mathrm{cn}^2(s|k^2)}. \quad (A.41)$$

Fig. A.2   Plots of the Frenet-Serret curvature (solid) and torsion (dashed) for $k = 0.5$. The dotted line shows the geodesic curvature $\kappa_g = 1$.

Figure A.2 shows the Frenet-Serret curvature (A.39) and the Frenet-Serret torsion (A.41) for the Seiffert spiral (A.36) for $k = 0.5$ (see Fig. A.1). As the point leaves the North Pole, the negative torsion causes it to drift westward (negative y) from the vertical circle on the (x, z)-plane.

A.3.4   *Frenet-Serret Formulas for Helical Path*

As a simple application of the Frenet-Serret formulas in the Frenet frame, we apply them to the helical path

$$\mathbf{r}(\theta) = a\,(\cos\theta\,\widehat{\mathbf{x}} + \sin\theta\,\widehat{\mathbf{y}}) + b\,\theta\,\widehat{\mathbf{z}} \quad (A.42)$$

parameterized by the angle θ. Here, the distance s along the helix is simply given as $s = c\theta$, where $c = \sqrt{a^2 + b^2} \equiv ds/d\theta$. Hence, the tangent unit vector is defined as

$$\hat{t} = \frac{d\mathbf{r}}{ds} = \cos\alpha \, (-\sin\theta \, \hat{x} + \cos\theta \, \hat{y}) + \sin\alpha \, \hat{z}, \qquad (A.43)$$

where $(a, b) \equiv (c\cos\alpha, c\sin\alpha)$ and $0 \le \alpha < \pi/2$ denotes the pitch of the helix (e.g., a circle is a helical path with pitch $\alpha = 0$). Next, the derivative of the tangent unit vector yields

$$\frac{d\hat{t}}{ds} = \frac{1}{c}\frac{d\hat{t}}{d\theta} = -\frac{\cos\alpha}{c} \, (\cos\theta \, \hat{x} + \sin\theta \, \hat{y}) \equiv \kappa \, \hat{n}, \qquad (A.44)$$

so that the normal unit vector is

$$\hat{n} = -(\cos\theta \, \hat{x} + \sin\theta \, \hat{y}) \qquad (A.45)$$

and the curvature is $\kappa = c^{-1}\cos\alpha$ (i.e., a circle of radius a, with pitch $\alpha = 0$, has a scalar curvature $\kappa = a^{-1}$). Lastly, the binormal vector is

$$\hat{b} = \hat{t} \times \hat{n} = \sin\alpha \, (-\sin\theta \, \hat{x} + \cos\theta \, \hat{y}) + \cos\alpha \, \hat{z}, \qquad (A.46)$$

so that its derivative yields

$$\frac{d\hat{b}}{ds} = \frac{\sin\alpha}{c} \, (\cos\theta \, \hat{x} + \sin\theta \, \hat{y}) \equiv -\tau \, \hat{n}, \qquad (A.47)$$

where the torsion is $\tau = c^{-1}\sin\alpha$. We can now easily verify that

$$\frac{d\hat{n}}{ds} = \frac{1}{c} \, (\sin\theta \, \hat{x} - \cos\theta \, \hat{y}) \equiv -\kappa \, \hat{t} + \tau \, \hat{b}.$$

A.4   Linear Algebra

The methods of Linear Algebra are introduced and used extensively in Chapters 7 and 8. A brief survey is presented here for convenience.

A fundamental object in linear algebra is the $m \times n$ matrix $\mathbf{A}$ with components (labeled A_{ij}) distributed on m rows ($i = 1, 2, ..., m$) and n columns ($j = 1, 2, ..., n$); for simplicity of notation, we write $\mathbf{A}_{(m \times n)}$ when we want to specify the *order* of the matrix $\mathbf{A}$ and we say that a matrix is *square* if $m = n$.

A.4.1   *Matrix Algebra*

We begin with a discussion of general properties of matrices and later focus our attention on square matrices (in particular 2×2 matrices). First, we can add (or subtract) two matrices only if they are of the same order; hence, the matrix $\mathsf{C} = \mathsf{A} \pm \mathsf{B}$ has components $C_{ij} = A_{ij} \pm B_{ij}$. Next, we can multiply a matrix A by a scalar a and obtain the new matrix $\mathsf{B} = a\,\mathsf{A}$ with components $B_{ij} = a\,A_{ij}$. Lastly, we introduce the *transpose* operation (denoted $^\top$): $\mathsf{A} \to \mathsf{A}^\top$ such that $(\mathsf{A}^\top)_{ij} \equiv A_{ji}$, i.e., the transpose of a $m \times n$ matrix is a $n \times m$ matrix. Note that the *column* vector

$$\mathbf{v} = \begin{pmatrix} v_1 \\ \vdots \\ v_n \end{pmatrix}$$

is a $n \times 1$ matrix while its transpose (a *row* vector) $\mathbf{v}^\top = (v_1, ..., v_n)$ is a matrix of order $1 \times n$. With this definition, we now introduce the operation of matrix multiplication

$$\mathsf{C}_{(m \times k)} = \mathsf{A}_{(m \times n)} \cdot \mathsf{B}_{(n \times k)},$$

where $\mathsf{C}_{(m \times k)}$ is a new matrix of order $m \times k$ with components

$$C_{ij} = \sum_{\ell=1}^{n} A_{i\ell}\,B_{\ell j}.$$

Note that the matrix multiplication

$$\mathbf{u}^\top \cdot \mathbf{v} = \sum_{i=1}^{n} u_i\,v_i = \mathbf{u} \cdot \mathbf{v}$$

coincides with the standard dot product of two vectors.

The remainder of this Section will now deal exclusively with square matrices. First, we introduce two important operations on square matrices: the determinant $\det(\mathsf{A})$ and the trace $\mathrm{Tr}(\mathsf{A})$ defined, respectively, as

$$\det(\mathsf{A}) \equiv \sum_{i=1}^{n} (-1)^{i+j}\,A_{ij}\,\mathrm{ad}_{ij} \equiv \sum_{j=1}^{n} (-1)^{i+j}\,A_{ij}\,\mathrm{ad}_{ij}, \quad \text{(A.48)}$$

$$\mathrm{Tr}(\mathsf{A}) \equiv \sum_{i=1}^{n} A_{ii}, \quad \text{(A.49)}$$

where ad_{ij} denotes the determinant of the *reduced* matrix obtained by removing the ith-row and j^{th}-column from A and the index j is fixed in the first expression in Eq. (A.48), while the index i is fixed in the second expression. Next, we say that the matrix A is invertible if its determinant $\Delta \equiv \det(\mathsf{A})$ does not vanish and we define the inverse A^{-1} with components

$$(\mathsf{A}^{-1})_{ij} \equiv \frac{(-1)^{i+j}}{\Delta} \, \mathrm{ad}_{ji},$$

which, thus, satisfies the identity relation

$$\mathsf{A} \cdot \mathsf{A}^{-1} = \mathsf{I} = \mathsf{A}^{-1} \cdot \mathsf{A},$$

where I denotes the $n \times n$ identity matrix. Note, here, that the matrix multiplication

$$\mathsf{A} \cdot \mathsf{B} \neq \mathsf{B} \cdot \mathsf{A}$$

of two matrices A and B is generally not commutative (for $\mathsf{A}, \mathsf{B} \neq \mathsf{I}$). For example, if

$$\mathsf{A} = \begin{pmatrix} a_1 & a_2 \\ a_3 & a_4 \end{pmatrix} \text{ and } \mathsf{B} = \begin{pmatrix} b_1 & b_2 \\ b_3 & b_4 \end{pmatrix},$$

then

$$\mathsf{A} \cdot \mathsf{B} = \begin{pmatrix} a_1 b_1 + a_2 b_3 & a_1 b_2 + a_2 b_4 \\ a_3 b_1 + a_4 b_3 & a_3 b_2 + a_4 b_4 \end{pmatrix}$$

and

$$\mathsf{B} \cdot \mathsf{A} = \begin{pmatrix} b_1 a_1 + b_2 a_3 & b_1 a_2 + b_2 a_4 \\ b_3 a_1 + b_4 a_3 & b_3 a_2 + b_4 a_4 \end{pmatrix} \neq \mathsf{A} \cdot \mathsf{B}.$$

Lastly, fundamental properties of a square $n \times n$ matrix A are discussed in terms of its eigenvalues $(\lambda_1, ..., \lambda_n)$ and eigenvectors $(\mathbf{e}_1, ..., \mathbf{e}_n)$ which satisfy the eigenvalue equation

$$\mathsf{A} \cdot \mathbf{e}_i = \lambda_i \, \mathbf{e}_i, \tag{A.50}$$

for $i = 1, ..., n$. Here, the determinant and the trace of the $n \times n$ matrix A are expressed in terms of its eigenvalues $(\lambda_1, ..., \lambda_n)$ as

$$\det(\mathsf{A}) = \lambda_1 \times ... \times \lambda_n \text{ and } \mathrm{Tr}(\mathsf{A}) = \lambda_1 + ... + \lambda_n.$$

In order to continue our discussion of this important problem, we now focus our attention on 2×2 matrices.

A.4.2 Eigenvalue Analysis of a 2 × 2 Matrix

Consider the 2×2 matrix

$$M = \begin{pmatrix} a & b \\ c & d \end{pmatrix}, \qquad (A.51)$$

where (a, b, c, d) are arbitrary real (or complex) numbers and introduce the following two matrix *invariants*:

$$\Delta \equiv \det(M) = ad - bc \quad \text{and} \quad \sigma \equiv \operatorname{Tr}(M) = a + d, \qquad (A.52)$$

which denote the determinant and the trace of matrix M, respectively.

A.4.2.1 Eigenvalues of M

The *eigenvalues* λ and *eigenvectors* e of matrix M are defined by the eigenvalue equation

$$M \cdot e = \lambda e. \qquad (A.53)$$

This equation has nontrivial solutions only if the determinant of the matrix $M - \lambda I$ vanishes (where I denotes the 2×2 identity matrix). This vanishing determinant yields the characteristic quadratic polynomial:

$$\det(M - \lambda I) = (a - \lambda)(d - \lambda) - bc \equiv \lambda^2 - \sigma \lambda + \Delta = 0, \quad (A.54)$$

and the eigenvalues $\lambda_\pm$ are obtained as the roots of this characteristic polynomial:

$$\lambda_\pm = \frac{\sigma}{2} \pm \sqrt{\frac{\sigma^2}{4} - \Delta} = \frac{(a + d)}{2} \pm \sqrt{\frac{(a - d)^2}{4} + bc}. \qquad (A.55)$$

Here, we note that the matrix invariants (σ, Δ) are related to the eigenvalues $\lambda_\pm$:

$$\lambda_+ + \lambda_- \equiv \sigma \quad \text{and} \quad \lambda_+ \cdot \lambda_- \equiv \Delta. \qquad (A.56)$$

Lastly, the eigenvalues are said to be *degenerate* if $\lambda_+ = \lambda_- \equiv \sigma/2$, i.e.,

$$\Delta = \frac{\sigma^2}{4} \quad \text{or} \quad bc = -\left(\frac{a - d}{2}\right)^2.$$

A.4.2.2 *Eigenvectors of* M

Next, the eigenvectors $e_\pm$ associated with the eigenvalues $\lambda_\pm$ are constructed from the eigenvalue equations $M \cdot e_\pm = \lambda_\pm e_\pm$, which yield the general solutions

$$e_\pm \equiv \begin{pmatrix} 1 \\ \mu_\pm \end{pmatrix} \epsilon_\pm, \qquad (A.57)$$

where $\epsilon_\pm$ denotes an arbitrary constant and

$$\mu_\pm = \frac{\lambda_\pm - a}{b} = \frac{c}{\lambda_\pm - d}. \qquad (A.58)$$

The normalization of the eigenvectors $e_\pm$ ($|e_\pm| = 1$), for example, can be achieved by choosing

$$\epsilon_\pm = \frac{1}{\sqrt{1 + (\mu_\pm)^2}}.$$

We note that the eigenvectors $e_\pm$ are not automatically orthogonal to each other (i.e., the dot product $e_+ \cdot e_-$ may not vanish). Indeed, we find

$$e_+ \cdot e_- = \epsilon_+ \epsilon_- (1 + \mu_+ \mu_-), \qquad (A.59)$$

where

$$1 + \mu_+ \mu_- = 1 + \frac{1}{b^2} (\lambda_+ - a)(\lambda_- - a) = 1 - \frac{c}{b},$$

whose sign is indefinite. This relation guarantees, however, that a symmetric matrix (with $b = c$) has orthogonal eigenvectors.

If the matrix M is not symmetric, by using the Gram-Schmidt orthogonalization procedure, we may construct two orthogonal vectors (e_1, e_2):

$$\left. \begin{array}{l} e_1 = \alpha\, e_+ + \beta\, e_- \\[2mm] e_2 = \gamma\, e_+ + \delta\, e_- \end{array} \right\}, \qquad (A.60)$$

where the coefficients $(\alpha, \beta, \gamma, \delta)$ are chosen to satisfy the orthogonalization condition $e_1 \cdot e_2 \equiv 0$; it is important to note that the vectors (e_1, e_2) are not themselves eigenvectors of the matrix M. For example, we may choose $\alpha = 1 = \delta$, $\beta = 0$, and

$$\gamma = - \left(\frac{e_+ \cdot e_-}{|e_+|^2} \right),$$

which corresponds to choosing $e_1 = e_+$ and constructing e_2 as the component of e_- that is orthogonal to e_+.

Lastly, we point out that any two-dimensional vector $\mathbf{u}$ may be decomposed in terms of the eigenvectors $\mathbf{e}_\pm$:

$$\mathbf{u} = \sum_{i=\pm} u_i \, \mathbf{e}_i \equiv \sum_{i=\pm} \left(\frac{\mathbf{u} \cdot \mathbf{e}_i}{|\mathbf{e}_i|^2} \right) \mathbf{e}_i, \qquad (A.61)$$

where we assumed, here, that the eigenvectors are orthogonal to each other. Furthermore, the *transformation* $\mathsf{M} \cdot \mathbf{u}$ generates a new vector

$$\mathbf{v} = \mathsf{M} \cdot \mathbf{u} = \sum_{i=\pm} u_i \, \mathsf{M} \cdot \mathbf{e}_i \equiv \sum_{i=\pm} v_i \, \mathbf{e}_i,$$

where the components of $\mathbf{v}$ are $v_i \equiv u_i \, \lambda_i$.

A.4.2.3 *Inverse of Matrix* M

The matrix (A.51) has an inverse, denoted M^{-1}, if its determinant Δ does not vanish. In this nonsingular case, we easily find

$$\mathsf{M}^{-1} = \frac{1}{\Delta} \begin{pmatrix} d & -b \\ -c & a \end{pmatrix}, \qquad (A.62)$$

so that $\mathsf{M}^{-1} \cdot \mathsf{M} = \mathbf{I} = \mathsf{M} \cdot \mathsf{M}^{-1}$. The determinant of M^{-1}, denoted Δ', is

$$\Delta' = \frac{d\,a - b\,c}{\Delta^2} \equiv \frac{1}{\Delta} = \frac{1}{\lambda_+ \cdot \lambda_-},$$

while its trace, denoted σ', is

$$\sigma' = \frac{d+a}{\Delta} \equiv \frac{\sigma}{\Delta} = \frac{1}{\lambda_+} + \frac{1}{\lambda_-}.$$

Hence, the eigenvalues of the inverse matrix (A.62) are

$$\lambda'_\pm \equiv \frac{1}{\lambda_\pm} = \frac{\lambda_\mp}{\Delta},$$

and its eigenvectors $\overline{\mathbf{e}}_\pm$ are identical to $\mathbf{e}_\pm$ since

$$\mathsf{M} \cdot \mathbf{e}_\pm = \lambda_\pm \, \mathbf{e}_\pm \quad \rightarrow \quad \mathbf{e}_\pm = \lambda_\pm \, \mathsf{M}^{-1} \cdot \mathbf{e}_\pm,$$

and

$$\mathsf{M}^{-1} \cdot \mathbf{e}_\pm = \lambda_\pm^{-1} \, \mathbf{e}_\pm \equiv \lambda'_\pm \, \mathbf{e}_\pm.$$

We note that once the inverse M^{-1} of a matrix M is known, then any inhomogeneous linear system of equations of the form $\mathsf{M} \cdot \mathbf{u} = \mathbf{v}$ may be solved as $\mathbf{u} \equiv \mathsf{M}^{-1} \cdot \mathbf{v}$.

A.4.2.4 *Special Case I: Real Symmetric Matrix*

A real matrix is said to be symmetric if its transpose, denoted $M^\top$ (i.e., $M_{ij}^\top = M_{ji}$), satisfies the identity $M^\top = M$, which requires that $c = b$ in Eq. (A.51). In this case, the eigenvalues are automatically real

$$\lambda_\pm = \left(\frac{a+d}{2}\right) \pm \sqrt{\left(\frac{a-d}{2}\right)^2 + b^2},$$

and the associated eigenvectors (A.57), which are defined with

$$\mu_\pm = -\left(\frac{a-d}{2b}\right) \pm \sqrt{1 + \left(\frac{a-d}{2b}\right)^2},$$

are automatically orthogonal to each other ($e_+ \cdot e_- = 0$) since $\mu_+ \mu_- \equiv -1$.

A.4.2.5 *Special Case II: Rotation Matrix*

Another special matrix is given by the rotation matrix

$$R = \begin{pmatrix} \cos\theta & \sin\theta \\ -\sin\theta & \cos\theta \end{pmatrix}, \tag{A.63}$$

with determinant $\det(R) = 1$ and trace $\mathrm{Tr}(R) = 2\cos\theta = \exp(i\theta) + \exp(-i\theta)$. The rotation matrix (A.63) is said to be *unitary* since its transverse $R^\top$ is equal to its inverse $R^{-1} = R^\top$ (which is possible only if its determinant is one).

The eigenvalues of the rotation matrix (A.63) are $\exp(\pm i\theta)$ and the eigenvectors are

$$e_\pm = \begin{pmatrix} 1 \\ \pm i \end{pmatrix}.$$

Note that the rotation matrix (A.63) can be written as $R = \exp(i\theta\,\sigma)$, where the matrix

$$\sigma = \begin{pmatrix} 0 & -i \\ i & 0 \end{pmatrix}$$

(also known as the Pauli *spin* matrix σ_2) satisfies the properties $\sigma^{2n} = I$ and $\sigma^{2n+1} = \sigma$ and, thus, we find

$$\exp(i\theta\,\sigma) = \sum_{n=0}^\infty \frac{(i\theta)^n}{n!}\,\sigma^n = \cos\theta\,I + i\sin\theta\,\sigma = R.$$

Note that the time derivative of the rotation matrix (A.63) satisfies the property

$$R^{-1} \cdot \dot{R} = i\dot\theta\,\sigma = \begin{pmatrix} 0 & \dot\theta \\ -\dot\theta & 0 \end{pmatrix}.$$

Lastly, we note that the rotation matrix (A.63) can be used to *diagonalize* a real symmetric matrix (e.g., the inertia tensor)

$$M = \begin{pmatrix} a & b \\ b & d \end{pmatrix},$$

by constructing the new matrix

$$\overline{M} = R \cdot M \cdot R^\top = \begin{pmatrix} \overline{a} & \overline{b} \\ \overline{b} & \overline{d} \end{pmatrix}, \tag{A.64}$$

where

$$\overline{a} = a + (d - a) \sin^2 \theta - b \sin 2\theta,$$
$$\overline{b} = b \cos 2\theta - \frac{1}{2}(d - a) \sin 2\theta,$$
$$\overline{d} = d - (d - a) \sin^2 \theta + b \sin 2\theta.$$

Next, by setting the non-diagonal element $\overline{b} \equiv 0$ (assuming that $d > a$), we obtain

$$\tan 2\theta = \frac{2b}{d - a} \rightarrow \begin{cases} \cos 2\theta = (d - a)/\sqrt{(a - d)^2 + 4b^2} \\[2mm] \sin 2\theta = 2b/\sqrt{(a - d)^2 + 4b^2} \end{cases},$$

where $\sigma = a + d$ and $\Delta = ad - b^2$ denote the trace and determinant of M, respectively. Hence $\overline{M}$ becomes a diagonal matrix

$$\overline{M} = \begin{pmatrix} \lambda_- & 0 \\ 0 & \lambda_+ \end{pmatrix}, \tag{A.65}$$

where the diagonal components are

$$\lambda_\pm = \frac{\sigma}{2} \pm \frac{1}{2}\sqrt{\sigma^2 - 4\Delta}.$$

Note that, since $\sigma = \overline{a} + \overline{d} = a + d = \lambda_+ + \lambda_-$ and $\Delta = \overline{a}\overline{d} - \overline{b}^2 = ad - b^2 = \lambda_+ + \lambda_-$, the trace and determinant of $\overline{M}$ are the same as that of M, i.e., the trace and determinant of any real symmetric matrix are invariant under the *congruence* transformation (A.64).

A.5 Numerical Analysis

The nonlinear ordinary differential equations obtained in this course are often impossible to solve analytically in terms of known mathematical functions. Since numerical software is often readily available to students (e.g.,

Mathematica, Maple, or Matlab), translating physical equations into dimensionless equations is a useful skill to acquire.

For example, the physical equation for the pendulum

$$\ddot{\theta} + \omega_g^2 \sin\theta = 0$$

can be translated into the dimensionless equation

$$\theta''(\tau) + \sin\theta(\tau) = 0,$$

where $\omega_g = \sqrt{g/\ell}$ and $\theta(\tau)$ is a function of the dimensionless time $\tau = \omega_g t$. The great advantage of this dimensionless formulation is that the pendulum problem can be solved for all possible values of ω_g. The dimensionless pendulum can thus be solved by using the initial conditions $\theta(0)$ and $\theta'(0)$ determined from the dimensionless energy equation

$$\epsilon = (\theta')^2/2 + (1 - \cos\theta).$$

Hence, by choosing ϵ and the initial angle θ_0, we can determine the initial velocity $\theta_0' \equiv \pm\sqrt{2[\epsilon - (1 - \cos\theta_0)]}$.

Note that it is often preferable to adopt a Hamiltonian representation when numerically integrating equations of motion. This means that, instead of solving k second-order ordinary differential equations (ODEs), we are solving $2k$ first-order ODEs. For example, for the pendulum problem, we numerically solve Hamilton's equations $\theta'(\tau) = p(\tau)$ and $p'(\tau) = -\sin\theta(\tau)$, which allows us to easily plot the orbits of the pendulum in terms of the phase-space coordinates (θ, p).

When we consider nonlinear coupled equations such as Eqs. (2.53) and (2.54), which describe the motion on the surface of an inverted cone of apex angle α, it is desirable to choose a *clock* frequency needed to define a dimensionless time. By introducing $\omega_g = \sqrt{(g/s_0)}\cos\alpha$ and $\tau = \omega_g t$, we obtain the dimensionless equations

$$\sigma'' = -1 + \frac{1}{\sigma^3} \quad \text{and} \quad \theta' = \frac{1}{\sigma^2 \sin\alpha},$$

where $\sigma \equiv s/s_0$ becomes the distance normalized to

$$s_0 = \left(\frac{p_\theta^2}{m^2 g \sin^2\alpha \cos\alpha}\right)^{\frac{1}{3}}.$$

Note that, while several (physical) parameters appear in Eq. (2.54), the normalized equation $\sigma'' + 1 = \sigma^{-3}$ contains no dimensionless parameters at all, while the equation for θ' only requires that the cone angle α (which can even be absorbed in a new definition of θ).

The following sample Mathematica code (v 4.0) was written to generate solutions of the problem of constrained motion on the surface of a cone and to create Fig. (2.9).

```
s0 = 3
p0 = 0
q0 = 0

a = π/8

t0 = 0.0
t1 = 20.0

solution1 = NDSolve[{
    s'[t] == p[t],
    p'[t] == -1 + 1/s[t]^3,
    q'[t] == 1/(Sin[a] s[t]^2),
    s[0] == s0,
    p[0] == p0,
    q[0] == q0 },
    {s[t],p[t],q[t]}, {t,t0,t1}]

plot1 = ParametricPlot[Evaluate[{s[t] Sin[a] Cos[q[t]], s[t] Sin[a] Sin[q[t]]}
    /. solution1], {t,t0,t1}, AxesLabel → { "x-axis", "y-axis"}]

plot2 = ParametricPlot[Evaluate[{s[t] Sin[a] Cos[q[t]], s[t] Cos[a]}
    /. solution1], {t,t0,t1}, AxesLabel → { "x-axis", "z-axis"}]
```

Note here that we are numerically solving 3 first-order ODEs: $\sigma' = p$, $p' = -1 + 1/\sigma^3$, and $\theta' = 1/(\sigma^2 \sin\alpha)$. The command plot1 generates the "top view" of Fig. 2.9 while plot2 generates the "side view". If we need to change the parameters or initial conditions for our numerical solution, we can create a new solution #, e.g., solution 2 = NDSolve[{$\cdots$}].

Appendix B

Elliptic Functions and Integrals*

The Jacobi and Weierstrass elliptic functions [14] used to be part of the standard mathematical arsenal of physics students [20]. They appear as solutions of many important problems in classical mechanics: the motion of a planar pendulum (Jacobi), the motion of a force-free asymmetric top (Jacobi), the motion of a spherical pendulum (Weierstrass), and the motion of a heavy symmetric top with one fixed point (Weierstrass). The problem of the planar pendulum, in fact, can be used to construct the general connection between the Jacobi and Weierstrass elliptic functions. The easy access to mathematical software by physics students suggests that they might reappear as useful tools in the undergraduate curriculum.[1]

B.1 Jacobi Elliptic Functions

B.1.1 *Definitions and Notation*

We begin our introduction of elliptic functions with the more familiar Jacobi elliptic functions [16]. The Jacobi elliptic function $\operatorname{sn}(z \,|\, m)$ is defined in terms of the inverse-function formula

$$
\begin{aligned}
z &= \int_0^\varphi \frac{d\theta}{\sqrt{1 - m \sin^2 \theta}} \\
&= \int_0^{\sin \varphi} \frac{dy}{\sqrt{(1 - y^2)(1 - m y^2)}} \\
&\equiv \operatorname{sn}^{-1}(\sin \varphi \,|\, m),
\end{aligned}
\tag{B.1}
$$

---

[1] An extended version of this Appendix can be found at arxiv.org/abs/0711.4064 and additional material was published in Eur. J. Phys. **30**, 729-750 (2009) and Commun. Nonlinear Sci. Numer. Simulat. **18**, 511-518 (2013).

where the modulus m is a positive number and the amplitude φ varies from 0 to 2π. From this definition, we easily check that $\mathrm{sn}^{-1}(\sin\varphi\,|\,0) = \sin^{-1}(\sin\varphi) = \varphi$. The solution to the differential equation

$$\left(\frac{dy}{dz}\right)^2 = \left(1 - y^2\right)\left(1 - m\,y^2\right) \tag{B.2}$$

is expressed in terms of the Jacobi elliptic function

$$y(z) = \begin{cases} \mathrm{sn}(z|m) & \text{(for } m < 1), \\[2ex] m^{-1/2}\,\mathrm{sn}\left(m^{1/2}\,z\,|\,m^{-1}\right) & \text{(for } m > 1). \end{cases} \tag{B.3}$$

By using the transformation $y = \sin\varphi$, the Jacobi differential equation (B.2) is also written as

$$\left(\frac{d\varphi}{dz}\right)^2 = 1 - m\,\sin^2\varphi, \tag{B.4}$$

and the solution to this equation is $\varphi(z) = \sin^{-1}[\mathrm{sn}(z|m)]$ for $m < 1$.

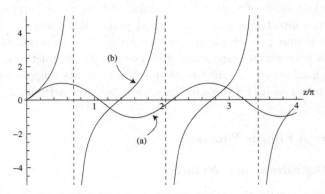

Fig. B.1   Plots of (a) $\mathrm{sn}(z|m)$ and (b) $-i\,\mathrm{sn}(iz|m)$ for $m = 1/16$ showing the real and imaginary periods $4\,K(m)$ and $4\,i\,K'(m)$.

The function $\mathrm{sn}(z|m)$ has a purely-real period $4\,K$, where the quarter-period K is defined as

$$K \equiv K(m) = \int_0^{\pi/2} \frac{d\theta}{\sqrt{1 - m\,\sin^2\theta}} \tag{B.5}$$

and a purely-imaginary period $4\,iK'$, where the quarter-period K' is defined as (with the complementary modulus $m' \equiv 1 - m$)

$$i\,K' \equiv i\,K(m') = i\int_0^{\pi/2} \frac{d\theta}{\sqrt{1 - m'\,\sin^2\theta}}. \tag{B.6}$$

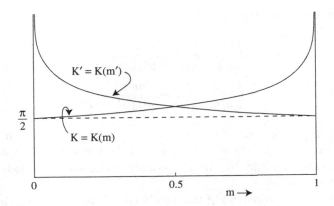

Fig. B.2   Plots of the quarter periods $K = K(m)$ and $K' = K(m') = K(1 - m)$.

Figure B.1 shows plots of sn z and $-i\,\mathrm{sn}(iz)$ for $m = 1/16$, which exhibit both a real period and an imaginary period. Note that, while the Jacobi elliptic function sn z alternates between -1 and $+1$ for real values of z (with zeroes at $2n\,K$), it also exhibits singularities for imaginary values of z at $(2n+1)\,iK'$ $(n = 0, 1, ...)$. Furthermore, as $m \to 0$ (and $m' \to 1$), we find $K \to \pi/2$ (or $4\,K \to 2\pi$) and $|K'| \to \infty$ (see Fig. B.2), and so sn $z \to \sin z$ becomes singly-periodic.

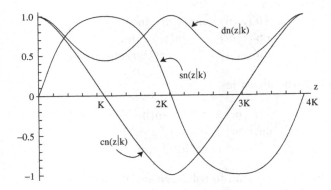

Fig. B.3   Plots of $\mathrm{sn}(z|m)$, $\mathrm{cn}(z|m)$, and $\mathrm{dn}(z|m)$ from $z = 0$ to $4\,K(m)$ for $m = 0.81$.

The additional Jacobi elliptic functions $\mathrm{cn}(z\,|\,m)$ and $\mathrm{dn}(z\,|\,m)$ are defined from the integrals

$$z = \int_{\text{cn}(z|m)}^{1} \frac{dy}{\sqrt{(1-y^2)\,(m'+m\,y^2)}}, \tag{B.7}$$

$$= \int_{\text{dn}(z|m)}^{1} \frac{dy}{\sqrt{(1-y^2)\,(y^2-m')}}, \tag{B.8}$$

with the properties $\text{cn}\,z \equiv \text{cn}(z|m) = \cos\varphi$, $\text{dn}\,z \equiv \text{dn}(z|m) = \sqrt{1-m\sin^2\varphi}$, and $\text{sn}^2 z + \text{cn}^2 z = 1 = \text{dn}^2 z + m\,\text{sn}^2 z$. The Jacobi elliptic functions $\text{cn}\,z$ and $\text{dn}\,z$ are also doubly-periodic with periods $4\,K$ and $4i\,K'$ (see Fig. B.3).

The following properties of the Jacobi elliptic functions $(\text{sn}, \text{cn}, \text{dn})$ are useful. First, we find the limits:

$$\begin{pmatrix} \text{sn}(z|0) \\ \text{cn}(z|0) \\ \text{dn}(z|0) \end{pmatrix} = \begin{pmatrix} \sin z \\ \cos z \\ 1 \end{pmatrix} \tag{B.9}$$

and

$$\begin{pmatrix} \text{sn}(z|1) \\ \text{cn}(z|1) \\ \text{dn}(z|1) \end{pmatrix} = \begin{pmatrix} \tanh z \\ \text{sech}\,z \\ \text{sech}\,z \end{pmatrix}. \tag{B.10}$$

Next, we find the derivatives with respect to the argument z:

$$\left. \begin{array}{l} \text{sn}'(z|m) = \text{cn}(z|m)\,\text{dn}(z|m) \\ \text{cn}'(z|m) = -\,\text{sn}(z|m)\,\text{dn}(z|m) \\ \text{dn}'(z|m) = -\,m\,\text{cn}(z|m)\,\text{sn}(z|m) \end{array} \right\}, \tag{B.11}$$

and, if $m > 1$, the identities:

$$\left. \begin{array}{l} \text{sn}(z|m) = m^{-1/2}\,\text{sn}(m^{1/2}\,z|m^{-1}) \\ \text{cn}(z|m) = \text{dn}(m^{1/2}\,z|m^{-1}) \\ \text{dn}(z|m) = \text{cn}(m^{1/2}\,z|m^{-1}) \end{array} \right\}. \tag{B.12}$$

We now turn our attention to solving a simple physical problem that highlights the periodic properties the Jacobi elliptic functions (B.1) and (B.7)-(B.8). Already, we have seen how the problems of the planar pendulum in Sec. 3.5.4 and the force-free asymmetric top in Sec. 7.2.3 can be solved simply and explicitly in terms of the Jacobi elliptic functions $(\text{sn}, \text{cn}, \text{dn})$.

B.1.2   *Motion in a Quartic Potential*

We look at particle orbits in the (dimensionless) quartic potential $U(x) = 1 - x^2/2 + x^4/16$. Here, the turning points for $E \equiv e^2 = U(x)$ are

$$\left.\begin{array}{c} \pm 2\,\sqrt{1+e} \quad (\text{for } e > 1) \\[2mm] 0 \text{ and } \pm\sqrt{8} \ (\text{for } e = 1) \\[2mm] \pm 2\,\sqrt{1 \pm e} \quad (\text{for } e < 1) \end{array}\right\}. \tag{B.13}$$

Each orbit is solved using the initial condition $x_0 = 2\,\sqrt{1+e}$ with the initial velocity $\dot{x}_0 < 0$:

$$\begin{aligned} t(x) &= -\int_{2\sqrt{1+e}}^{x} \frac{dy}{\sqrt{2\,(e^2-1) + y^2\,(1 - y^2/8)}} \\[2mm] &= -\int_{2\sqrt{1+e}}^{x} \frac{\sqrt{8}\,dy}{\sqrt{[4\,(e+1) - y^2]\,[y^2 + 4\,(e-1)]}} \\[2mm] &= \frac{1}{\sqrt{e}} \int_{0}^{\Phi(x)} \frac{d\varphi}{\sqrt{1 - m\,\sin^2\varphi}}, \end{aligned} \tag{B.14}$$

where $m \equiv (1 + e)/2e$ while we used the trigonometric substitution $y = 2\,\sqrt{1+e}\,\cos\varphi$ with

$$\Phi(x) \equiv \cos^{-1}\left[\frac{x}{2\sqrt{1+e}}\right] \tag{B.15}$$

to obtain the last expression in Eq. (B.14). The Jacobi elliptic solutions obtained from Eq. (B.14) are shown in Fig. B.4 for the orbit (a), with $e > 1$, the separatrix orbit (b), with $e = 1$, and the orbit (c), with $e < 1$.

For $e > 1$ (i.e., $m < 1$), corresponding to orbit (a) in Fig. B.4, we use Eq. (B.1) to find

$$\sin\Phi(x) = \text{sn}(\sqrt{e}\,t|m) = \sqrt{1 - \frac{x^2(t)}{4\,(1+e)}},$$

which yields the phase-portrait coordinates $(x, \dot{x})$:

$$\left.\begin{array}{l} x(t) = 2\,\sqrt{1+e}\ \text{cn}(\sqrt{e}\,t|m) \\[3mm] \dot{x}(t) = -2\,\sqrt{e\,(1+e)}\ \text{sn}(\sqrt{e}\,t|m)\,\text{dn}(\sqrt{e}\,t|m) \end{array}\right\}, \tag{B.16}$$

where the velocity $\dot{x}(t)$ is obtained by using Eq. (B.11). For $e = 1$ (i.e., the separatrix orbit with $m = 1$), corresponding to orbit (b) in Fig. B.4, the phase-portrait coordinates become

$$\left.\begin{array}{l} x(t) = \sqrt{8}\ \text{sech}\,t \\[3mm] \dot{x}(t) = -\sqrt{8}\ \text{sech}\,t\ \tanh t \end{array}\right\}, \tag{B.17}$$

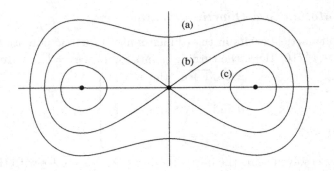

Fig. B.4   Phase portait for orbits (B.16)-(B.18) of the quartic potential $U(x) = 1 - x^2/2 + x^4/16$ for (a) $e > 1$, (b) $e = 1$ (separatrix), and (c) $e < 1$.

where the limits (B.10) were applied to Eq. (B.16). Lastly, for $e < 1$ (i.e., $m > 1$), corresponding to orbit (c) in Fig. B.4, we apply the relations (B.12) on Eq. (B.16) to obtain

$$\left.\begin{array}{l} x(t) = 2\sqrt{1+e}\ \mathrm{dn}(\tau\,|m^{-1}) \\[2mm] \dot{x}(t) = -\sqrt{8}\ e\ \mathrm{sn}(\tau\,|m^{-1})\,\mathrm{cn}(\tau\,|m^{-1}) \end{array}\right\}, \qquad (B.18)$$

where $\tau = t\,\sqrt{(1+e)/2}$. The orbits (B.16)-(B.18) are combined to yield the phase portrait for the quartic potential shown in Fig. B.4.

B.2   Weierstrass Elliptic Functions

B.2.1   *Definitions and Notation*

The Weierstrass elliptic function $\wp(z; g_2, g_3)$ is defined as the solution of the differential equation [18]

$$\begin{aligned} (ds/dz)^2 &= 4\,s^3\ -\ g_2\,s\ -\ g_3 \\ &\equiv 4\,(s-e_1)\,(s-e_2)\,(s-e_3). \end{aligned} \qquad (B.19)$$

Here, (e_1, e_2, e_3) denote the roots of the cubic polynomial $4s^3 - g_2\,s - g_3$ (such that $e_1 + e_2 + e_3 = 0$), where the invariants g_2 and g_3 are defined in terms of the cubic roots as

$$\left.\begin{array}{l} g_2 = -4\,(e_1\,e_2 + e_2\,e_3 + e_3\,e_1) = 2\,(e_1^2 + e_2^2 + e_3^2) \\[3mm] g_3 = 4\,e_1\,e_2\,e_3 \end{array}\right\}, \qquad (B.20)$$

and $\Delta = g_2^3 - 27\,g_3^2$ is the modular discriminant. Since physical values for the constants g_2 and g_3 are always real (and $g_2 > 0$), then either all three roots are real or one root (say e_a) is real and we have a conjugate pair of complex roots (e_b, e_b^*) with $\mathrm{Re}(e_b) = -e_a/2$. The applications of Weierstrass elliptic functions are analyzed in terms of four different cases based on the signs of $(g_3, \Delta) = [(-,-),(-,+),(+,-),(+,+)]$, with two special cases $(g_3 \neq 0, \Delta = 0)$ and $(g_3 = 0, \Delta > 0)$.

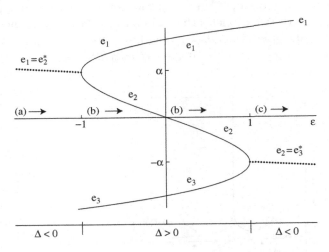

Fig. B.5 Cubic roots (e_1, e_2, e_3) as a function of $\epsilon \equiv (3/g_2)^{3/2}\,g_3$ with fixed value g_2, where $\alpha \equiv \sqrt{g_2/12}$ and $\Delta = g_2^3\,(1 - \epsilon^2)$. The three roots satisfy $e_1 + e_2 + e_3 = 0$.

In general, the roots (e_1, e_2, e_3) of the cubic polynomial on the right side of Eq. (B.19) can be expressed in terms of the parameters $\alpha \equiv \sqrt{g_2/12}$ and $\epsilon \equiv (3/g_2)^{3/2}\,g_3 \equiv -\cos\varphi$ as

$$\begin{pmatrix} e_1 \\ e_2 \\ e_3 \end{pmatrix} \equiv 2\,\alpha \begin{pmatrix} \cos[(\varphi - \pi)/3] \\ \cos[(\varphi + \pi)/3] \\ -\cos(\varphi/3) \end{pmatrix}, \qquad (B.21)$$

and the discriminant is

$$\Delta = g_2^3 - 27\,g_3^2 = g_2^3\,(1 - \epsilon^2). \qquad (B.22)$$

These cubic roots are shown in Fig. B.5 as a function of ϵ for $\alpha = 1/2$ (i.e., $g_2 = 3$); the polynomial $4s^3 - g_2 s - g_3$ is positive (and ds/dz is real) to the left of the curve and negative (and ds/dz is imaginary) to the right of the curve. The three cubic roots (as shown in Fig. B.5) are connected smoothly in the complex φ-plane

$$\varphi = \begin{cases} -i\,\psi & (\psi \geq 0,\ \epsilon \leq -1) \\ \phi & (0 \leq \phi \leq \pi,\ -1 \leq \epsilon \leq 1) \\ \pi + i\,\psi & (\psi \geq 0,\ \epsilon \geq 1) \end{cases} \qquad (\text{B.23})$$

Here, for $\epsilon \leq -1$, the imaginary phase $\varphi \equiv -i\,\psi$ (with $\psi \geq 0$) yields the complex-conjugate roots $e_1 = a - ib = e_2^*$ (with $b > 0$) and the real root $e_3 = -2a < -1$; for $-1 \leq \epsilon \leq 1$, the real phase $\varphi \equiv \phi$ (with $0 \leq \phi \leq \pi$) yields three real roots $e_1 > e_2 > e_3$; and for $\epsilon \geq 1$, the complex phase $\varphi = \pi + i\,\psi$ (with $\psi \geq 0$) yields the real root $e_1 = 2a > 1$ and the complex-conjugate roots $e_2 = -a - ib = e_3^*$ (with $b > 0$). Note that $e_2 = 0$ (and $e_1 = \sqrt{3}\,a = -e_3$) for $g_3 = 0$ (i.e., $\phi = \pi/2$); this case is called the *lemniscatic* case.

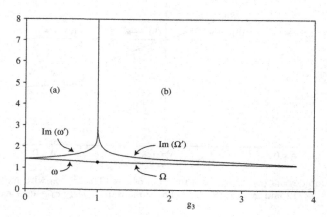

Fig. B.6 Plots of (a) ω and $\text{Im}(\omega')$ for $0 < g_3 < 1$ and $g_2 = 3$ (i.e., $\Delta > 0$) and (b) Ω and $\text{Im}(\Omega')$ for $g_3 > 1$ and $g_2 = 3$ (i.e., $\Delta < 0$). Note that $\omega'(g_2, 0) = i\,\omega(g_2, 0)$, $\Omega(g_2, 1) = \omega(g_2, 1)$, and both (Ω, Ω') decrease to zero as g_3 becomes infinite.

For $0 < \epsilon < 1$ (i.e., $\Delta > 0$), $\wp(z)$ has two different periods $2\,\omega$ and $2\,\omega'$ along the real and imaginary axes, respectively, with the half-periods ω and ω' defined as

$$\omega(g_2, g_3) = \int_{e_1}^{\infty} \frac{ds}{\sqrt{4s^3 - g_2\,s - g_3}}, \qquad (\text{B.24})$$

$$\omega'(g_2, g_3) = i \int_{-\infty}^{e_3} \frac{ds}{\sqrt{|4s^3 - g_2\,s - g_3|}}. \qquad (\text{B.25})$$

The plots of $\omega(g_2, g_3)$ and $\omega'(g_2, g_3)$ are shown in Fig. B.6 for $g_2 = 3$ as functions of g_3. Note that for $g_3 = 0$ (with $e_1 = -e_3$ and $e_2 = 0$), we find that $\omega' \equiv i\,\omega$ while $|\omega'|$ approaches infinity as g_3 approaches one.

Table B.1   Cubic Roots (e_1, e_2, e_3) and Half Periods $(\omega_1, \omega_2, \omega_3)$ for $g_2 = 3$ $(\alpha = 1/2)$.

(g_3, Δ)	e_1	e_2	e_3	ω_1	ω_2	ω_3
$(-,-)$	$a - ib$	$a + ib$	$-2a < -1$	$|\Omega'| + i\Omega/2$	$-|\Omega'| + i\Omega/2$	$-i\Omega$
$(-,+)$	$d > 0$	$c - d > 0$	$-c < 0$	$|\omega'|$	$i\omega - |\omega'|$	$-i\omega$
$(+,+)$	$c > 0$	$d - c > 0$	$-d < 0$	ω	$-\omega - \omega'$	ω'
$(+,-)$	$2a > 1$	$-a - ib$	$-a + ib$	Ω	$-\Omega/2 - \Omega'$	$-\Omega/2 + \Omega'$

For $\epsilon > 1$ (i.e., $\Delta < 0$), on the other hand, $\wp(z)$ has two different periods 2Ω and $2\Omega'$ along the real and imaginary axes, respectively, with the half-periods Ω and Ω' defined as

$$\Omega(g_2, g_3) = \int_{e_1}^{\infty} \frac{ds}{\sqrt{4s^3 - g_2 s - g_3}}, \tag{B.26}$$

$$\Omega'(g_2, g_3) = i \int_{-\infty}^{e_1} \frac{ds}{\sqrt{|4s^3 - g_2 s - g_3|}}. \tag{B.27}$$

The plots of $\Omega(g_2, g_3)$ and $\Omega'(g_2, g_3)$ are shown in Fig. B.6 for $g_2 = 3$ as functions of g_3. Note that $\omega(g_2, 1) = \Omega(g_2, 1)$, $|\Omega'|$ approaches infinity as g_3 approaches one, and that both Ω and Ω' approach zero as g_3 approaches infinity.

Table B.1 shows the cubic roots $e_i = (e_1, e_2, e_3)$, defined by Eq. (B.21), and the half periods $\omega_i = (\omega_1, \omega_2, \omega_3)$, defined as

$$\omega_i(g_2, g_3) \equiv \int_{e_i}^{\infty} \frac{ds}{\sqrt{4s^3 - g_2 s - g_3}}$$

$$= \int_{e_i}^{\infty} \frac{ds}{2\sqrt{(s - e_1)(s - e_2)(s - e_3)}}. \tag{B.28}$$

The cubic roots and half periods satisfy the following properties:

$$\left. \begin{array}{l} \wp(\omega_i) = e_i \\ \wp(z + \omega_i) = e_i + (e_i - e_j)(e_i - e_k)[\wp(z) - e_i]^{-1} \\ \wp(z + 2\omega_i) = \wp(z) \end{array} \right\}, \tag{B.29}$$

where $i \neq j \neq k$ so that $\wp(\omega_i + \omega_j) = e_k$. Figure B.7 shows the plots of $\wp(z + \omega_2)$ and $\wp(z + \omega_3)$ for one complete period from $z = 0$ to $2\omega_1$, which clearly satisfies the identities (B.29).

The Weierstrass elliptic function $\wp(z; g_2, g_3)$ obeys the homogeneity relation

$$\wp\left(\lambda z; \lambda^{-4} g_2, \lambda^{-6} g_3\right) = \lambda^{-2} \wp(z; g_2, g_3), \tag{B.30}$$

where $\lambda \neq 0$. By choosing $\lambda = -1$, for example, we readily verify that the Weierstrass elliptic function has even parity, i.e., $\wp(-z; g_2, g_3) =$

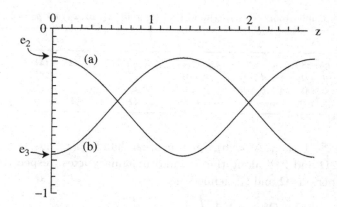

Fig. B.7   Plots of (a) $\wp(z+\omega_2)$ and (b) $\wp(z+\omega_3)$ for $g_2 = 3$ and $g_3 = 0.5$ (with $\epsilon = 0.5$) over one complete period from 0 to $2\omega_1$. Note that $\wp(\omega_j) = e_j$ for $j = 2$ or 3 and $\wp(\omega_i + \omega_j) = e_k$, for $i = 1$ and $(j, k) = (2, 3)$ or $(3, 2)$.

$\wp(z; g_2, g_3)$. On the other hand, for $\lambda = i$, we find that the half-period assignments for $g_3 < 0$ in Table B.1 are based on the relation

$$\wp(z; g_2, g_3) = -\wp(iz; g_2, |g_3|). \tag{B.31}$$

For example, for $-1 < g_3 < 0$ (and $\Delta > 0$), we find for $\wp(\omega_1; g_2, g_3)$:

$$\wp(|\omega'|; g_2, -|g_3|) = -\wp(\omega'; g_2, |g_3|) = -(-d) = d,$$

which corresponds exactly to $e_1 = d$ found in Table B.1 for the case $(g_3, \Delta) = (-, +)$.[2]

In general, the connections between the half-periods

$$(\omega_1^+, \omega_2^+, \omega_3^+) \rightarrow (\omega_1^-, \omega_2^-, \omega_3^-)$$

and the cubic roots $(e_1^+, e_2^+, e_3^+) \rightarrow (e_1^-, e_2^-, e_3^-)$ as g_3 changes sign from positive $(+)$ to negative $(-)$ are found in Table B.1 to be

$$\left.\begin{array}{c} (\omega_1^-, \omega_2^-, \omega_3^-) \equiv (-i\,\omega_3^+, -i\,\omega_2^+, -i\,\omega_1^+) \\[2mm] (e_1^-, e_2^-, e_3^-) \equiv (-e_3^+, -e_2^+, -e_1^+) \end{array}\right\}. \tag{B.32}$$

Once again, these connections follow a non-standard convention. For example, according to the standard convention [18] for the case $(g_3, \Delta) = (+, -)$,

---

[2]The reader should be warned that only the case $(g_3, \Delta) = (+, +)$ in Table B.1 follows the standard mathematical convention [18]. The convention for the remaining cases $(g_3, \Delta) = [(-, -), (-, +), (+, -)]$ in Table B.1 are based on the output of Mathematica, on which Eq. (B.21) and the Weierstrass path (B.23) are based, and the convention adopted for ω_2 satisfies the condition $\omega_1 + \omega_2 + \omega_3 = 0$.

the root e_2 is real (while $e_1^* = e_3$) and the corresponding half-period ω_2 is also real (in contrast to the convention adopted in Table B.1). The connections (B.32) shown in Table B.1 are simply based on the smooth dependence of the cubic roots on the single parameter ϵ (for fixed g_2). These connections enable us to describe consistent orbital dynamics in several problems in classical mechanics.

B.2.2 *Motion in a Cubic Potential*

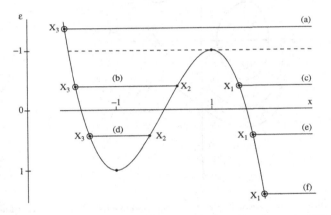

Fig. B.8 Cubic-potential energy levels $E = x - x^3/3$ showing orbits (a) $E > 2/3$ (unbounded orbits; $\epsilon < -1$), (b) and (c) $0 < E < 2/3$ (bounded and unbounded orbits; $-1 < \epsilon < 0$), (d) and (e) $-2/3 < E < 0$ (bounded and unbounded orbits; $0 < \epsilon < 1$), and (f) $E \leq -2/3$ (unbounded orbit; $\epsilon \geq 1$).

We have already seen that the solution of the sleeping top problem can be written in terms of the Weierstrass elliptic function (Sec. 7.4.3). As an additional physical problem, we consider particle orbits in a (dimensionless) cubic potential $U(x) = x - x^3/3$. Here, the cubic-potential orbits $x(t)$ are solutions of the differential equation

$$\dot{x}^2 = 2\left(E - x + \frac{x^3}{3}\right)$$
$$\equiv \frac{2}{3}(x - x_1)(x - x_2)(x - x_3), \qquad (B.33)$$

and the turning points (x_1, x_2, x_3) are shown in Fig. B.8 (with $x_1 + x_2 + x_3 = 0$). By writing $x(t) = 6\, s(t)$, Eq. (B.33) is transformed into the standard Weierstrass elliptic equation (B.19), where the invariants are $g_2 = 1/3$ and

$g_3 = -E/18$, so that $\epsilon \equiv -3E/2$. Note that bounded orbits exist only for $-1 < \epsilon < 1$ (i.e., $\Delta > 0$).

Fig. B.9 Plots of $\dot{x}(t)$ versus $x(t)$ for cubic potential (B.33) show bounded and unbounded orbits: Orbit (a) $E > 2/3$ ($\epsilon < -1$ and $\Delta < 0$); orbits (b)-(e) $-2/3 < E < 2/3$ ($-1 < \epsilon < 1$ and $\Delta > 0$); and orbit (f) $E < -2/3$ ($\epsilon > 1$ and $\Delta < 0$). The dotted lines are the bounded and unbounded separatrix orbits for $E = 2/3$ and circles denote particle positions at $t = 0$.

The cubic-potential solution for Eq. (B.33) is

$$x(t) = 6 \, \wp(t + \gamma), \tag{B.34}$$

where the constant γ is determined from the initial condition $x(0)$. Figure B.9 shows the orbits (a)-(f) associated with initial conditions identified by a circle and a qualitative description of these orbits is summarized in Table B.2. Note that the turning points $x_i = 6 \, e_i$ ($i = 1, 2, 3$) are simply related to the standard cubic roots e_i. Lastly, the separatrix solution is obtained from orbit (b) as E approaches $2/3$ and the period $2 \, |\omega'|$ becomes infinite.

Lastly, we note that the imaginary time range for orbit (a) takes into account the relation (B.31) since $g_3 < 0$ for this orbit. In addition, the connections (B.32) allow us to describe the orbits (a)-(f) in Figs. B.8 and B.9 (and Table B.2) smoothly as the single (energy) parameter ϵ is varied.

Table B.2 Bounded and Unbounded Orbits in a Cubic Potential (see Figs. B.8 and B.9).

Orbit	Energy	Time Range	Constant γ	Period
(a)	$E > 2/3$	$-i\Omega < t < i\Omega$	$-i\Omega$	Unbounded
(b)	$0 < E < 2/3$	$0 < t < 2\lvert\omega'\rvert$	$-i\omega$	$2\lvert\omega'\rvert$
(c)	$0 < E < 2/3$	$-\lvert\omega'\rvert < t < \lvert\omega'\rvert$	$\lvert\omega'\rvert$	Unbounded
(d)	$-2/3 < E < 0$	$0 < t < 2\omega$	ω'	2ω
(e)	$-2/3 < E < 0$	$-\omega < t < \omega$	ω	Unbounded
(f)	$E < -2/3$	$-\Omega < t < \Omega$	Ω	Unbounded

B.3   Connection between Elliptic Functions

In this Section, we return to the planar pendulum problem of Sec. 3.5.4 to establish a connection between the Jacobi and Weierstrass elliptic functions. First, we write $z = 1 - \cos\theta$ (i.e., $0 < z < 2$) and transform Eq. (3.51) into the cubic-potential equation

$$(z')^2 = 2z(2-z)(\epsilon - z), \tag{B.35}$$

with roots at $z = 0, 2$ and $\epsilon \equiv E/(mg\ell)$. When $\epsilon < 2$, the motion is periodic between $z = 0$ and $z = \epsilon$, while the motion is periodic between $z = 0$ and $z = 2$ for $\epsilon > 2$. We recover the standard Weierstrass differential equation (B.19) by setting

$$z(\tau) = 2\wp(\tau + \gamma) + \mu, \tag{B.36}$$

where $\mu \equiv (\epsilon + 2)/3$ and the constant γ is determined from the initial condition $z(0)$.

The root corresponding to $z = 0$ is labeled $e_c = -\mu/2$, the root corresponding to $z = 2$ is labeled $e_b = 1 - \mu/2$, and the root corresponding to $z = \epsilon$ is labeled $e_a = \mu - 1$ and we easily verify that $e_a + e_b + e_c = 0$ (see Fig. B.10). The Weierstrass invariants are $g_2 = 1 + 3(\mu - 1)^2$ and $g_3 = \mu(\mu - 1)(\mu - 2)$, and the modular discriminant is $\Delta = \epsilon^2(2 - \epsilon)^2 \geq 0$.

The planar pendulum is now discussed in terms of 4 cases labeled (a)-(d) in Fig. B.10. For cases (a) and (b), where $2/3 < \mu < 4/3$ (i.e., $0 < \epsilon < 2$), we find

$$e_3 = -\mu/2 < e_2 = \mu - 1 < e_1 = 1 - \mu/2,$$

so that

$$\left.\begin{array}{l} \kappa = (e_1 - e_3)^{1/2} = 1 \\[2mm] m = (e_2 - e_3)/(e_1 - e_3) = (3\mu - 2)/2 = \epsilon/2 < 1 \end{array}\right\}. \tag{B.37}$$

For cases (c) and (d), where $\mu > 4/3$ (i.e., $\epsilon > 2$), we find

$$e_3 = -\mu/2 < e_2 = 1 - \mu/2 < e_1 = \mu - 1,$$

so that

$$\left.\begin{array}{l} \kappa = (e_1 - e_3)^{1/2} = (\epsilon/2)^{1/2} > 1 \\[2mm] m = (e_2 - e_3)/(e_1 - e_3) = 2/\epsilon < 1 \end{array}\right\} . \qquad \text{(B.38)}$$

Figure B.11 shows a plot of g_3 as a function of the parameter ϵ, which can be used with the information presented in Table B.1 to describe the motion of the planar pendulum in terms of the Weierstrass elliptic function.

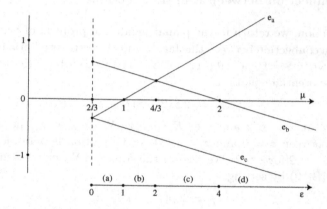

Fig. B.10   Plots of the cubic roots (e_a, e_b, e_c) as functions of $\mu = (\epsilon + 2)/3$. The cases (a)-(d) are discussed in the text. Note that for cases (a) and (b), or $\epsilon < 2$, we find $e_c < e_a < e_b$, while for cases (c) and (d), or $\epsilon > 2$ we find $e_c < e_b < e_a$. The bounded motion of the planar pendulum $(-1 \le z \le 1)$ occurs between the two lowest cubic roots: $e_c < e_a$ (for $\epsilon < 2$) or $e_c < e_b$ (for $\epsilon > 2$).

We first consider case (a), where $0 < \epsilon < 1$ (i.e., $2/3 < \mu < 1$ and $g_3 > 0$), the periodic motion is bounded between $e_3 = -\mu/2$ (i.e., $z = 0$) and $e_2 = \mu - 1 < 0$ (i.e., $z = \epsilon$). Using the initial condition $z(0) = 0$, we find that $\wp(\gamma) = -\mu/2 \equiv e_3$ which implies that $\gamma = \omega'$ (see ω_3 in Table B.1 for $g_3 > 0$ and $\Delta > 0$). The Weierstrass solution of the planar pendulum for $0 < \epsilon < 1$ is therefore

$$z(\tau) = 2\,\wp(\tau + \omega') + \mu, \qquad \text{(B.39)}$$

with the period of oscillation $2\,\omega$.

For case (b), where $1 < \epsilon < 2$ (i.e., $1 < \mu < 4/3$ and $g_3 < 0$), the periodic motion is bounded between $e_3 = -\mu/2$ (i.e., $z = 0$) and $e_2 = \mu - 1 < 0$ (i.e.,

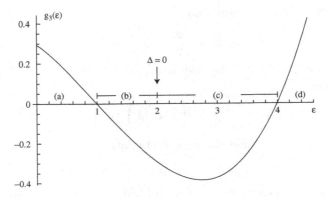

Fig. B.11   Plot of the Weierstrass invariant g_3 as a function of ϵ. For case (a), $g_3 > 0$ and $\Delta > 0$; for cases (b) and (c), $g_3 < 0$ and $\Delta \geq 0$; and for case (d) $g_3 > 0$ and $\Delta > 0$.

$z = \epsilon$). Using the initial condition $z(0) = 0$, we find that $\wp(\gamma) = -\mu/2 \equiv e_3$ which implies that $\gamma = -i\,\omega$ (see ω_3 in Table B.1 for $g_3 < 0$ and $\Delta \geq 0$). The Weierstrass solution of the planar pendulum for $1 < \epsilon < 2$ is

$$z(\tau) = 2\,\wp(\tau - i\omega) + \mu, \tag{B.40}$$

with the period of oscillation $2\,|\omega'|$. As expected, when $\epsilon \to 2$ (i.e., $\Delta \to 0$ and $m \to 1$), the period $2\,|\omega'|$ approaches infinity as we approach the separatrix.

For case (c), where $2 < \epsilon < 4$ (i.e., $4/3 < \mu < 2$ and $g_3 < 0$), the periodic motion is bounded between $e_3 = -\mu/2$ (i.e., $z = 0$) and $e_2 = 1 - \mu/2$ (i.e., $z = 2$), with the period of oscillation $2\,|\omega'|$. Using the initial condition $z(0) = 0$, we find that $\wp(\gamma) = -\mu/2 \equiv e_3$ which implies that $\gamma = -i\,\omega$ (see ω_3 in Table B.1 for $g_3 < 0$ and $\Delta \geq 0$) and thus the Weierstrass solution of the planar pendulum for $2 < \epsilon < 4$ is again given by Eq. (B.40). Note that the separatrix solution ($\epsilon = 2$) is represented by orbits (b) and (c) as $|\omega'| \to \infty$.

Lastly, for case (d), where $\epsilon > 4$ (i.e., $\mu > 2$ and $g_3 > 0$), the periodic motion is bounded between $e_3 = -\mu/2$ (i.e., $z = 0$) and $e_2 = 1 - \mu/2$ (i.e., $z = 2$). Using the initial condition $z(0) = 0$, we find that $\wp(\gamma) = -\mu/2 \equiv e_3$ which implies that $\gamma = \omega'$ (see ω_3 in Table B.1 for $g_3 > 0$ and $\Delta > 0$) and thus the Weierstrass solution of the planar pendulum for $\epsilon > 4$ is again given by Eq. (B.39).

We conclude our discussion of the planar pendulum by using the Jacobi and Weierstrass solutions of this problem to establish a general relation between these elliptic functions. First, we use the Jacobi elliptic solution

(3.57) for the case $\epsilon < 2$ and find

$$z(\tau) = 2 \sin^2 \frac{\theta}{2} = 2m \operatorname{sn}^2(\kappa\tau \,|\, m), \qquad (B.41)$$

where $m = \epsilon/2$ and $\kappa = 1$. By comparing Eqs. (B.39) and (B.41), we obtain the relation

$$\wp(\tau + \omega_3) = -\frac{\mu}{2} + \left(\frac{3}{2}\mu - 1\right) \operatorname{sn}^2\left(\tau \,\bigg|\, \frac{3}{2}\mu - 1\right),$$

which is just one example of the general relation

$$\wp(\tau + \omega_3) \equiv e_3 + (e_2 - e_3) \operatorname{sn}^2(\kappa\tau \,|\, m). \qquad (B.42)$$

Next, using the Jacobi elliptic solution (3.64) for $\epsilon > 2$, we find

$$z(\tau) = 2 \operatorname{sn}^2\left(\sqrt{\epsilon/2}\,\tau \,\big|\, 2/\epsilon\right) = 2 \operatorname{sn}^2\left(m^{1/2}\tau \,\big|\, m^{-1}\right), \qquad (B.43)$$

and thus we recover once again the relation (B.42).

Appendix C

Noncanonical Hamiltonian Mechanics*

Modern formulations of Hamiltonian mechanics [1] rely on the use of non-canonical phase-space coordinates and the methods of differential geometry [5]. The purpose of this Appendix is to present a brief introduction to the noncanonical single-particle Hamiltonian mechanics. We also present an application of the (canonical) Hamiltonian perturbation method to the problem of the perturbed simple harmonic oscillator.

C.1 Differential Geometry

Differential k-forms

$$\omega_k = \frac{1}{k!} \, \omega_{i_1 i_2 \ldots i_k} \, \mathrm{d}z^{i_1} \wedge \mathrm{d}z^{i_2} \wedge \cdots \wedge \mathrm{d}z^{i_k}$$

are fundamental objects in the differential geometry of n-dimensional space (with coordinates $\mathbf{z}$), where the components $\omega_{i_1 i_2 \ldots i_k}$ are antisymmetric with respect to interchange of two adjacent indices since the wedge product $\wedge$ is skew-symmetric (i.e., $\mathrm{d}z^a \wedge \mathrm{d}z^b = -\mathrm{d}z^b \wedge \mathrm{d}z^a$) with respect to the exterior derivative d (which has properties similar to the standard derivative d).

Note that the exterior derivative $\mathrm{d}\omega_k$ of a differential k-form (or k-form for short) ω_k is a $(k+1)$-form. For example, the exterior derivative of a 0-form f is defined as

$$\mathrm{d}f \equiv \partial_a f \, \mathrm{d}z^a, \tag{C.1}$$

and, thus, $\mathrm{d}f$ is a differential 1-form; note that its components are the components of the gradient ∇f. Next, the exterior derivative of a 1-form Γ is a 2-form: $\mathrm{d}\Gamma \equiv \mathrm{d}\Gamma_b \wedge \mathrm{d}z^b = \partial_a \Gamma_b \, \mathrm{d}z^a \wedge \mathrm{d}z^b$, which, as a result of the skew-symmetry of the wedge product $\wedge$, may be expressed as

$$d\Gamma = \frac{1}{2}\left(\partial_a\Gamma_b - \partial_b\Gamma_a\right)dz^a \wedge dz^b$$

$$\equiv \frac{1}{2}\,\omega_{ab}\,dz^a \wedge dz^b, \qquad (C.2)$$

where $\omega_{ab} = -\omega_{ba}$ denotes the antisymmetric components of the 2-form $\omega \equiv d\Gamma$.

An important difference between the exterior derivative d and the standard derivative d comes from the property that $d^2\omega_k = d(d\omega_k) \equiv 0$ for any k-form ω_k. Indeed, for a 0-form, we find

$$d^2 f = \partial^2_{ab}f\,dz^a \wedge dz^b = 0,$$

since $\partial^2_{ab}f$ is symmetric with respect to interchange $a \leftrightarrow b$ while $\wedge$ is antisymmetric. For a 1-form, we find

$$d^2\Gamma = \frac{1}{3!}\left(\partial_a\omega_{bc} + \partial_b\omega_{ca} + \partial_c\omega_{ab}\right)dz^a \wedge dz^b \wedge dz^c = 0,$$

which vanishes since $\partial_a\omega_{bc} + \partial_b\omega_{ca} + \partial_c\omega_{ab} \equiv 0$ vanishes identically since $\omega \equiv d\Gamma$.

A k-form ω_k is said to be *closed* if its exterior derivative is $d\omega_k \equiv 0$, while a k-form ω_k is said to be *exact* if it can be written in terms of a $(k{-}1)$-form Γ_{k-1} as $\omega_k \equiv d\Gamma_{k-1}$. Poincaré's Lemma states that all closed k-forms are exact (as can easily be verified), while its converse states that all exact k-forms are closed. For example, the infinitesimal volume element in three-dimensional space with curvilinear coordinates $\mathbf{u} = (u^1, u^2, u^3)$ and Jacobian $\mathcal{J}$: $\Omega \equiv \mathcal{J}(\mathbf{u})\,du^1 \wedge du^2 \wedge du^3$ is a closed 3-form since $d\Omega = 0$. Hence, according to the converse of Poincaré's Lemma, there exists a 2-form σ such that $\Omega \equiv d\sigma$, where $\sigma \equiv \frac{1}{2}\,\epsilon_{ijk}\,\sigma^k(\mathbf{u})\,du^i \wedge du^j$ defines the infinitesimal area 2-form, with the Jacobian defined as $\mathcal{J} \equiv \partial\sigma^i(\mathbf{u})/\partial u^i$.

We now introduce the inner-product operation involving a vector field $\mathbf{v}$ and a k-form ω_k, denoted as $\mathbf{v}\cdot\omega_k$, which produces a $(k-1)$-form. For example, for a 1-form, it is defined as $\mathbf{v}\cdot\Gamma = v^a\,\Gamma_a$ while for a 2-form, it is defined as

$$\mathbf{v}\cdot\omega \equiv \frac{1}{2}\left(v^a\,\omega_{ab}\,dz^b - \omega_{ab}\,v^b\,dz^a\right) = v^a\,\omega_{ab}\,dz^b.$$

Note that $d(\mathbf{v}\cdot\Omega) = \mathcal{J}^{-1}\partial_a(\mathcal{J}\,v^a)\,\Omega \equiv (\nabla\cdot\mathbf{v})\,\Omega$, which can be used to derive the divergence of any vector field expressed in arbitrary curvilinear coordinates.

C.2   Lagrange and Poisson Tensors

The Poincaré-Cartan one-form [1] (Jules Henri Poincaré, 1854-1912; Élie-Joseph Cartan, 1869-1951) is expressed in canonical phase-space coordinates $(\mathbf{q}, \mathbf{p})$ as

$$\Gamma_c \equiv \frac{\partial L}{\partial \dot{q}^i} \, dq^i - H \, d\sigma = p_i \, dq^i - H \, d\sigma, \qquad (C.3)$$

where σ represents the Hamiltonian orbit parameter. The Lagrange one-form (a generalization of the Poincaré-Cartan one-form) is expressed in terms of general noncanonical phase-space coordinates z^a as

$$\Gamma \equiv \Lambda_a \, dz^a - H \, d\sigma, \qquad (C.4)$$

where

$$\Lambda_a \equiv \frac{\partial L}{\partial \dot{\mathbf{q}}} \cdot \frac{\partial \mathbf{q}}{\partial z^a}.$$

The two-form $\boldsymbol{\omega} \equiv d\gamma$ is written as

$$\boldsymbol{\omega} = \frac{1}{2} \omega_{ab} \, dz^a \wedge dz^b - dH \wedge d\sigma, \qquad (C.5)$$

where the components of the Lagrange two-form are

$$\omega_{ab} \equiv \frac{\partial \Lambda_b}{\partial z^a} - \frac{\partial \Lambda_a}{\partial z^b}. \qquad (C.6)$$

The phase-space Euler-Lagrange equation for the coordinate z^a is obtained by the contraction $\delta z^a \cdot \boldsymbol{\omega} \equiv 0$, which yields

$$\omega_{ab} \frac{dz^b}{d\sigma} = \frac{\partial H}{\partial z^a}. \qquad (C.7)$$

The noncanonical Hamilton's equations are obtained from Eq. (C.7) provided the antisymmetric Lagrange matrix $\boldsymbol{\omega}$ (with components ω_{ab}) can be inverted. This inversion condition is represented by $\det(\boldsymbol{\omega}) \neq 0$. The inverse of the Lagrange matrix yields the Poisson matrix $\mathsf{J} \equiv \boldsymbol{\omega}^{-1}$ with components J^{ab} that satisfy the condition $J^{ab} \omega_{bc} \equiv \delta^a{}_c$. The noncanonical Hamilton's equation (C.7) for z^a is therefore written as

$$\frac{dz^a}{d\sigma} = J^{ab} \frac{\partial H}{\partial z^b} \equiv \{z^a, H\}, \qquad (C.8)$$

where we introduced the antisymmetric Poisson bracket (Siméon-Denis Poisson, 1781-1840)

$$\{F, G\} \equiv \frac{\partial F}{\partial z^a} J^{ab} \frac{\partial G}{\partial z^b}. \qquad (C.9)$$

The antisymmetry of the Poisson matrix guarantees the antisymmetry of the Poisson bracket $\{G, F\} = -\{F, G\}$. An important property of the Poisson bracket is that it must satisfy the Jacobi condition (expressed in terms of three arbitrary functions F, G, and H)

$$\{F, \{G, H\}\} + \{G, \{H, F\}\} + \{H, \{F, G\}\} = 0, \qquad (C.10)$$

which can be expressed in terms of the components of the Poisson matrix as

$$J^{ad} \partial_d J^{bc} + J^{bd} \partial_d J^{ca} + J^{cd} \partial_d J^{ab} = 0. \qquad (C.11)$$

Note that when canonical coordinates are used (for which the Poisson components J^{ab} are either ± 1 or 0), the Jacobi identity (C.10) is trivially satisfied. The condition (C.11) can also be expressed in terms of the Lagrange components ω_{ab} as

$$\partial_a \omega_{bc} + \partial_b \omega_{ca} + \partial_c \omega_{ab} = 0,$$

which is trivially satisfied since the Lagrange components are defined by Eq. (C.6).

As a simple example of noncanonical Hamiltonian mechanics, we consider the Poincaré-Cartan one-form written in terms of the eight-dimensional noncanonical coordinates $z^a = (x^\mu, p_\mu) = (ct, \mathbf{x}; -w/c, \mathbf{p})$:

$$\Gamma = \left(p_\mu + \frac{e}{c} A_\mu\right) dx^\mu = \left(\mathbf{p} + \frac{e}{c} \mathbf{A}\right) \cdot d\mathbf{x} - (w + e\Phi) dt, \qquad (C.12)$$

where the energy coordinate w is canonically conjugate to time t. The Lagrange two-form

$$\omega \equiv d\Gamma = dp_\mu \wedge dx^\mu + \frac{e}{2} F_{\mu\nu} dx^\mu \wedge dx^\nu \equiv \frac{1}{2} \omega_{ab} dz^a \wedge dz^b \quad (C.13)$$

is expressed in terms of the Faraday electromagnetic tensor components $F_{\mu\nu} \equiv \partial_\mu A_\nu - \partial_\nu A_\mu$. From the two-form (C.13), we construct the 8×8 antisymmetric Lagrange matrix

$$\boldsymbol{\omega} = \begin{pmatrix} (e/c)\mathsf{F} & -\mathbf{I} \\ \mathbf{I} & 0 \end{pmatrix},$$

which is composed of 4×4 block matrices. Its inversion yields the 8×8 antisymmetric Poisson matrix

$$\mathsf{J} \equiv \boldsymbol{\omega}^{-1} = \begin{pmatrix} 0 & \mathbf{I} \\ -\mathbf{I} & (e/c)\mathsf{F} \end{pmatrix},$$

from which we obtain the noncanonical Poisson bracket

$$\{F, G\} = \left(\frac{\partial F}{\partial x^\mu} \frac{\partial G}{\partial p_\mu} - \frac{\partial F}{\partial p_\mu} \frac{\partial G}{\partial x^\mu}\right) + \frac{e}{c} F^{\mu\nu} \frac{\partial F}{\partial p^\mu} \frac{\partial G}{\partial p_\nu}. \qquad (C.14)$$

When we combine this Poisson bracket with the Hamiltonian $H = p_\mu\, p^\mu/2m$, we obtain Hamilton's equations of motion

$$\dot{z}^a \;=\; \{z^a,\, H\} \;=\; \begin{cases} \dot{x}^\mu = v^\mu \\[6pt] \dot{p}_\mu = (e/c)\, F_{\mu\nu}\, v^\nu \end{cases}$$

which describe the relativistic motion of a particle in an electromagnetic field [7].

Lastly, an important property of Hamilton's equations is that the equations satisfy the Liouville Theorem

$$\frac{\partial}{\partial z^a}\left(\mathcal{J}\, \frac{dz^a}{d\sigma} \right) \;=\; 0, \tag{C.15}$$

which can also be expressed in terms of the Liouville identities $\partial_a(\mathcal{J}\, J^{ab}) = 0$, where the Jacobian $\mathcal{J}$ is the determinant of the matrix $\partial(\mathbf{q}, \mathbf{p})/\partial\mathbf{z}$. For the Poisson bracket (C.14), the Liouville identities are $\partial F_{\mu\nu}/\partial p_\mu \equiv 0$. Lastly, the Liouville Theorem is trivially satisfied in the case of canonical Hamilton's equations.

C.3   Hamiltonian Perturbation Theory

Hamiltonian methods offer powerful tools in perturbation theory. For example, when an exact dynamical invariant is destroyed by a small perturbation, Hamiltonian perturbation methods can be used to construct an adiabatic invariant that is preserved to arbitrary order in the perturbation amplitude.

We consider, for example, the unperturbed (canonical) Hamiltonian $H_0 = p^2/2 + q^2/2$ for a simple harmonic oscillator with unit mass and unit frequency. By introducing the transformation to *action-angle* coordinates $\mathbf{z} = (J, \theta)$, where $q = \sqrt{2\,J}\,\sin\theta$ and $p = \sqrt{2\,J}\,\cos\theta$, we readily find that the new unperturbed Hamiltonian $K_0(\mathbf{z}) \equiv H_0(q(\mathbf{z}), p(\mathbf{z})) = J$ is independent of the angle θ. Hence, the new unperturbed Hamilton's equations are $\dot{J}_0 = -\partial K_0/\partial\theta \equiv 0$ and $\dot{\theta}_0 = \partial K_0/\partial J = 1$. The action variable is therefore an invariant of the unperturbed Hamiltonian system.

We now introduce the perturbation Hamiltonian $\epsilon\, H_1(q, p) = -\epsilon\, q^4/24$ in the original simple-harmonic-oscillator Hamiltonian system (where ϵ appears as an ordering parameter), which is translated into the new perturbation Hamiltonian

$$\epsilon\, K_1(J, \theta) \;=\; -\frac{\epsilon}{6}\, J^2 \sin^4\theta. \tag{C.16}$$

We note here that in order for the perturbation to be considered small (i.e., $|K_1| < K_0$), we require that $\epsilon < 6/J_{\max}$ during the evolution of the perturbed system. Because the new Hamiltonian $K \equiv K_0 + \epsilon K_1$ now depends on the angle variable θ, the action variable J is no longer invariant, i.e., $\dot{J} = -\epsilon \partial K_1/\partial\theta \neq 0$. The purpose of Hamiltonian perturbation theory is to construct new action-angle coordinates $\overline{z} = (\overline{J}, \overline{\theta})$ in terms of which the transformed Hamiltonian

$$\overline{K}(\overline{z}) = K(J(\overline{z}), \theta(\overline{z})) \tag{C.17}$$

is independent of the new angle variable $\overline{\theta}$ up to arbitrary orders in ϵ.

Since the transformation $(J, \theta) \to (\overline{J}, \overline{\theta})$ we seek is canonical, it may be expressed in the following form

$$\overline{z}^a = z^a + \epsilon\{S_1,\, z^a\} + \epsilon^2 \left(\{S_2,\, z^a\} + \frac{1}{2}\left\{S_1,\, \{S_1,\, z^a\}\right\} \right) + \cdots, \tag{C.18}$$

where the functions $(S_1, S_2, \cdots)$ are said to *generate* the canonical transformation, and the action-angle canonical Poisson bracket is $\{F, G\} = \partial_\theta F \, \partial_J G - \partial_J F \, \partial_\theta G$. The new Hamiltonian, on the other hand, is expressed in terms of these generating functions as

$$\overline{K} = K - \epsilon\{S_1,\, K\} - \epsilon^2 \left(\{S_2,\, K\} + \frac{1}{2}\left\{S_1,\, \{S_1,\, K\}\right\} \right) + \cdots, \tag{C.19}$$

which ensures the scalar-invariance property (C.17) is satisfied. Note that the *direct* transformation approach used here is different from the standard perturbation analysis based on mixed-variable generating functions [7]. The main advantage of the direct approach is that it can easily be generalized to arbitrary orders in the perturbation parameter ϵ.

When the original Hamiltonian

$$K = K_0(J) + \epsilon K_1(J, \theta) + \epsilon^2 K_2(J, \theta) + \cdots \tag{C.20}$$

is expanded in powers of ϵ with each perturbation term $K_n(J, \theta)$ $(n \geq 1)$ expressed as an explicit function of θ, the transformed Hamiltonian (C.19) is also expressed as an expansion in powers of ϵ

$$\overline{K} = \overline{K}_0(\overline{J}) + \epsilon \overline{K}_1(\overline{J}) + \epsilon^2 \overline{K}_2(\overline{J}) + \cdots \tag{C.21}$$

By inserting Eqs. (C.20)-(C.21) in Eq. (C.19), we obtain the following expressions up to second order in ϵ:

$$\overline{K}_0 = K_0, \tag{C.22}$$

$$\overline{K}_1 = K_1 - \{S_1, K_0\} = K_1 - \frac{\partial S_1}{\partial \theta}, \tag{C.23}$$

$$\overline{K}_2 = K_2 - \{S_2, K_0\} - \{S_1, K_1\} + \frac{1}{2}\left\{S_1, \{S_1, K_0\}\right\}$$

$$= K_2 - \frac{\partial S_2}{\partial \theta} - \overline{K}_1' \frac{\partial S_1}{\partial \theta} - \frac{1}{2}\left\{S_1, \frac{\partial S_1}{\partial \theta}\right\}. \tag{C.24}$$

At zeroth order, we easily find $\overline{K}_0 = \overline{J}$. At first order, we impose the condition that $\overline{K}_1$ is independent of $\overline{\theta}$ by $\overline{\theta}$-averaging the right side of Eq. (C.23), which yields

$$\overline{K}_1(\overline{J}) \equiv \langle K_1(\overline{J}, \overline{\theta})\rangle, \tag{C.25}$$

while the $\overline{\theta}$-dependent part $\widetilde{K}_1 \equiv K_1 - \langle K_1\rangle$ is cancelled by choosing S_1 such that $\partial S_1/\partial \overline{\theta} \equiv \widetilde{K}_1$. Likewise, the transformed second-order Hamiltonian is defined as

$$\overline{K}_2 \equiv \langle K_2\rangle - \frac{1}{2}\frac{\partial}{\partial \overline{J}}\left\langle\left(\frac{\partial S_1}{\partial \overline{\theta}}\right)^2\right\rangle = \langle K_2\rangle - \frac{1}{2}\frac{\partial\langle(\widetilde{K}_1)^2\rangle}{\partial \overline{J}}, \tag{C.26}$$

while the second-order generating function S_2 is chosen to cancel all explicit $\overline{\theta}$-dependence on the right side of Eq. (C.24).

We note that since the new action variable

$$\overline{J} = J + \epsilon\frac{\partial S_1}{\partial \theta} + \epsilon^2\left(\frac{\partial S_2}{\partial \theta} + \frac{1}{2}\left\{S_1, \frac{\partial S_1}{\partial \theta}\right\}\right) + \cdots \tag{C.27}$$

is expressed in terms of a truncated asymptotic series in powers of ϵ, it is not an exact dynamical invariant, i.e., $\dot{\overline{J}} = \mathcal{O}(\epsilon^{n+1})$ if the new Hamiltonian $\overline{K} = \overline{K}_0 + \cdots + \epsilon^n \overline{K}_n$ is truncated at order ϵ^n. Hence, the new action variable (C.27) called an *adiabatic* invariant [13].

Returning to the perturbation term (C.16), for example, we find

$$\langle K_1\rangle = -\frac{J^2}{16} \quad \text{and} \quad \widetilde{K}_1 = -\frac{J^2}{48}(\cos 4\theta - 4\cos 2\theta) \equiv \frac{\partial S_1}{\partial \theta},$$

so that, up to first order in ϵ, the new Hamiltonian is

$$\overline{K} = \overline{J} - \frac{\epsilon}{16}\overline{J}^2, \tag{C.28}$$

while the new action-angle variables are

$$\overline{J} = J - \frac{\epsilon}{48}J^2(\cos 4\theta - 4\cos 2\theta) \quad \text{and} \quad \overline{\theta} = \theta + \frac{\epsilon}{96}J(\sin 4\theta - 8\sin 2\theta).$$

Fig. C.1 Plots of the old action $J(t)$ and the new action $\overline{J}(t)$ (with first-order corrections) corresponding to the initial conditions $q(0) = 2$ and $p(0) = 0$ with perturbation parameter $\epsilon = 0.25$.

If we now write $\overline{J}$ in terms of the original coordinates (q, p), we find

$$\overline{J} = \frac{1}{2}\left(p^2 + q^2\right) - \frac{\epsilon}{192}\left(5\,q^4 - 6\,p^2 q^2 - 3\,p^4\right), \tag{C.29}$$

and we easily verify that $\dot{\overline{J}} = \mathcal{O}(\epsilon^2)$, with $\dot{q} = p$ and $\dot{p} = -q + \epsilon\,q^3/6$.

Figure C.1 shows plots of the old action $J(t) = p^2(t)/2 + q^2(t)/2$ and the new action $\overline{J}(t)$, given by Eq. (C.29), as functions of time t for $\epsilon = 0.25$. Note that since $1 < J \le 2$ during its time evolution, then $\epsilon = 0.25 < 6/J_{\max} = 3$ satisfies the condition of applicability of Hamiltonian perturbation theory. One can clearly see that, even for a large value of the perturbation parameter ϵ, the new action $\overline{J}$ shows much smaller oscillations than the old action J. One could further reduce the oscillations in the new action $\overline{J}$ by proceeding to second order in the Hamiltonian perturbation analysis [see Eq. (C.27)], which requires us to evaluate the generating function S_2 in Eq. (C.24).

Lastly, we note that the simplicity of the new Hamilton's equations of motion

$$\dot{\overline{J}} = -\frac{\partial\overline{K}}{\partial\overline{\theta}} = 0 \quad\text{and}\quad \dot{\overline{\theta}} = \frac{\partial\overline{K}}{\partial\overline{J}} = 1 - \frac{\epsilon}{8}\overline{J} \equiv \overline{\Omega}, \tag{C.30}$$

implies that the old action-angle variables can be evaluated explicitly as functions of time by inverting the transformation (C.18):

$$J(t) = \overline{J} + \frac{\epsilon}{48}\,\overline{J}^2\left[\cos 4(\overline{\theta}_0 + \overline{\Omega}t) - 4\cos 2(\overline{\theta}_0 + \overline{\Omega}t)\right], \tag{C.31}$$

$$\theta(t) = \overline{\theta}_0 + \overline{\Omega}t - \frac{\epsilon}{96}\,\overline{J}\left[\sin 4(\overline{\theta}_0 + \overline{\Omega}t) - 8\sin 2(\overline{\theta}_0 + \overline{\Omega}t)\right]. \tag{C.32}$$

By extension, the old coordinates

$$q(t) = \sqrt{2J(t)}\,\sin\theta(t) \quad\text{and}\quad p(t) = \sqrt{2J(t)}\,\cos\theta(t) \tag{C.33}$$

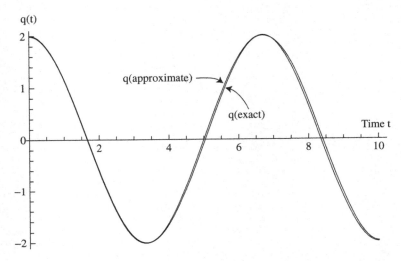

Fig. C.2 Plots of exact solution $q(t)$ of the perturbed Hamiltonian problem (C.34) and the approximate solution $q(t) = \sqrt{2J(t)}\sin\theta(t)$, with $(J(t), \theta(t))$ given by Eqs. (C.31)-(C.32), for the same initial condition $(q(0), p(0)) = (2, 0)$ and $\epsilon = 0.25$.

have also been solved explicitly as functions of time. Hamiltonian perturbation theory has therefore allowed us to solve explicitly the Hamilton's equations

$$\dot{q} = p \quad \text{and} \quad \dot{p} = -q + \frac{\epsilon}{6}\, q^3 \qquad \text{(C.34)}$$

for small enough values of the perturbation parameter ϵ. It is important to note that the solution (C.33), with $(J(t), \theta(t))$ given by Eqs. (C.31)-(C.32), starts to deviate from the true solution of Eq. (C.34) for times of order ϵ^{-n} when the Hamiltonian perturbation analysis has been carried out up to order ϵ^n. For example, Fig. C.2 shows that the approximate solution (C.33) begins to deviate from the exact solution at a time close to $\epsilon^{-1} = 4$. Note that the deviation of the approximate solution oscillates around the exact solution and the amplitude of the deviation depends on the initial conditions (i.e., how well the perturbation condition $\epsilon < 6/J_{\max}$ is satisfied).

Bibliography

[1] Arnold, V. I., *Mathematical Methods of Classical Mechanics, 2nd ed.* (Springer-Verlag, 1989).

[2] Basdevant, J.-L., *Variational principles in physics* (Springer, 2006).

[3] Born, M., and Wolf, E., *Principles of Optics, 5th ed.* (Pergamon Press, 1975).

[4] Dugas, R., *A History of Mechanics* (Dover, 1988).

[5] Flanders, H., *Differential Forms with Applications to the Physical Sciences* (Dover, 1989).

[6] Fox, C., *An Introduction to the Calculus of Variations* (Dover, 1987).

[7] Goldstein, H., Poole, C., and Safko, J., *Classical Mechanics, 3rd ed.* (Addison Wesley, 2002).

[8] Gradshteyn, I. S., and Ryzhik, I. M., *Table of Integrals, Series, and Products (Seventh Ed.)* (Academic Press, 2007).

[9] Gray, C. G., Karl, G., and Novikov, V. A., *The four variational principles of mechanics*, Ann. Phys. **251**, 1-25 (1996).

[10] Gray, C. G., Karl, G., and Novikov, V. A., *From Maupertuis to Schroedinger: Quantization of classical variational principles*, Am. J. Phys. **67**, 959-961 (1999).

[11] Gray, C. G., Karl, G., and Novikov, V. A., *Progress in classical and quantum variational principles*, Rep. Prog. Phys. **67**, 159-208 (2004).

[12] Lanczos, C., *The Variational Principles of Mechanics, 4th ed.* (Dover, 1970).

[13] Landau, L. D., and Lifshitz, E. M., *Mechanics, 3rd ed.* (Pergamon Press, 1976).

[14] Lawden, D. F., *Elliptic Functions and Applications* (Springer-Verlag, 1989).

[15] Marion, J. B., and Thornton, S. T., *Classical Dynamics of Particles and Systems, 4th, ed.* (Harcourt, 1995).

[16] W. P. Reinhardt and P. L. Walker, *Jacobian Elliptic Functions*, in NIST Handbook of Mathematical Functions (Cambridge University Press, Cambridge, 2010), Chap. 22.

[17] Nesbet, R. K., *Variational principles and methods in theoretical physics and chemistry* (Cambridge University Press, 2003).

[18] W. P. Reinhardt and P. L. Walker, *Weierstrass Elliptic and Modular Functions*, in NIST Handbook of Mathematical Functions (Cambridge University

Press, Cambridge, 2010), Chap. 23.

[19] Tracy, E. R., Brizard, A. J., Richardson, A. S., and Kaufman, A. N., *Ray Tracing and Beyond: Phase space methods in plasma wave theory* (Cambridge University Press, Cambridge, 2014).

[20] Whittaker, E. T., *A Treatise on the Analytical Dynamics of Particles and Rigid Bodies, 4th ed.* (Dover, 1944).

[21] Yourgrau, W., and Mandelstam, S., *Variational Principles in Dynamics and Quantum Theory* (Dover, 1968).

Index

Brachistochrone problem, 13

Center-of-mass frame
angular momentum, 68
definition, 67
Lagrangian, 68
reduced mass, 68
Central-force potential
finite potential well, 122
isotropic simple harmonic
oscillator, 122
Lagrangian, 106
orbit equation, 107
orbit integral, 109
radial force equation, 106
Christoffel symbol, 9, 27, 28, 52
Collision kinematics
CM frame, 137
LAB frame, 134
LAB-CM connections, 138
Continuous Lagrangian systems
Euler's fluid equations, 250
general field theory, 243
Noether Theorem, 245
Schroedinger equation, 248
transverse waves, 241

d'Alembert-Lagrange equation, 44

Elliptic functions
Jacobi, 95, 96, 196, 275, 287
Weierstrass, 88, 99, 280, 287

Energy method
bounded and unbounded orbits, 87
cubic potential, 285
Morse potential, 89
pendulum, 93
pendulum
libration orbits, 94
rotation orbits, 96
separatrix orbit, 96
quartic potential, 279
simple harmonic oscillator, 88
turning points, 86, 114
Euler's First Equation, 5, 6, 18
Euler's Second Equation, 7, 8, 19, 50
Euler-Fermat ray equation, 29
Euler-Lagrange equations, 50, 52, 53,
55, 56, 58, 60, 64–66, 68, 77, 83,
165, 200, 219, 222, 242, 245, 248,
251
Euler-Maupertuis equation, 39
Eulerian rigid-body dynamics
Euler equations, 189
torque-free asymmetric top, 194
torque-free symmetric top, 191

Frenet-Serret formulas, 23, 34, 38,
154, 169, 170, 259
Functional variation
first-order, 4
second-order, 5

Hamiltonian mechanics

adiabatic invariance, 297
canonical equations, 78
Hamilton-Jacobi equation, 40, 78, 81
noncanonical coordinates, 291
noncanonical Poisson bracket, 293
perturbation theory, 295
ray optics, 81
Historical Names
Bernoulli, 13
Bouger, 24
Cartan, 293
Coriolis, 164
d'Alembert, 42
de Broglie, 40
Descartes, 38
Euler, 3
Fermat, 15
Feynman, 82
Foucault, 172
Hamilton, 78
Huygens, 29
Jacobi, 6
Kepler, 113
Lagrange, 3
Maupertuis, 38
Newton, 53
Noether, 61
Poincaré, 293
Poisson, 293
Routh, 65
Rutherford, 146
Schroedinger, 82
Snell, 15

Inertia tensor
body frame, 186
continuous mass distribution, 184
discrete mass distribution, 181
principal moments of inertia, 185
principle-axes theorem, 183

Jacobi equation, 6, 28
Jacobi-Euler equation, 6
Jacobi-Euler-Fermat equation, 29

Kepler Problem
equations of motion, 113
Kepler problem
First Law, 115
Hohmann transfer orbits, 130
Kepler's Equation, 129
perturbed orbit, 120
Second Law, 116
Third Law, 117
unbounded orbits, 117

Lagrangian mechanics
configuration space, 53
constrained motion, 49
constrained variations, 11
curvilinear coordinates, 52
degree of freedom, 47
four-step method, 51
frozen degrees of freedom, 59
gauge transformation, 63
generalized coordinates, 43, 47
holonomic constraint, 48
Lagrange multiplier, 11, 46
Lagrange's δ-operator, 5, 43
Routh reduction, 65
Lagrangian rigid-body dynamics
Eulerian angles, 197
Eulerian angular velocities, 198
Symmetric top, 199
Symmetric top with one fixed point
nutation, 206
Routh-Lagrange equations, 204
sleeping top, 209, 231
Laplace-Runge-Lenz vector, 119
Legendre conditions, 5
Legendre transformation, 78, 80
Light
law of reflection, 16
light geodesics, 27
optical length, 17
ray curvature, 23
ray torsion, 23
Snell's Law of refraction, 17, 19
wavefront representation, 29

Motion in an electromagnetic field
  gauge invariance, 84
  Hamiltonian, 84
  Lagrangian, 83

Noether Theorem, 24, 61, 65, 245,
  248, 252
Normal-mode analysis
  coupled linear oscillators, 226
  coupled nonlinear oscillators, 228
  Lissajous figures, 227
  one degree of freedom, 223
  two degrees of freedom, 224
Numerical analysis, 66, 102, 103, 120,
  174, 194, 272

Potential-well scattering
  hard potential, 143
  hard surface, 152
  soft potential, 150

Rayleigh dissipation function, 73
Rotating frames
  angular momentum, 166
  centrifugal acceleration, 164
  Coriolis acceleration, 164, 174
  Coriolis drifts, 170

Coriolis inertia circle, 169
Foucault pendulum, 174
Lagrangian, 164
Magnus force, 178
time derivative, 162
Rotation matrix, 186
Rutherford scattering
  Kepler potential, 145
  modified potential, 147

Scattering cross section
  geometry, 140
  LAB-CM relations, 142
  total cross section, 141
Stability of equilibrium points
  bead on a rotating hoop, 219
  circular orbits in central-force
    potentials, 220
  sleeping top, 231

Variational Principles
  Fermat, 40
  Hamilton, 44, 45, 50
  Jacobi, 39
  Maupertuis, 38
  Stationary Phase, 81
Virial Theorem, 111

Printed in the United States
By Bookmasters